ENGINEERING TECHNOLOGY AND APPLICATIONS

PROCEEDINGS OF THE 2014 INTERNATIONAL CONFERENCE ON ENGINEERING
TECHNOLOGY AND APPLICATIONS (ICETA 2014), TSINGTAO, CHINA, 29–30 APRIL 2014

Engineering Technology and Applications

Editors

Fun Shao

Digital Library Department, Library of Huazhong University of Science and Technology, China

Wise Shu

Arts College, Hubei Open University, China

Tracy Tian

Bos'n Academic Service Centre, China

CRC Press
Taylor & Francis Group
Boca Raton London New York Leiden

CRC Press is an imprint of the
Taylor & Francis Group, an **informa** business

A BALKEMA BOOK

CRC Press/Balkema is an imprint of the Taylor & Francis Group, an informa business

© 2014 Taylor & Francis Group, London, UK

Typeset by MPS Limited, Chennai, India
Printed and bound in Great Britain by CPI Group (UK) Ltd, Croydon, CR0 4YY.

Published by: CRC Press/Balkema
                P.O. Box 11320, 2301 EH Leiden, The Netherlands
                e-mail: Pub.NL@taylorandfrancis.com
                www.crcpress.com – www.taylorandfrancis.com

ISBN: 978-1-138-02705-3 (Hardback)
ISBN: 978-1-315-73751-5 (ebook PDF)

Table of contents

Mechanical engineering

Engineering Technology and Applications – Shao, Shu & Tian (Eds)
© 2014 Taylor & Francis Group, London, ISBN 978-1-138-02705-3

Preface

Engineering Technology and Applications contains the contributions presented at the 2014 International Conference on Engineering Technology and Applications (ICETA 2014, Tsingtao, China, 29–30 April 2014). The book is divided into three main topics:

– Civil and environmental engineering
– Electrical and computer engineering
– Mechanical engineering

Considerable attention is also paid to big data, cloud computing, neural network algorithms and social network services. The book will be invaluable to professionals and academics in civil, environmental, electrical, computer and mechanical engineering.

Engineering Technology and Applications – Shao, Shu & Tian (Eds)
© 2014 Taylor & Francis Group, London, ISBN 978-1-138-02705-3

Acknowledgements

It must be appreciated that the proceedings can be successfully published. Thanks to the authors who contributed their papers to the conference which consisted of this book. Thanks to the E-science Website of Chinese Academy of Science, which provided a platform to release the conference and relevant information. Of course, the organizing committee is one of the most important parts and all members made great efforts to ensure the quality of papers. Finally, thanks to all editors at Bos'n Academic Service Centre and CRC Press.

Engineering Technology and Applications – Shao, Shu & Tian (Eds)
© 2014 Taylor & Francis Group, London, ISBN 978-1-138-02705-3

Committees

PROGRAM CHAIRS

Dr. Max Lee, Union University, United States of America
Dr. Shee Wung, The Chinese University of Hong Kong

GENERAL CHAIRS

Prof. Hong T. Yu, Huazhong University of Science and Technology, China
Dr. Fun Shao, Huazhong University of Science and Technology, China

TECHNICAL COMMITTEES

Prof. Efu Liu, China Pharmaceutical University, China
Dr. Long Tan, Xiangya School of Medicine, Central South University, China
PhD. Chen Chen, Huazhong University of Science and Technology, China
PhD. Fang Xiao, Wuhan National Laboratory for Optoelectronics, China
Prof. Zuo Tang, Central South University, China
PhD. Yhong Zhao, Huazhong University of Science and Technology, China
PhD. Yuan Liu, Technology Center of Foton Motor Co., Ltd., China
Dr. Bifa Zhu, Hubei University of traditional Chinese Medicine, China
PhD. Fanze Hua, Huazhong University of Science and Technology, China
PhD. Lifeng Wong, Beijing University of Posts & Telecommunications, China
Dr. Jian Wang, The Third Research Institute of Ministry of Public Security, China

SECRETARY GENERAL

Mi Tian, Bos'n Academic Service Centre

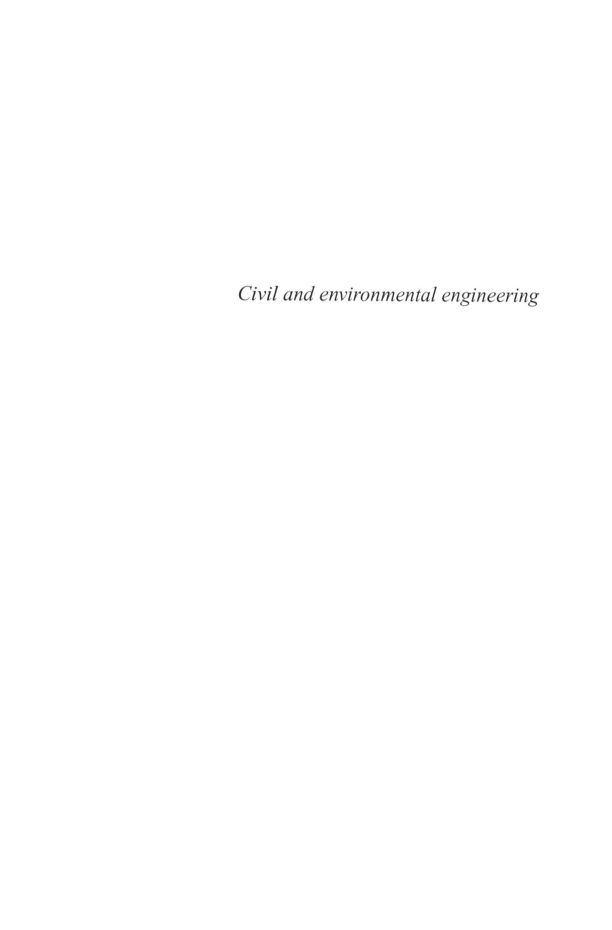

Civil and environmental engineering

Engineering Technology and Applications – Shao, Shu & Tian (Eds)
© *2014 Taylor & Francis Group, London, ISBN 978-1-138-02705-3*

Study on solidifying or stabilizing heavy metal pollution sludge with multiple additives and the leaching experiment

Mingming Li
Tianjin Research Institute for Water Transport Engineering, Tianjin, China

Dianjun Zuo & Yuting Zhang
Tianjin Research Institute for Water Transport Engineering, Tianjin, China
College of Civil and Transportation Engineering, Hohai University, Nanjing, Jiangsu, China

Xiaoyu An
Tianjin Research Institute for Water Transport Engineering, Tianjin, China

ABSTRACT: With sludge in the estuary of Dagu drainage canal in Tianjin as the study object and cement, bentonite, quicklime and coal ash as additives to solidify or stabilize sludge produced by heavy metal pollution, the paper, using the orthogonal method to design the experiment, studies the unconfined compressive strength after the solidification and stabilization of the heavy metal sludge and the leaching effect of heavy metals in the leaching experiment. The results indicate that among the four additives, cement has the greatest effect on the unconfined compressive strength of the solidified body, coming up successively are quicklime, coal ash and bentonite. By making a comprehensive summary of the unconfined compressive strength and leaching experiments above, the paper draws the conclusion that the best additive ratio of curing agents to treat heavy metal pollution sludge in the estuary of Dagu drainage canal, i.e. cement 8 (wt%), bentonite 4 (wt%), quicklime 3 (wt%) and coal ash 2 (wt%), not only solidifies the sludge and stabilizes heavy metals but also achieves maximum economic benefits.

Keywords: multiple additive; heavy metal; stabilization/solidification; unconfined compressive strength; leaching experiment

1 INTRODUCTION

Dagu drainage canal is one of the two major drainage rivers in Tianjin. Its estuary is the main drainage watercourse in the south of Haihe River. Heavy metals and organic pollutants precipitate in the downstream of the canal in the form of particulates through physical and chemical processes, especially in the estuary, which has caused poor water quality. Now the area has been seriously polluted and the ecosystem here damaged, which has seriously hampered the development of local economy and affected the safety of agricultural irrigation areas[1,2].

Currently, the solidification/stabilization technology has been widely used in the disposal of heavy metal pollution sites and landfill of solid waste[3,4]. Compared with other remedial technologies (such as chemical treatment and bioremediation), the technology has many advantages, such as low cost, convenient construction, high intensity of foundation soil after treatment and low biodegradation. Use a mixing pile and other construction machines to mix and stir heavy metal pollution soil and cement as well as other additives in situ to wrap the pollutants, absorb heavy metals and change their forms, so as to prevent them from spreading to the surrounding environment and thereby transform harmful substances into stable solid materials acceptable by the environment. The treatment method can not only solve the problem of heavy metal pollution, but also make solidified/stabilized soil highly strong and resistant to deformation, thereby meeting construction requirements. Owing to these advantages, the solidification/stabilization technology of cement and other additives has been more widely used in the recovery of heavy metal pollution sites in foreign countries.

For the comprehensive improvement project of Dagu drainage canal, the author has conducted a preliminary study on the effect of cement, bentonite, quicklime and coal ash as curing agents. DERMATAS D[5−7] has shown that the shear strength and compressive strength of soil and other indexes have increased significantly over time, and the concentration of Cr, Cu and Pb meet relevant requirements after heavy metal pollution soil is treated with cement and other curing agents. Xue Yongjie[8−10] studied that coal ash and quicklime have a good solidification/stabilization effect on sludge containing cadmium, copper, zinc, mercury and other heavy metals. Fushimi H et al[11,12] found that bentonite is a kind of highly active inorganic ash clay material. It, with a strong water-absorbing and ion adsorption capacity, is widely used in water purification. Relevant studies have shown that heavy metals Zn, Cu and Pb will be removed by over 80% if adding 0.1% of bentonite into 10 mg/L ion solution. To find the best additive ratio of these curing agents, the author uses the orthogonal experiment to conduct the compressive strength test and toxicity leaching experiment on solidified bodies with different additive ratios to evaluate their compression resistance and solidification effect, which provides technical and theoretical support for the safe treatment of heavy metal pollution sludge and resource utilization[13,14].

2 EXPERIMENT PROGRAM

2.1 Unconfined compressive strength experiment

2.1.1 Experimental materials and devices
Experimental sludge is collected from Tianjin Dagu drainage canal. It's dark black or dark brown, with an obvious stink. The sample is stored in a plastic bucket for later use. The main physical indexes of original soil are listed in Table 1.

The instrument used in the compressive strength test is an unconfined compression tester, and the standard adopted is the Soil Engineering Test Method GB T50123-1999.

2.1.2 Test procedure and step
The experiment involves a number of factors and each factor has many changes. If using the traditional test method, it will cost much time, so the author uses the orthogonal test[15−17].

The test uses the L25_5_6 standard orthogonal table and takes the unconfined compressive strength of the sample as the index. The following four factors are used in the test, including cement, bentonite, quicklime and coal ash. The 5 levels of each factor are shown in Table 2.

Table 1. Physical parameters of original soil used in this study.

Moisture [%]	pH	Organic substances [%]	Heavy metals [mg/kg^{-1}]			
			Cr	Cu	Zn	Pb
59.98	7.93	0.81	552	131	384	63

Table 2. Factors and levels of orthogonal test.

Factor	Level Content [wt %]				
Cement	A	2	4	6	8	10
Bentonite	B	1	2	3	4	5
Quicklime	C	1	2	3	4	5
Coal ash	D	1	2	3	4	5

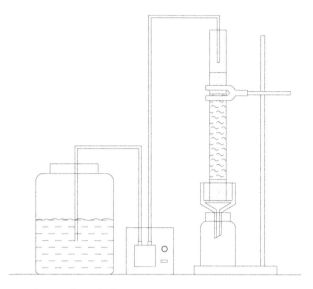

Figure 1. Leaching experiment schematic diagram.

Test steps: mix the original soil with the additives in a certain proportion and stir in a cement paste mixer uniformly, then put a mold with the specification of 50×100 mm into the mixer and demould after 24 hours. Maintain the mold in a curing oven for 7 days at a temperature of 20°C and humidity of 95%. Determine the unconfined compressive strength after curing.

2.2 Heavy metal leaching experiment

2.2.1 Experimental materials and devices

The experimental device is shown in Figure 1, including a reservoir tank, plastic tube, peristaltic pump, straight glass, filter paper, Buchner funnel, collection bottle and holder.

After the compressive strength test, select several formulas with higher compressive strength as the sample to be used in the heavy metal toxicity leaching experiment. Water used in the experiment is deionized water and sand is pure quartz sand with a size of 20 mesh.

2.2.2 Test procedure and step

Crush the sample and sieve to select powder with a size of 30 mesh for use. As shown in Figure 1, to achieve uniform water distribution, we continually add the pure quartz sand into the bottom of the filter tube with a diameter of 5 cm, with the thickness reaching 5 cm. Then add the sample with the thickness reaching 20 cm, and fill the pure quartz sand again with the thickness reaching 5 cm. Seal the top of the filter tube with plastic wrap and leave a hole in the middle to insert the hose which connects to the creep flow pump. Connect the bottom of the filter tube to the Buchner funnel and prepare four layers of filter paper. The collection bottle is placed under the funnel. The flow rate of the pump is 500 ml/d, continuing leaching for 7 days.

3 EXPERIMENTAL RESULT AND ANALYSIS

3.1 Analysis of compressive strength

To ensure the solidified body of heavy metal pollution sludge can be used as engineering soil, it's necessary to determine its unconfined compressive strength. Generally, when the compressive strength reaches 100 kPa, the solidified body will meet general engineering requirements. The

Table 3. Results of orthogonal test.

Test No.	Factor (Additive ratio wt%)				Compressive strength (kPa)
	Cement A	Bentonite B	Quick C	Coal ash D	
1	2	1	1	1	6.50
2	2	2	2	2	3.47
3	2	3	3	3	10.47
4	2	4	4	4	10.34
5	2	5	5	5	11.48
6	4	1	2	3	18.96
7	4	2	3	4	22.59
8	4	3	4	5	25.56
9	4	4	5	1	52.79
10	4	5	1	2	22.68
11	6	1	3	5	34.14
12	6	2	4	1	71.20
13	6	3	5	2	58.13
14	6	4	1	3	27.04
15	6	5	2	4	43.14
16	8	1	4	2	63.58
17	8	2	5	3	82.24
18	8	3	1	4	73.92
19	8	4	2	5	40.86
20	8	5	3	1	67.41
21	10	1	5	4	180.75
22	10	2	1	5	94.97
23	10	3	2	1	73.03
24	10	4	3	2	242.05
25	10	5	4	3	116.35
K_1	8.452	60.786	45.022	54.186	
K_2	28.516	54.894	35.892	77.982	
K_3	46.730	48.222	75.332	51.012	
K_4	65.602	74.616	57.406	66.148	
K_5	141.430	52.212	77.078	41.402	
R	132.978	26.394	41.186	36.580	

results of orthogonal test on unconfined compressive strength are shown in Table 3, and the effect curve of unconfined test is in Figure 2.

We can know from R value in Table 3 that, cement has the greatest effect on the unconfined compressive strength of the solidified body, coming up successively are quicklime, coal ash and bentonite. Since $K_5 > K_4 > K_3 > K_2 > K_1$, when the content of cement reaches 10%, the compressive strength of the sample significantly improves. The second is quicklime. Since $K_5 \approx K_3 > K_4 > K_1 > K_2$, there is little difference between the third level and fifth level. Select the third level (content 3%) for the subsequent test. The third is coal ash. Since $K_2 > K_4 > K_1 > K_3 > K_5$, the second level (content 5%) has greater impact. Bentonite has the weakest impact. Since $K_4 > K_1 > K_2 > K_5 > K_3$, the fourth level (content 4%) has greater impact.

In summary, cement has obvious impact on the unconfined compressive strength of the solidified body while quicklime, coal ash and bentonite have weaker impact.

Figure 2 shows that the effect value of cement improves with the increasing proportion, which is consistent with the test result of the unconfined compressive strength. And the cusps of the effect curve (vantage points) are in the position of Level 4 of bentonite and Level 2 of coal ash, which indicates the optimal level combination of bentonite and coal ash–B4D2, i.e. the best additive ratio of the curing agent: bentonite accounts for 4 (wt%) and coal ash for 2 (wt%). From Table 2, when cement is less than 8 (wt%), the unconfined compressive strength is less than 100 kPa, which cannot meet general engineering requirements, so we increase the percentage of cement and determine the percentage of quicklime (i.e. 5%) in the supplementary experiment of unconfined compressive strength. The test arrangement and result is shown in Table 4.

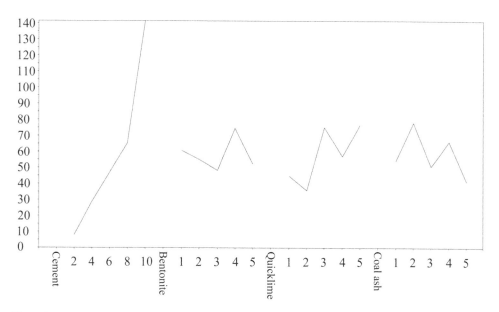

Figure 2. Relationships between the factors and strength solidified body of the orthogonal test.

Table 4. A supplementary experiment of unconfined compressive strength of solidified body.

Factor (Additive ratio wt%)				
A	B	C	D	Compressive strength (kPa)
8	4	5	2	227.45
10	4	5	2	266.63
8	4	3	2	212.36
13	4	3	2	474.12
17	4	3	2	703.70
20	4	3	2	1348.88

From Table 2 and 4, we learn that: when bentonite, quicklime and coal ash respectively account for 4 (wt%), 3 (wt%) and 2 (wt%), the undefined compressive strength of the solidified body improves greatly with the increasing percentage of cement. And when the percentage reaches 20%, the strength still does not reach the upper limit, which is determined by the nature of cement itself. Meanwhile, because of the high cost of cement, increasing the percentage of cement will greatly increase the cost of sludge treatment, which will consequently leads to low economic benefit. As for the additive ratio of curing agents, I will make a comprehensive analysis according to the subsequent leaching test and develop an optimal solution.

3.2 Heavy metal leaching analysis

We conduct a leaching experiment on the solidified body prepared with six kinds of curing agents in Table 3, solidified body $A_5B_4C_3D_2$ and the original soil, and measure the change rule of the concentration of heavy metals Cr, Cu, Zn and Pb (see Figure 3~6).

From the curve trends in Figure 3~6, we can see that the concentration of heavy metals gradually decreases and gets stable with time, of which the concentration of Cr in the leaching solution of solidified body 8.4.3.2 is the lowest and lower than that in the original soil. This indicates that when using cement as the curing agent, although the compressive strength of pollution soil will improve

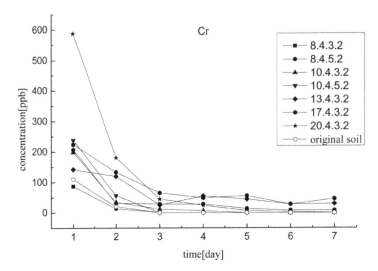

Figure 3. The concentration of heavy metal element Cr versus time. Note: 8.4.3.2 in the figures means that cement, bentonite, quicklime and coal ash respectively account for 8%, 4%, 3% and 2% in the solidified body, the same below.

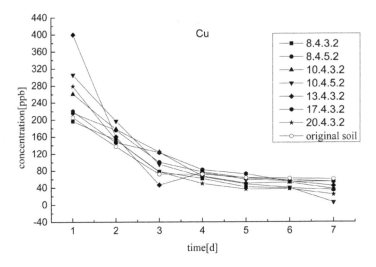

Figure 4. The concentration of heavy metal element Cu versus time.

greatly with the increasing percentage of cement, the content of Cr in soil will also increase as there is much Cr contained in a large amount of cement, which may cause secondary pollution and high soil pollution control costs but poor economic benefit.

Figure 4 shows the concentration of Cu in the leaching solution during the first three days is higher than that in the original soil, but lower than that during the later phase. Comparing these curves and values, we can know that the additive ratio of 8.4.3.2 has the best effect on the curing of Cu; Figure 5 shows the concentration of Zn in the leaching solution is lower than that in the original soil, which indicates that the additive ratio has a better effect on the curing of Zn; Figure 6 shows the concentration of Pb in the leaching solution is lower than that in the original soil, but it doesn't change obviously. The absolute value is greater than 9.2 ppb, showing the additive ratio has a weaker effect on the curing of Pb.

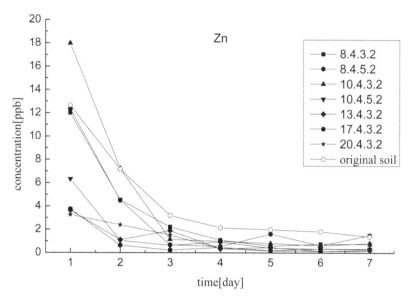

Figure 5. The concentration of heavy metal element Zn versus time.

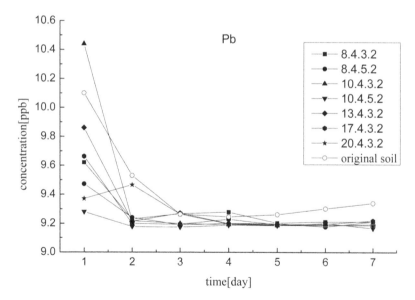

Figure 6. The concentration of heavy metal element Pb versus time.

In the figures above, some curves show a low–high–low trend. The reason is that at the beginning of leaching, a lot of free heavy metal ions in soil are leached, which leads to the first peak. With the continuous leaching, soluble heavy metals not completely solidified/stabilized start leaching after dissolved, resulting in the second leaching peak, but which is much lower than the first peak. After a certain period, the leaching amount gets stabilized. The change trend of elements in the leaching solution is related to their nature and mode of occurrence.

In summary, the additive ratio of 8.4.3.2 (i.e. cement percentage of 8 (wt%), bentonite 4 (wt%), quicklime 3 (wt%) and coal ash 2 (wt%)) has the best stabilizing effect on heavy metals. In addition, since the unconfined compressive strength of pollution soil needs to meet engineering requirements, therefore, the solution can meet relevant requirements and achieve the best economic benefit.

4 CONCLUSION AND RECOMMENDATION

(1) Through the unconfined compressive strength test and leaching experiment, it's concluded that the best additive ratio of curing agents for treating heavy metal pollution sludge in Dagu estuary is that cement accounts for 8 (wt%), bentonite 4 (wt%), quicklime 3 (wt%) and coal ash 2 (wt%). With this ratio, we can solidify sludge/stabilize heavy metals and obtain the best economic benefit at the same time.
(2) The treatment of heavy metals in sludge is affected by many factors, so we shall take into account of different factors, in order to achieve the best effect. Because of different forms of heavy metals and the complexity of removal, it's necessary to combine different methods together, which will be one of the future trends.

REFERENCES

[1] Zhang, Y.X., Wang, L.T., Wang, X., Shi, Z.L. & Duan, K. Design scheme on landfill site of dredged sediment from Dagu river. *China Water & Waste Water*, 22(8) 46–50.
[2] Guang-Hong Wu, Rui-Xian Su, Wan-Qing Li, Hong-Qi Zheng, Source and enrichment of heavy metals in sewage-irrigated area soil of dagu sewage discharge channe, J. Environmental science, 29(6) 1693–1698.
[3] Yan-Jun Du, Fei Jin, Song-Yu Liu, Lei Chen, Fan Zhang, Review of stabilization/solidification technique for remediation of heavy metals contaminated lands, J. Rock and Soil Mechanics, 32(1) 116–124.
[4] Fang Wu, Ai-Ping Luo, Nan Li, Experimental study on solidification and soaking test of sewage sludge containing heavy metal ion, J. Techniques and Equipment for Environmental Pollution Control, 4(12) 40–43.
[5] Chun-Yang Yin, Shaaban M G, Mahmud H B, Chemical stabilization of scrap metal yard contaminated soil using ordinary Portland cement: Strength and leachability aspects, J. Building and Environment, 42(2) 794–802.
[6] Dermatas D, Xiao-Gguang Meng, Utilization of fly ash for stabilization/solidification of heavy metal contaminated soils, J. Engineering Geology, 70(3) 377–394.
[7] Fu-Sheng Zha, Long Xu, Ke-Rui Cui, Strength characteristics of heavy metal contaminated soils stabilized/
solidified by cement, J. Rock and Soil Mechanics, 33(3) 652–656.
[8] Yong-Jie Xue, Shu-Jing Zhu, Hao-Bo Hou, Experimental study of quicklime-ash solidification of heavy metal contaminated soil, J. Coal Ash China, 19(3) 10–12.
[9] Xue-Feng Wang, Gui-Fen Zhu, Hui-Yong Zhang, Study of the effect of amend regents on the combined pollution of heavy-metals, J. Journal of Henan Normal University (Natural Science), 32(3) 133–136.
[10] Xiao-Lei Jia, Pan-Yue Zhang, Guang-Ming Zeng, Effect of coal fly ash addition on cement stabilization/solidification of heavy metal contaminated sediments, J. Journal of Safety and Environment, 10(5) 50–53.
[11] Fushimi H., On the adsorption on removal phenomenon and recovery technique of heavy metal ions by use of clay minerals, R. Tokyo: Department of Physics and chemical Science, Waseda University, 1980.
[12] Wei Zhu, Cheng Lin, Lei Li, T. Ohki, Solidification/stabilization (S/S) of sludge using dditive, J. Environmental Science, 28(5) 1020–1025.
[13] Hu-Yuan Zhang, Bao Wang, Xing-Ling Dong, Lei Feng, Zhi-Ming Fan, Leachability of heavy metals from solidified sludge, J. Science in China(Series E:Technological Sciences), 52(7) 1906–1912.
[14] Qi-Jun Yu, Shigeyoshi Nagataki, Jin-Mei LIN. Solidification of municipal solid waste incineration fly ash with cement and its leaching behaviors of heavy metals, J. 18(1) 55–60.
[15] Zhong-An Xu, Tian-Bao Wang, Chang-Ying Li, Li-Yan Bao, Qing-Mei Ma, Brief introduction to the orthogonal test design, J. Sci/Tech Information Development & Economy, 12(5) 48–50.
[16] Rui-Jiang Liu, Ye-Wang Zhang, Chong-Wei Wen, Jian Tang. Study on the design and analysis methods of orthogonal experiment, J. Experimental Technology and Management, 27(9) 52–55.
[17] Yu-Feng Yuan, Analysis of Multi-target Orthogonal Experiment, J. Journal of Hubei Automotive Industries, 19(4) 53–56.

Engineering Technology and Applications – Shao, Shu & Tian (Eds)
© *2014 Taylor & Francis Group, London, ISBN 978-1-138-02705-3*

Preparation and properties of CNT/Sn-Pb composite coating

Chunhua Li & Chen Chen
Baoding Vocational and Technical College, Baoding, Hebei, China

ABSTRACT: The effect of different amounts of carbon nanotubes (CNTs) on the properties of Sn-Pb composite coating is investigated by preparing a CNT/Sn-Pb composite coating through composite electroplating process and putting pre-acidized CNTs into the plating bath. The result indicates that: CNTs eliminated impurities and aggregate structure after mixed acid ultrasonic treatment and improved its dispersibility; the thickness of the coating decreases with the increase of the amount of CNTs; the hardness of the coating reaches its peak when the amount of CNTs is 2 g/L; the friction factor first decreases and then increases with the amount of CNTs; and the friction factor reaches its valley and provides the best wear resistance when the amount of CNTs is 2 g/L.

Keywords: carbon nanotubes; acid treatment; composite coating; friction and wear behavior

1 INTRODUCTION

A sliding bearing is a bearing that works under sliding friction. The part that supports the bearing is the journal. The accessory associated with the journal is the bush. The antifriction material cast over the surface of the bush is liner. The bushing and lining materials are collectively termed sliding bearing materials. An important part of modern machinery, sliding bearings are heavily vulnerable to damage during their service. As modern machinery progresses towards high speed, low depletion and light weight[1–3] to contribute to a green, pollution-free society, bearing materials are also calling for better behavior. How to improve the service performance of bearing materials or the research on new bearing materials has become the hot spot of the present science community. Conventional bearing alloys are supplied either as hard matrix + soft particles or as soft matrix + hard particles, both of which have their merits and demerits. The former provides very good bearing capacity and fatigue resistance but poor in wear and embedding behavior while the latter provides good friction resistance but are poor in bearing and hot fatigue performance. A popular material widely used for auto and internal combustion engines is Babbitt metal, a tin- or lead-base soft alloy with good antifriction performance, small friction factor and sound plastic, elastic and anticorrosion behavior. The only problem with this metal is that it has fairly low bearing capacity. Many researchers have tried to stiffen Babbitt metal through experiments to improve its bearing and fatigue performances while retaining its sound friction behavior[4–5].

In 1991, Professor Iijima of Japan's NEC discovered multiwall carbon nanotubes (MWNTs), which soon became the hot spot for the material science community for its excellent mechanical, physical and chemical behavior and novel structure[6]. In this paper, a CNT/Sn-Pb composite coating was prepared by adding acidized carbon nanotubes (CNTs) into the Sn-Pb alloy using composite electro-deposition process, and the behavior of the resultant composite coating was examined in hopes of providing theoretical and technical support for the practical application of composite coatings.

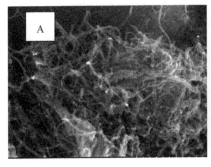 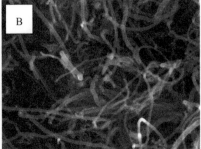

A. Morphology of untreated CNTs   B. Morphology of treated CNTs

Figure 1.   SEM images of CNTs.

2   PRETREATMENT OF CNTS

2.1   *Pretreatment process*

Under composite electro-plating, CNTs often appear like aggregate structures. This prevents them from distributing evenly in the composite coating, thereby limiting the uniformity and compactness of the coating and making it impossible to use all the functions expected from the CNT coating. The only solution is to minimize the aggregation of CNTs and increase their uniformity and dispersibility in the coating.

The pretreatment process used in our study was supersonic treatment of the CNTs in a sulfuric-nitric mixture. During the test, multiwall carbon nanotubes (MWNTs) 5–15 μm in length, 60–100 nm in diameter were used. 200 ml concentrated sulfuric solution and 200 concentrated nitric solution, respectively, and 2 g CTNs were weighed and mixed together in a three-neck flask. A reflux condenser was placed into the flask, which was then placed in an ultrasonic bath and treated under normal temperature for 10 h. After that, the CNTs were moved into a funnel over a Buchner flask that was covered by 0.45 μm microporous membrane. The mixture was filtered by a circulating vacuum pump and scrubbed in deionized water until neutral before it was moved into a clean breaker and dried in a vacuum drying box for 5 h under 80°C.

2.2   *Pretreatment result*

Treated and untreated CNTs were observed under a scanning electron microscope (SEM).

From Fig. 1A, the untreated CNTs are stained with impurities on the surface and tangle with each other showing heavy aggregation. This is explained by: 1) the presence of amorphous carbon, graphite particles and catalyzers containing carbon nano particles of five-membered or seven-membered cycles on the surface of CNTs, which limit the behavior and application of CNTs[7–8]; and 2) the fairly high surface energy and large curvature intrinsic to CNTs, which result in aggregation of CNTs and reduce their dispersibility, effective aspect ratio and thereby their performance.

Fig. 1B shows the morphology of the acidized CNTs, over the surface of which there are notably fewer impurities. The reason lies in the fact that the impurities typically comprise carbon nano particles and amorphous carbon. The former are a polyhedron containing numerous five-memered cyclic structures that can be oxidized by the high oxidation property of concentrated sulfuric acid and concentrated nitric acid. The latter is a multilayered structure with a lot of unsaturated dangling bonds that are also oxidized. The acid oxidation reduced the impurities on the surface of the CNTs.

From Fig. 1B, the CNTs are also less aggregated and the tangling among the tubes lessened. This is because there are five- or seven-membered cycles on the wall and roof of the CNTs on which carbon atoms are highly energetic and therefore easily oxidized. During the acidization, the large-curvature part of the CNTs where more five or seven-memered cycles had gathered was oxidized by oxygen atoms released from the mixed acid. When this reached a certain point, the end was cut off. The acid penetrated into the CNT layers. The irregularly structured layers of the outer and inner walls were oxidized. As it was oxidized, the tube wall became thinner and thinner. When this reached a certain point, the weak portion of the CNTs broke, and the aggregated CNTs broke up into shorter, smaller-curvature, open-ended tubes.

This demonstrates that acidization not only reduces the surface impurities of CNTs, but also improves the aggregation of the tubes, thereby resulting in better performance of the tubes.

3 PREPARING THE COMPOSITE PLATING

3.1 *Experimental materials and solution formula*

Sheet copper was used as the matrix for the test. Reagents used in the test included fluoboric acid HBF_4, boric H_3BO_3, tin powder, Lead oxide yellow PbO, basic copper carbonate $Cu_2(OH)_2CO_3$, concentrated sulfuric acid, concentrated nitric acid, alcohol, acetone, deionized water, all analytically pure, and lead electrode and CNTs.

The plating solution was formulated by: stannous fluoborate $Sn(BF_4)_2$: 85 g/L; Lead fluoroborate $Pb(BF_4)_2$: 40 g/L; fluoboric acid HBF_4: 120 g/L; boric acid H_3BO_3: 25 g/L; resorcinol: 1.5 g/L; gelatin: 3 g/L and a proper amount of dispersant.

3.2 *Electroplating process*

First, the sheet copper was pretreated: grinding $\rightarrow$ degreasing $\rightarrow$ derusting to remove rust $\rightarrow$ ultrasonic degreasing and scrubbing. After the plating solution was prepared and mixed up, the acidized CNTs, together with some surfactant were added into the solution and then shaken in an ultrasonic bath for 30 min.

After all the CNTs were soaked, they were electro-plated, during which the plating solution was stirred by a magnetic stirrer at a constant speed. Next, the electroplated test sample was washed in deionized water and dried with an air dryer.

Cathodic current density: $2 A/dm^2$; plating duration: 10~20 min; anode: lead plate. 0, 1, 2, 3 and 4 g/L CNTs were added.

4 EXPERIMENTAL RESULT AND ANALYSIS

4.1 *Amount of CNTs vs coating thickness*

Fig. 2 shows the relation between the amount of CNTs and the coating thickness. From this curve, the Sn-Pb coating thickness was 11.5 μm maximum before CNTs were added and declined as the amount of CNTs increased. This is because CNTs are a refractory material which, when suspended in the plating solution during the electro-plating process, prevent the syndeposition of the Sn-Pb ions. Furthermore, the increase in the CNTs also result in aggregation, which further prevents the Sn-Pb ions from depositing on the coating surface.

4.2 *Amount of CNTs vs coating hardness*

Fig. 3 shows the relation between the amount of CNTs and the coating hardness. From this curve, the coating hardness first increased and then decreased with the amount of CNTs increased, and reached its peak, 45.5 HV when the amount of CNTs was 2 g/L. CNTs in the coating help reduce the

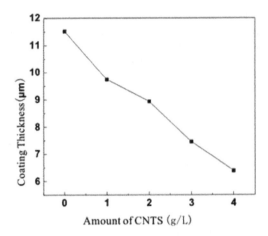

Figure 2.   Relation between the amount of CNTs and the coating thickness.

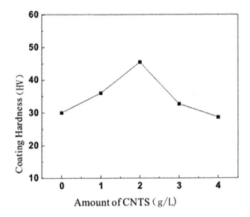

Figure 3.   Relation between the amount of CNTs and the coating hardness.

size of the particles in the coating, and the combination of the coating and the CNTs increases the hardness of the coating. When a limited amount of CNTs is added, the CNTs are able to distribute in the coating to prevent dislocation movement in the coating, thereby strengthening the coating[9]. When the amount of CNTs reaches 2 g/L or more, the aggregation of the CNTs will prevent them from distributing in the coating and thereby reduce their ability to increase dispersibility. This explains when the coating hardness would increase first and then decrease as the amount of CNTs increased.

4.3   *Friction and wear behavior of the CTN/Sn-Pb composite coating*

4.3.1   *Friction and wear testing*

The friction and wear test was conducted on a MPX-2000 friction and wear tester. The upper and lower rings were processed in the way illustrated in Fig. 4. The matrix for the test sample was brass for the upper ring and 45# steel for the lower ring. The upper ring sample was composite electro-deposited in plating solutions containing different amount of CNTs at the current density of 2 A/dm^2. Then the prepared CNT/Sn-Pb composite coating sample was tested for friction and wear behavior.

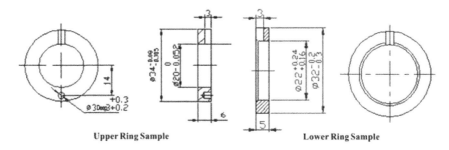

Upper Ring Sample   Lower Ring Sample

Figure 4. Sketch of the test sample.

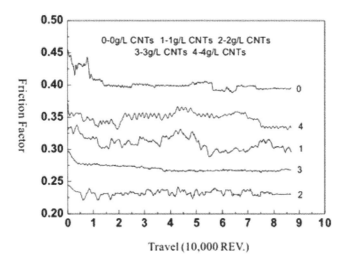

Figure 5. Relation between the amount of CNTs and the friction factor.

Parameters used for the wear and friction testing: speed: 500 r/min; travel: 9000 rev; lubricant: 10# engine oil.

4.3.2 *Amount of CNTs vs friction factor*

Fig. 5 shows the relation between the friction factor and the amount of CNTs. From this curve, the Sn-Pb coating had the largest friction factor when the amount of CNTs was 0 or when no CNT was added, suggesting that its wear resistance is the lowest when no CNT was added, and decreased notably when CNTs were added, suggesting that the CNTs in the composite coating helped increase its wear resistance. The curve also shows that the friction factor of the coating was the lowest when 2 g/L CNTs were added.

The increased wear resistance of the composite coating as a result of CNTs is mainly attributable to the high strength, tenacity and elastic modulus intrinsic to CNTs. Adding CNTs into the Sn-Pb coating will help increase its strength as well as bearing capacity. Furthermore, as CNTs will detach from the matrix during the friction and the detached CNTs will cover over the wear surface of the coating, their self-lubrication will reduce the friction factor and consequently the wear resistance[10−11]. The friction factor is the lowest when the content of CNTs is 2 g/L because, when the amount of CNTs is too small, the detached CNTs are not able to cover the wear surface fully. When the amount increases to 2 g/L, the detached CTNs are able to do so, bringing the coating to its peak wear resistance level. When more CNTs are added, they won't be able to distribute in the coating evenly and their self-lubrication won't work well[12−13], resulting in higher friction factors.

In a word, a composite coating provides better wear resistance when it contains CNTs than when it doesn't, and has the best wear resistance when 2 g/L CNTs are added into the coating.

5 CONCLUSIONS

(1) Before they are pretreated, the excellent performance of CNTs is limited by the excessive impurities contained in them and the heavy aggregation of the tubes themselves. Mixed-acid ultrasonic treatment can help reduce the presence of impurities as well as the aspect ratio of the CNTs, resulting in less aggregation and higher dispersion.
(2) During the electro-plating process, the coating hardness first increased and then decreased as the amount of CNTs increased, and reached its peak, 45.5 HV, when 2 g/L CNTs were added. The composite coating containing CNTs displayed better wear resistance than the one containing no CNTs. The friction factor first decreased and then increased as the amount of CNTs increased, and reached its valley, i.e. provides the best wear resistance, when 2 g/L CNTs was added.

REFERENCES

[1] Whitney, W.J. (1995) An Advanced Aluminum-Tin-Silicon Engineer Bearing Alloy and its Performance Characteristics [A]. SAE Tech [C]. Warrendale USA, 950–953.
[2] Whitney W.J. An Advanced Aluminum-Tin-Silicon Engine Bearing Alloy. Foreign Internal Combustion Engine, 2000(2): 58–62.
[3] Whitney W.J. Aluminum-base bearings and bushings for long-lived engines. Foreign Locomotive & Rolling Stock Technology, 2001(2): 27–30.
[4] Shi Y.S. Development Trends of Antifriction Material of Crankshaft Bush of Internal Combustion Engine. Internal Combustion Engine Parts, 2003(6): 39–41.
[5] Xie X.H. Material and manufacturing process of foreign engine cranshaft bushings. Foreign Locomotive & Rolling Stock Technology, 1999(3): 1–5.
[6] S. Iijima. Helical microtubules of graphitic carbon [J]. Nature, 1991, 35(4): 56–58.
[7] Jia Z.J. Research on carbon nanotube/polymer composites [D]. (PhD dissertation). Beijing: Tsinghua University, 1999.
[8] Cao M.S., Cao C.B., Xu J.Q. Nano-materials Science [M]. Harbin: Harbin Engineering University Press, 2002.
[9] Chen X.H., Zhang G., Chen C.S. et al. Tribological Behavior of Electroless Ni-P-Carbon Nanotube Comoposite Coating [J]. Journal of Inorganic Materials, 2003, 18(6): 1320–1324.
[10] Han G., Chen W.Q., Xia J.B. et al. Friction and Wear Behavior of Electroless Wear-Resistant and Self-Lubricating Ni-P Composite Coatings [J]. Tribology, 2004, 24(3): 216–218.
[11] Chen W.X., Gan H.Y., Tu J.P. et al. Friction and Wear Behavior of Ni-P-Carbon Nanotubes Electroless Composite Coating [J]. Tribology, 2002, 22(4): 241–244.
[12] Jiang J.L., Dai J.F., Yuan X.M. et al. Fabrication and Tribological Behavior of Carbon Nanotube/Al Matrix Composites [J]. Tribology, 2007, 27(3): 119–222.
[13] Tu J.P., Zhou T.Z., Wang L.Y. et al. Friction and wear behavior of Ni-based carbon nanotubes composite coatings [J]. Journal of Zhejiang University, 2004, 38(7): 931–934.

Engineering Technology and Applications – Shao, Shu & Tian (Eds)
© *2014 Taylor & Francis Group, London, ISBN 978-1-138-02705-3*

A study on polycrystalline material microstructure based on 3D simulation analysis

Jing Xu
Northeast Petroleum University at Qinhuangdao, Qinhuangdao Hebei, China

ABSTRACT: The microstructure of polycrystalline materials determines the macro-mechanical properties of the materials, and the mechanical properties and service capability of materials are the most concerned issue by the users, therefore, a corresponding scientific and reasonable 3D reconstruction of the polycrystalline material microstructure will be of considerable significance to social development. In this paper, the author analyzed the micro-structural features of the poly-crystalline materials' composition and studied the rationality when Laguerre model being applied in the 3D reconstruction of the material microstructure so as to provide a space structure basis for the further study on the mechanical properties and service capability of materials. In this paper, the author first discussed the mathematical principles for the fitting of the Laguerre map and Laguerre model, dissected the feasibility when the model being applied in the microstructure study on polycrystalline materials, then according to the various anisotropic characteristics displayed by the materials, the author proposed a simulation method for the orientation of composition which constitutes the texture of the polycrystalline materials; finally, with the five algorithms given in the Laguerre model, the author worked out the corresponding 3D simulation images of the polycrys-talline material microstructures, upon which the author further elaborated the applicable fields and the characteristics of the five models.

Keywords: polycrystalline material microstructure, Laguerre model, 3D reconstruction, conversion of coordinates

1 INTRODUCTION

As to the polycrystalline materials, people are inclined to focus on the rendered macro-mechanical properties and service capability, yet the macro-mechanical properties are determined by the microstructure of material, therefore, it's necessary to study the microstructure of materials. When it comes to the study on polycrystalline material microstructure, we shall, on the one hand, commence from the means of detection, on the other hand, explore from the perspective of simulation tech-nique, because the improved simulation technique will facilitate the research of material mechanical properties on a more scientific and convenient basis; as for the study on 3D simulation design and 3D reconstruction over polycrystalline material microstructure, many a scholar and researcher has made great contributions, by the efforts of whom the performance exploration of polycrys-talline materials is propelled much further than expected; for instance, Cao Zhiyuan et al. (2008) studied how to calculate the macro-response of functionally gradient plate based directly on the provided material component distribution and the net-shaped microstructure graph on the prepara-tion stage, thus he proposed the mechanical quantity 3D distributional pattern of the functionally gradient plates under different boundary conditions and the deformation of the macroscopic equal-stress diagram caused by the local abrupt change in conventional structure[1]; Zhu Xiaoyan (2011) introduced the measurement and analysis basis of crystalline material texture, the 3D orientation analysis as well as the interference of Friendel's law, she also introduced the maximum entropy principle in ODF analysis, thus overcoming the interference of Friendel's law and making quantitative texture closer to the actual conditions[2]; Zhang Bin et al. (2013) used software to conduct corresponding numerical calculation and analysis over the mesomechanics of the typical volume elements from the polycrystalline material microstructure[3].

Based on the previous studies, the author further probed into the 3D simulation model on polycrystalline material microstructure and proposed the Laguerre model algorithms in this paper, upon which the author expected to work out scientific 3D simulation images of material microstructure, thus realizing the purpose to provide more applicable models for material microstructure mechanical analysis.

2 LAGUERRE MODEL ANALYSIS ON THE COMPOSITION OF POLYCRYSTALLINE MATERIALS

In the long course of practice and theoretical investigation, we come to understand the performance of chemicals relies not only on their chemical composition but also, to a certain degree, on the microstructure of the materials. In reality, people anticipate obtaining chemical materials with lengthened service time and they expect the study on materials could scientifically reveal the service duration and performance of given materials.

The study on polycrystalline material composition can be comprehensively conducted from three aspects, namely, its geometric structure, interface and the simulation of its topology. With a view to making this study more accessible and scientific, the author built a few Laguerre models for this paper and applied them to the structural analysis over the polycrystalline material composition. In this chapter, the author firstly introduced the theoretical basis of Laguerre map, and elaborated the fitting principles of Laguerre model, and finally presented the readers with corresponding Laguerre maps under different coefficients of variation.

2.1 Mathematical theory for Laguerre map

The geometric features and topology of polycrystalline material microstructure largely resemble that of the cell body structure when generated from the Laguerre algorithms, in case a rational yardstick is used and the statistical average property, as the microstructure of a given crystal can reach, can be obtained, then various geometric models of material microstructures which are highly resembling the various compositions in the organization of the material will be worked out.

In order to build a scientific and reasonable geometric model for the material microstructure, a d dimension Euclid space is defined, and each point P_i in the space is given a weight value w_i, and the geometric model in the space is right constructed by organically connecting the weighted point set.

The space power distance L from the defined point p to the weighted point p_i can be obtained by the following computing formula (1):

$$L = d(p, p_i)^2 - w_i \tag{1}$$

In formula (1), $p \notin P, p_i \in P$, if the collection constituted by the weighted points with power distance to the weighted point p_i not more than that of any other random weighted point is named the associated Laguerre unit cell of p_i, the mathematical expression of the collection is shown in formula (2) as follows:

$$C(p_i, w_i) = \bigcup_{j \neq i} \left\{ x \in R^d \middle| d(x, p_i)^2 - w_i \leq d(x, p_j)^2 - w_i, \forall (p_j, w_j) \in P \right\} \tag{2}$$

As can be known from the defined mathematical formula (2) for Laguerre unit cell, the form of unit cell is subject not only to the space position of the point set, but to the weight distributing as

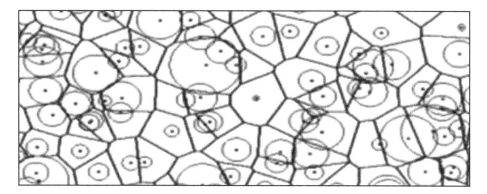

Figure 1. A 2D Laguerre map based on 2D random distribution of weighted points.

well. As to the weighted points with negative value, we can add the absolute value of the largest negative weight in the point set to each weighted point, thus to convert the negative weighted points into non-negative ones, the two-dimension weighted points exhibit a two-dimensional disc shape in Laguerre map while the 3D weighted points exhibit a 3D sphere. Fig. 1 reveals a 2D Laguerre map with randomly distributed weight values.

2.2 Theoretical analysis over the Laguerre model fitting

In size distribution, the crystal grains of polycrystalline materials usually take on logarithmic normal distribution or gamma distribution, the crystal grain is a polyhedron, the sides of which are normally less than 14, and each facade is an approximate pentagon. Under the influence of preparation technology and processing method, the geometric characteristics of the microstructure will change accordingly. Therefore, if we attempt to obtain a microstructure with identical performance index as that of the real material, we have to carry out model fitting first.

Each point in the Laguerre map, as is introduced in the last section, is given a weight value, we can make a quantitative control of the space position and value size of the weighted point set according to the mathematical definition of Laguerre unit cell so as to generate a desirable Laguerre map that meets the geometric characteristic requirement of the real material microstructure. This is the process defined as "model fitting". In this paper, model fitting is realized by sphere packing at the weighted points, the sphere packing herein mentioned refers to the practice to fill in spheres with certain volume fractions within a defined space, and the spheres cannot overlap mutually; then the coordinates of each sphere and their radius squared will be taken as a weighted point $P_i(p_i, r_i^2)$, when sphere packing finished, the collection of spheres constitutes a new weighted point set with specified features, and the regularity and geometric characteristics of Laguerre unit cell generated by such point set structures are closely related to the construction parameters of sphere packing. The parameters reflecting the geometric characteristics of sphere packing consist of the numbers of spheres, volume fraction v_v and coefficient of variation cv, among which the volume fraction refers to the quotient of the standard deviation and the mean value of sphere volume.

The model fitting process of the Laguerre unit cell volume against logarithmic normal distribution refers to the practice to carry on fitting over the volume of each cell body according to logarithmic normal distribution within the volume area V, herein N spheres under logarithmic normal distribution shall be constructed, and they occupy an overall space volume v_v, and the probability density function is expressed in formula (3) as follows:

$$g(v_s) = \frac{1}{\sqrt{2\pi}\sigma v_s} e^{-\frac{(\log v_s - \mu)}{2\sigma^2}}, v_s \geq 0 \tag{3}$$

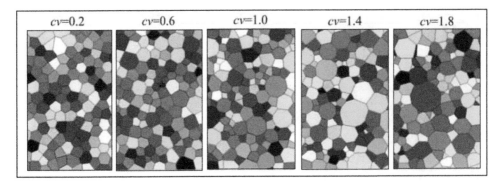

Figure 2.   2D polycrystalline material Laguerre maps with identical volume fraction but at different variation coefficients.

In formula (3), the mean value $\bar{\upsilon}_s$ and the standard deviation σ_s of the logarithmic normal distribution can be expressed with an expected value μ and the standard deviation σ in formula (4) as follows:

$$\bar{\upsilon}_s = \exp\left(\mu + \frac{\sigma^2}{2}\right), \sigma_s = \sqrt{\exp\left(\sigma^2 - 1\right)\exp\left(2\mu + \sigma^2\right)} \tag{4}$$

The quotient of the average sphere volume $\bar{\upsilon}_s$ and the standard deviation σ_s of their volume distribution equals the variation coefficient of volume $c\upsilon$, among which $\bar{\upsilon}_s$ can also be expressed in formula (5) with the relations of V, υ_v and N:

$$\bar{\upsilon}_s = \frac{V \cdot \upsilon_v}{N} \tag{5}$$

In this way, we can acquire the expected value μ and standard deviation σ in normal distribution expressed with parameter υ_v and $c\upsilon$, which can be shown in formula (6) as follows:

$$\mu = \frac{1}{2}\log\left[\frac{V \cdot \upsilon_v^2}{N^2\left(c\upsilon^2 + 1\right)}\right], \sigma = \sqrt{\log\left(c\upsilon^2 + 1\right)} \tag{6}$$

2.3   *Laguerre map at different coefficients of variation*

As can be known from the analysis in the last two sections, the average volume of cell body $\bar{\upsilon} = V/N$ and parameters υ_v and $c\upsilon$ are mutually independent, yet the change of sphere υ_v and $c\upsilon$ may influence the standard deviation and geometric characteristics of the cell body volume. Via the large quantity of models generated and the corresponding statistical analysis, we can obtain the correspondence among the υ_v, $c\upsilon$ and the characteristic parameters of cell body, Fig. 2 shows us the 2D polycrystalline material microstructure geometric model when volume fraction being set 0.68 at variation coefficients of 0.2, 0.6, 1.0, 1.4 and 1.8 respectively.

3   3D SIMULATION ANALYSIS ON POLYCRYSTALLINE MATERIAL MICROSTRUCTURE

3.1   *Anisotropy analysis on polycrystalline materials*

The microstructure of polycrystalline materials has a great influence on their performance. To predict the microcosmic failure behavior, a major method would be building a relational model

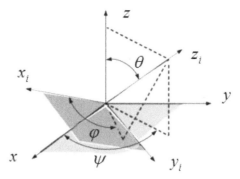

Figure 3. The defined sketch map of crystal orientation under the reference frame and the local coordinates.

between the polycrystalline material microstructure and its performance. At a meso-scale, the microstructure of polycrystalline materials take on the feature of local unevenness and anisotropy, which result in the anisotropy in mechanical properties and mesoscopic stress field of the materials. In order to carry on orientation simulation of the polycrystalline material microstructure's composition, the author applied a random number generator to simulate the orientation distribution of composition in an Euler space according to a specified rule.

For the orientation of crystal grain, we may use a reference frame to adjust one set of coordinate axes thus to determine the rotation of crystal lattice, in the coordinate system, θ is defined as the polar angle, ψ the azimuth angle and φ the rotation angle. As is shown in Fig. 3, the three interval angles are used to describe the spatial orientation of crystal grains, as for the i randomly oriented crystal grain, the relation between the local coordinate $(\alpha_i, \beta_i, \gamma_i)$ and the reference frame (α, β, γ) can be described in formula (7), the rotation matrix $\mathbf{R}$ in formula (7) is a 3*3 matrix comprised of parameters θ, ψ and φ, which is shown in formula (8) as follows:

$$[\cos\alpha_i \quad \cos\beta_i \quad \cos\gamma_i]^T = \mathbf{R}[\cos\alpha \quad \cos\beta \quad \cos\gamma]^T \tag{7}$$

$$\mathbf{R} = \begin{bmatrix} \cos\theta\cos\psi\cos\varphi - \sin\psi\sin\varphi & \cos\theta\sin\psi\cos\varphi + \cos\psi\sin\varphi & -\sin\theta\cos\varphi \\ -\cos\theta\cos\psi\sin\varphi - \sin\psi\cos\varphi & -\cos\theta\sin\psi\sin\varphi + \cos\psi\cos\varphi & -\sin\theta\sin\varphi \\ \sin\theta\cos\psi & \sin\theta\sin\psi & \cos\theta \end{bmatrix} \tag{8}$$

When formula (7) and (8) are made simultaneous, we will realize the effect to describe the orientation of a crystal grain in one set of space coordinates under the global coordinate system, then with the help of randomizer, the crystal orientation simulation in polycrystalline material microstructure can be generated according to a certain statistical regular distribution.

3.2 Model analysis on polycrystalline material 3D simulation result

To realize the 3D simulation image for polycrystalline materials, corresponding design on Laguerre map is essential. By adjusting the sphere packing parameters in specified range, we can have the size of Laguerre unit cell in specified range reach the actual size of crystal grains. Likewise, by adjusting the average crystal size in different area of a representative volume element, we can construct the polycrystalline micro-fluctuation image reflecting local crystal and regional gradient change, and enable it to resemble the geometrical texture of the representative volume element in the polycrystalline matrix; based on the above-mentioned principles and by the control over the weighted point set, we may conduct conversion of coordinates and realize the 3D simulation of polycrystalline material microstructure.

The control over weighted point set can be carried out from two aspects: the control over space position and the distribution setting of weight value size. The space transformation of Laguerre

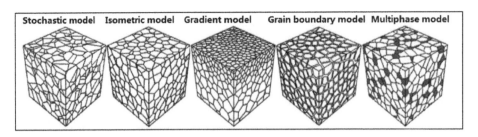

Figure 4.   The 3D simulation results of polycrystalline material microstructure constructed under different model principles.

unit cell is realized by spatial translation of the unit cell vertex or scale-transform; from the above two aspects and the space transformation model of Laguerre unit cell, we can obtain 3D simulation images of polycrystalline microstructure with different features. In this paper, the author presented the model building principles and simulation results for five different polycrystalline material microstructures, namely, the stochastic model, isometric model, gradient model, grain boundary model and the multiphase model.

Stochastic model is a Laguerre map generated from uniform distribution by setting the space position and weight value distribution of weighted points; in isometric model, the sphere packing algorithm is adopted to construct weighted point set, which is further taken as a seed point set of Laguerre map; gradient model is designed by making some grain sizes of polycrystalline materials distributed in gradient along a certain direction; grain boundary model is realized by space scale variation of the crystal vertex according to the average dimension scale of crystal boundary; when the polycrystalline composition is of multiple phases, and the mean crystal size, distribution characteristics and crystallographic orientations of the composition in each phase are also different, which result in the conversion of material performance, then we can resort to multiphase model, the multiphase model is designed with a method of mono-phase polycrystalline material microstructure model fitting to construct multiple groups of sphere collections with specified distribution, then all the spheres are packed to realize control over the space position of spheres, next is the Laguerre map construction to generate corresponding cell body structure, and the last step is unit cell identification, which is used to obtain the corresponding unit cell generated for each group of spheres to serve as each phase of the multiphase polycrystalline material microstructure, upon which the 3D simulation for multiphase polycrystalline material microstructure is thus realized. Fig. 4 shows us the 3D simulation results for the five models mentioned above.

According to Fig. 4, the 3D simulation image generated from the stochastic model has such characteristics as the shape of unit cell being irregular and that the unit cell size can only be measured by mean value, thus it is often restricted by other geometric parameters when used in polycrystalline material structure simulation; the 3D image of the final unit cell form generated from the isometric model may be reckoned that the crystal grains have the same growth velocity at each direction, when combined with the sphere packing point set, we can design the distribution pattern of grain size, numbers of crystal facades, crystal shape as well as other factors, thus realizing the model fitting of polycrystalline material microstructure; as can be seen from the 3D simulation image generated from gradient model, the grain size takes on gradient distribution towards a certain direction; by the 3D simulation image generated from grain boundary model, the scaling operation is completed pursuant to the principle that the crystal position in space differs from the scale mode, and all the geometric models of crystal are thus regenerated; as is shown from the 3D simulation image from the multiphase model, the composition of poly-crystal is comprised by multiple phases and the mean grain size, distribution pattern and crystallographic orientations of each phase are different, the simulation result renders this feature as well.

4 CONCLUSIONS AND REMARKS

1) The author analyzed the different microstructure of polycrystalline materials, built Laguerre models and realized the 3D simulation images for five types of polycrystalline material microstructure with such models and algorithm;
2) The author lucubrated the fitting principles of Laguerre map and Laguerre model, and provided theoretical basis for the 3D reconstruction of polycrystalline material microstructure;
3) Since part of the polycrystalline material grain may take on anisotropy characteristic, the author also provided an analytical approach to anisotropy and proposed the mutual transformation algorithm of rotation matrix against the relative coordinate system and reference frame;
4) In order to make 3D reconstruction for different polycrystalline material microstructures feasible, the author conducted theoretical analysis on stochastic model, isometric model, gradient model, grain boundary model and multiphase model respectively, and then discussed the characteristics of the 3D simulation images generated from the five models of the polycrystalline material microstructures;
5) The purpose of 3D simulation study on polycrystalline material microstructure is to probe into the mechanical response features of materials, upon which the performance of materials can be predicted. The contents discussed in this paper are a basic link, or preparatory link to the mechanical response study, so the author did not analyze the mechanical properties of materials. As is implied in the paper, the author suggests further study should extend to the mechanical property features generated from the polycrystalline material microstructure.

REFERENCES

[1] Cao, Z.Y., et al. (2008) Microelement method for scale-span analyses of materials with microstructure. *Journal on Numerical Methods and Computer Applications*, 29 (3), 186–196.
[2] Zhu, X.Y. (2011) The three-dimensional orientation analysis and development of crystalline material texture. *Xinjiang Youse Jinshu*, (5), 57–59.
[3] Zhang, B., et al. (2013) Polycrystal material microstructure simulation and numerical computation. *Journal of Jilin University*, 43(2), 368–375.
[4] Li, J.C., et al. (2009) Simulation of polycrystal material microstructures and calculation of its orientation distribution function. *Journal of Heilongjiang Institute of Science and Technology*. 19(5), 401–406.
[5] Yang, Z.G., et al. (2009) Computer reconstruction of polycrystalline material microstructure. *Journal of Engineering Graphics*, (1), 125–129.
[6] Liu, Y.Z., et al. (2008) Post-processing of finite element analysis for the 3d microstructure of polycrystalline materials. *Journal of Shandong University*, 38(2), 13–17.
[7] Feng, W., et al. (2009) Mechanistic and multiscale predictions for properties of polycrystalline materials. *Lanzhou: Lanzhou University*.
[8] Sun, Y., et al. (2011) Research progress in internal stress of single crystal materials. *Materials Review*, 25(8), 1–4.
[9] Li, X.H., et al. (2011) Research progress in three-dimensional simulation of material microstructure. *Materials Review*, 25, 245–248.
[10] Ren, H., et al. (2008) Mechanical response computation of three-dimensional polycrystalline material microstructures. *Journal of Lanzhou University of Technology*, 34(1), 1–5.
[11] Chen, B.M., et al. (2004) Programming for computer simulation of materials microstructure. *Journal of Lanzhou University of Technology*, 30(1), 172–174.

Engineering Technology and Applications – Shao, Shu & Tian (Eds)
© 2014 Taylor & Francis Group, London, ISBN 978-1-138-02705-3

Research on application of water conservancy project bidding decision model based on BP neural network

Ling Liu
Institute of Water Conservancy Engineering, Tianjin Agricultural University, Tianjin, China

Huiran Ji
Tianjin Binhai New Area Construction and Transport Bureau, Tianjin, China

Xiaoying Guo & Shuhong Sun
Institute of Water Conservancy Engineering, Tianjin Agricultural University, Tianjin, China

ABSTRACT: Currently bidding is the major way of construction project contracting. Bidding decisions are made by construction companies according to their advantages and characteristics of the target project so as to bid selectively. By this way, it can realize not only saving both human and financial resources but also improving the successful rate of the bid. According to terms of the contract established by Chinese Ministry of Water Resources in 2009 and actual competition environment of the domestic construction market, a water conservancy project bidding decision model is constructed based on BP neural network. It can solve the problem of nonlinear relationship between features and cost of water conservancy and hydropower project, which cannot effectively be solved by conventional bidding decision method. The built model is applied and tested. The results show that the model achieved the desired results and can be applied in the actual bidding decision, which has great realistic meaning for bidding decisions of water conservancy project.

Keywords: BP neural network, water conservancy project, bidding decision, decision model

1 INTRODUCTION

The bidding law of the People's Republic of China is implemented since January 1, 2000, which made project bidding contracting system be a mature and scientific project transaction method [1]. Since China's accession to the WTO, its domestic modernization construction has a rapid development and bidding plays an increasingly important role in its domestic economic activity. Contractors can obtain all kinds of projects by the way of bidding, so bidding decisions' correct or not, relating to their development foreground. If contractors cast every bid, it will inevitably lead to the resource waste of the enterprises, and produce adverse factors on the later development. In bidding for a project, contractor need to consider many factors, it is necessary to deeply research and accumulate large amounts of data and make scientific evaluation and analysis, in order to ensure the correctness of the bid decision. Water conservancy project bidding decision-making refers that the contractors decide whether to participate in a bidding competition through collecting information of this project and research the market actual situation. Before participating, it needs to know the bidding project's actual situation, and combined with the company strength to do the bidding strategy analysis. Conventional bidding decision analysis methods are limited by the problems of complex nonlinear relationship, however, the water conservancy project bidding decision-making based on BP neural network model has many advantages, such as strong learning ability and parallelism. It is an effective method for solving complex nonlinear problems in bidding [2], [3].

1.1 Bidding decision model review

Although bidding system has been a market economic system for more than one hundred years, research on bidding decision-making model is started relatively late, it was not until 1990 that Irtishad Ahmad first proposed a bidding decision model [4]. That model used decision analysis technique by comparing each two factors and obtained the weight of every factor and solved the problem of whether or not to bid. It considered four types of a total of 13 factors. In 2000, M. Wanous proposed a bidding model which based on parameters [5], the model identified affective decision parameters through face-to-face survey method, and applied these parameters to calculate the index rules of bidding, then made decisions. In 2003, Liu Erlie draw lessons from foreign research, based on the fuzzy set theory, proposed a multi-objective project bidding decision-making model [6]. That model can reduce the influence of uncertain factors in the decision making, and the fuzzy set theory conforms to human's thinking way.

1.2 Quotation decision model review

Tender offer theoretical research is started early, which can date back to Emblen's dissertation Competitive Bidding for Corporate Securities in 1944. Firedman proposed Firedman model in Operations Research in 1956. This is the first published tender offer research result. The model based on the hypothesis that each competitor's winning rate is mutually independent and mutually interfering to the contractors [7]. Since then, many scholars did a lot of work on the basis of Fierdman model, and achieved fruitful results. On this basis, the neural network algorithm is gradually applied to the bid decision-making model. In 2006, Mr. Zhu used BP neural network in the quote decision analysis of real estate [8]. In 2007, Chen Shouyu proposed fuzzy optimization neural network weights for BP model [9]. In 2009, Wang Aihua, Sun Jun, based on BP neural network, proposed engineering project management model flowchart in the highway construction project [10]. In 2013, Wang Bo standing on the point of view of the contractors, built the water conservancy project bidding decision model based on BP neural network, applied to the engineering practice and got many achieves.

2 BIDDING DECISION-MAKING EVALUATION INDEX SYSTEM

According to risk analysis in the Water Resources and Hydropower Engineering Construction Standard Bidding Document (2009), and current domestic competition situation of the water resources and hydropower engineering construction market, it is concluded that the main influences on water conservancy project to the contractors for bidding decision-making are: (1) the hardware condition of enterprise itself, (2) the professional qualifications, (3) similar project experience, (4) construction and management level of the enterprise, (5) site of the project, (6) project construction period, (7) project funding sources, (8) type and quantity of all kinds of mechanical equipment in the process of project construction of the. In addition, the strength and their relationship with the owners of competitive enterprises should also be taken into consideration and analysis. The workers' conditions and their familiarity with the project should also be considered as a condition [11], [12].

3 BP NEURAL NETWORK DECISION-MAKING MODEL

3.1 The establishment of BP neural network decision-making model

The water conservancy project bidding decision-making model was established based on BP neural network. Use network function's approximation ability, in view of complex non-linear function in the actual process do the mapping processing. Set risk factors in engineering as the input vector, and standardize processing before input it; all kinds of risk factors as the training sample to train the

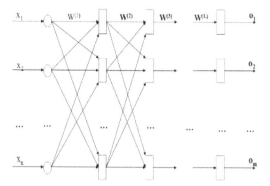

Figure 1. BP neural network structure.

network, the output vectors are corresponding to the final bid decision-making evaluation results. The design of the network model is based on the number of input and output nodes to determine the initial value of the hidden node data, then the network was trained, by increasing the number of hidden nodes more groups of training results, would be obtained then analyzed. The network structure as shown in Figure 1, the $X_1, X_2, \ldots, X_n$ are the input nodes, $O_1, O_2, \ldots, O_m$ are the output nodes, $W^{(1)}$, $W^{(2)}, \ldots, W^{(L)}$ are the connection weights between each layer. In the decision-making model of BP neural network, BP neural network model plays an important role similar as "expert", standardize all risk factor's sample data as the inputs of BP neural network decision model, set the bidding decision result as the output of the BP neural network model. After training, the neural network obtains the subjective judgment and evaluation knowledge to the bid evaluation, evaluation of knowledge, and the tendency to important indexes. The BP neural network has very strong learning ability.

3.2 BP neural network study process

BP neural network learning algorithm used in bidding decision-making is shown in Figure 2.

3.3 Training and testing of BP neural network decision-making model

(1) Forward propagation phase:

A. Take a sample (X_p, Y_p) from the sample set, Y_p is ideal output, X_p is the input of the network;
B. Calculate the corresponding actual output Op:

$$O_p = F_1(\ldots(F_2(F_1(X_p W^{(1)})W^{(2)})\ldots)W^{(L)})$$

(2) Backward propagation stage – error propagation stage:

A. Calculate the error of the actual output O_p and the corresponding ideal output Y_p;
B. Adjust the matrix with the method of minimizing error.
C. Measure the error of the sample p in the network with equation (1):

$$E_p = \frac{1}{2}\sum_{j=1}^{m}\left(y_{pj} - o_{pj}\right)^2 \tag{1}$$

D. Measure the error of the whole sample set in the network with equation (2):

$$E = \sum_{p} E_p \tag{2}$$

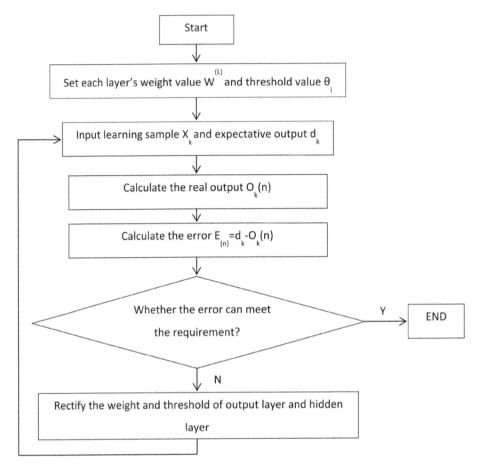

Figure 2. Flow chart of the BP neural network learning algorithm.

Table 1. Project's evaluation value.

| Projects | Basic index | | | | | | | | | | | | | | | | |
	B1	B2	B3	B4	C1	C2	C3	C4	D1	D2	D3	E1	E2	F1	F2	F3	F4
Optimal value	0	0	0	0	55	25	50	60	100	0	100	100	100	100	10	35	100
Worst value	25	45	30	100	0	85	0	0	0	100	0	0	0	0	50	100	0
A	20	24	28	15	30	28	20	18	68	58	50	66	42	65	8	50	64
B	10	23	20	45	50	20	46	31	75	30	80	55	48	68	12	45	68
C	12	22	30	53	45	30	35	30	60	35	45	20	46	31	9	60	35
D	30	42	38	29	35	25	25	25	66	42	65	38	29	35	14	42	45

Adopt the bidding decision-making model established in this paper to do bidding decision-making analysis and evaluation and choose 15 existing hydraulic engineering bidding projects. The previous 10 are training samples, these projects' related data as the input of the network training samples. The following 5 are check samples, these 5 projects' related data as the input of the check sample. Use Matlab software to do learning training of the neural network model. Bidding project evaluation value is as shown in Table 1, in this table, B, C, D, E, F represents the bidding decision-making factors. Through learning training, results are shown in Table 2. Input training sample data into the neural network and complete training, after training the network has higher recognition ability. Then check the network's training supervision with normalized training samples. Through

Table 2. BP neural network model output.

Results	Project 1	2	3	4	5	6	7	8	9	10	11	12	13	14	15
Expert evaluation	0.4201	0.5415	0.6726	0.3021	0.5522	0.7851	0.3219	0.5916	0.6012	0.2835	0.4428	0.3614	0.3972	0.6216	0.4821
actual output	0.4521	0.5308	0.6517	0.3189	0.5505	0.7724	0.3325	0.5823	0.6025	0.2911	0.4403	0.3628	0.3877	0.6311	0.4824

Table 3. Bidding project evaluation value.

Projects	B1	B2	B3	B4	C1	C2	C3	C4	D1	D2	D3	E1	E2	F1	F2	F3	F4
Optimal value	0	0	0	0	55	25	50	60	100	0	100	100	100	100	10	35	100
worst value	25	45	30	100	0	85	0	0	0	100	0	0	0	0	50	100	0
1	20	24	28	15	30	32	12	35	78	20	80	49	40	12	4	69	19
2	24	23	20	50	21	42	25	28	40	18	30	27	11	17	18	48	38
3	12	39	22	53	46	30	18	16	19	21	39	15	17	25	20	45	37
4	30	42	19	29	50	40	36	27	22	10	18	35	10	45	17	22	65
5	9	18	38	55	44	49	40	8	28	35	28	40	18	20	21	18	31
6	10	12	45	69	19	27	11	12	45	69	19	19	21	25	18	10	46
7	15	17	50	48	38	15	17	17	50	48	38	22	10	30	9	21	48
8	18	25	39	45	17	22	35	25	39	45	17	28	35	19	16	11	33
9	21	10	25	20	48	18	31	10	25	20	48	25	39	38	15	17	41
10	23	31	27	25	40	10	9	21	27	25	40	10	25	17	11	35	25
11	28	40	18	30	25	21	16	11	36	27	22	10	61	48	18	31	50
12	16	19	21	39	31	31	15	17	40	8	28	35	9	21	8	12	35
13	27	22	10	18	10	43	30	22	11	12	45	69	16	11	10	34	42
14	8	28	35	16	16	51	35	23	17	17	50	48	42	24	7	26	30
15	16	27	36	21	7	29	41	30	20	55	65	84	68	72	6	12	60

testing, compare and analyze the training results of the structure and actual expert structure, the general error of the model can meet the practical requirements. Results tally with the actual so that the model training is successful.

4 REAL CASE ANALYSES

A construction company pretend to make a biding to one of the four water conservancy project (A, B, C, D), using the method proposed in this paper. First of all, 10 experts with rich experience in some aspect of the bidding are selected from relevant decision making departments of the company to build an expert team. After several rounds of experts' discussion, each evaluation value of each bidding project is identified based on Table 1. The structure of the values is shown in Table 3, where i is assumed as one of the elements in the sample set of the training sample, then the jth fuzzy number of basic index is Z_{ij}. By standardized processing of the sample set of input data, the optimal value and the worst value of the jth basic index can be concluded as Z_{jB} and Z_{jW}. Do standardize processing of each index assessment value of the four projects (A, B, C, D), and then input them into trained BP neural network of the bidding decision-making model. After calculation of the model, the final index assessment value of the four project respectively are $I_A = 0.435$, $I_B = 0.785$, $I_C = 0.546$, $I_D = 0.718$. Therefore, the bidding risk of project B is smaller than others and can be bid. The conclusion is the same as actual decision made by expert's team.

5 CONCLUSION AND PROSPECT

(1) Conservancy project is analyzed and all kinds of typical engineering bidding information are collected in this paper to build the water conservancy project bidding decision model based on

BP neural network. The model can well solve complex nonlinear problems in biddings, which can provide important decision basis for the contractors.

(2) In the process of BP network training, a large number of practical data is very important. In the process of computing, the selection of network parameters and training samples as well as the determination of error value has a magnificent effect on the bidding decision.

(3) The bidding decision is a complicated process involving many factors, so the water conservancy project bidding decision model based on BP neural network, which has strong learning ability, can solve complex nonlinear problems in the biddings well.

(4) The evaluation results of water conservancy project bidding decision model built in the paper and experts evaluation results can match well, which shows the scientific and accuracy of the model.

Outlook: Only the initial weights and threshold of the input are optimized in the paper. The optimization of the network structure and weights should also be discussed in the next step. How to apply data mining technology to the decision-making model is the emphases and difficulties of future research. Neural network technology application in water conservancy project bidding decision model is the development trend of future bidding decisions.

REFERENCES

[1] People's Republic of China Ministry of Water Resources. Wang Yang: Deepen Reform and Improve the Mechanism to Vigorously Promote the Construction of Farmland Water Conservancy. Central Government Portal, 2013-10-24.

[2] Gao, J.J. (2013) Applied Research on the Model of Artificial Neural Network to Quickly Estimate the Project Cost of Highway. Shandong Jianzhu University, 24–39.

[3] Wang, B., Dun, X.C. & Li, Z.Y. (2013) Water conservancy projects bidding decision model and its application based on BP neural network. *Water Resources and Power*, 31(03), 131–134.

[4] Irtishad, A. (1990) Decision-support system for modeling Bid/No-Bid Decision problem. *Journal of Construction Engineering and Management*, 116(4), 457–466.

[5] Waous, M., Boussbaaine, A.H. (2000) To bid or not to bid: a parametric solution. *Construction Management & Economies*, 18(4), 457–466.

[6] Liu, E.L., Wang, J., & Luo, G. (2003) Project bidding decision based on fuzzy logic. *China Civil Engineering Journal*, 36(3), 57–63.

[7] Friedman, L.A. (1956) Competitive bidding strategy. *Operation Research*, 82(4), 104–112.

[8] Zhu, M.Q. (2006) The application of BP neural network in real estate investment risk analysis. *Sichuan Building Science*, 32(6), 243–246.

[9] Chen, S.Y. (1997) Multi-objective decision-making theory and application of neural network with fuzzy optimum selection. *Journal of Dalian University of Technology*, 37(6), 693–698.

[10] Wang, A.H. & Sun, J. (2009) Application of BP neural network in construction project management. *Construction Management Modernization*, 23(4), 304–309.

[11] People's Republic of China Ministry of Water Resources. The Current Situation of Water Conservancy and "Twelfth Five" Plan of Water Conservancy. Central Government Portal, 2011-10-12.

[12] The CPC Central Committee and State Council. The Decision on Speeding Up the Reform and Development of the Water Conservancy. Central Government Portal, 2010-12-31.

Engineering Technology and Applications – Shao, Shu & Tian (Eds)
© 2014 Taylor & Francis Group, London, ISBN 978-1-138-02705-3

The influence of calcium content on hydrogen chloride emission during Refuse Derived Fuel combustion

Mingming Li, Yue Zhao, Wenbin Pei, Xiaoqiang Liu & Wendong Ji
Tianjin Research Institute for Water Transport Engineering Tianjin, China

ABSTRACT: The chlorinated substances in Refuse Derived Fuel (RDF) will produce hydrogen chloride (HCl) gas during combustion. The HCl gas is not only harmful to human health, but corrodes furnace and boiler tube. In this study, quick lime (CaO) was added to RDF that will be incinerated to produce HCl and the influence of CaO in generation of HCl was investigated. During experiments, CaO added RDF was incinerated in the tube furnace and the contained in flue gas was measured by titration. The titration according to ASTM: E 776-87, Standard Test Method for Forms of Chlorine in Refuse-Derived Fuel. The results of the experiment about HCl emissions during RDF combustion showed that CaO can be control the HCl emissions. The removal efficiency of the HCl using CaO addition was approximately 60%–70%.

Keywords: Refuse Derived Fuel; hydrogen chloride; quick lime; tube furnace

1 INTRODUCTION

The disposal of municipal solid waste (MSW) has been change dgradually from landfill to incineration due to its advantages of volume reduction, thermal energy recovery and others. However, power generation efficiency in incineration of MSW is low and limited by incombustible components.

Refuse Derived Fuel (RDF) has a higher heating value and the better combustion performance than MSW due to a series of treatment processes, such as size reduction, separation, drying and densification. With higher heating value in the RDF product, better energy recovery efficiency could also be achieved from the engineering perspective[1].

One of the problems of the RDF is its high content of chlorine, both inorganic (i.e. sodium chloride: NaCl) and organic (i.e. Polyvinylchlorid: PVC). The raw gas during RDF combustion would contain not only the typical impurities of tar and NH_3 but also some chlorine both inorganic (HCl and chlorine gas) as organic (chlorinated organics). Chlorine reacts with other elements to form toxic materials that react with hydrogen to form HCl, or react with metals to form metal chlorides[2]. The calcium can be effected on HCl emission during RDF combustion.

So HCl must be removed from flue gases before its emission into the atmosphere, because, it is one of the most troublesome substances among acidic gases. That is why HCl is the most harmful emissions of RDF combustion.

The chlorine in RDF can be formed to HCl during combustion. CaO is a kind of sorbent that used to control the HCl emission. The investigation on the relationship between CaO addition and HCl emission during combustion is the mainly objective in this paper. So this paper has made a investigation about the influence of calcium content on HCl emission during RDF combustion. Based on the experimental, we can found out the relationship between CaO addition ratios and HCl emission during combustion. The experimental of the influence of combustion air supply ratios on HCl emission also had been tested to found out the relationship between them.

2 THEORY

2.1 *Generation of HCl by combustion*

The chlorinated substances in waste will produce HCl gas during combustion. The HCl gas will corrode furnace and the HCl gas will be formation of acid rain when it emissions in the air.

There are some papers about the characteristics of PVC thermal decomposition or combustion generates HCl and the characteristics of the generation of HCl by NaCl. According to the research, the emissions of HCl mainly come from the decomposition of the organic chlorine in waste. But also there are much of HCl be generated during the combustion of the inorganic chlorine in waste. Some papers show that the optimum temperature of the chlorine's removing by calcium is 873–973K. And calcium will be removing the chlorine during the combustion of RDF.

The main source of HCl is the NaCl and PVC in waste. The main reaction of NaCl is[3]:

$$NaCl + H_2O \rightarrow NaOH + HCl \tag{1}$$

$$2NaCl + H_2O + SO_2 \rightarrow Na_2SO_3 + 2HCl \tag{2}$$

$$2NaCl + H_2O + SO_3 \rightarrow Na_2SO_4 + 2HCl \tag{3}$$

$$2NaCl + H_2O + SiO_2 \rightarrow Na_2SiO_3 + 2HCl \tag{4}$$

The main reaction of PVC is:

$$PVC \Rightarrow L + HCl + R + HC \tag{5}$$

where L is condensable organics, R is solid coke, HC is volatile organic compounds. In the case of full combustion, the 50%–60% of NaCl in waste will be translated into HCl. The inorganic chlorine is solid form as NaCl when temperature below than 360. But when °C temperature higher than 360°C, NaCl and the other elements begin to generating $Al_2O_3 \cdot Na_2O_6 \cdot SiO_2$[4]. So the inorganic chlorine will be translated into HCl gas. According to the research, the conversion rate of NaCl translated into HCl is significantly increased by the presence of the Al_2O_3 and SiO_2.

The main reaction is:

$$2NaCl + H_2O + mSiO_2 \rightarrow Na_2O \bullet mSiO_2 + 2HCl \tag{6}$$

But when temperature higher than 800, HCl concentration started to 800°C decreased and gaseous NaCl concentration started to decreased, The reason is the reaction of NaCl translated into HCl gas will be adversed when temperature higher than 800 and $\Delta G > 0$.

The main reaction is[5]:

$$4NaCl(g) + 2SO_2(g) + O_2(g) + H_2O(g) \rightarrow 2NaSO_4(cr,l) + 4HCl(g) \tag{7}$$

2.2 *General chemistry of CaO and HCl reaction*

The most often used solids for HCl capture is CaO[6~9]. In this paper, CaO had been used as the sorbent that captured HCl during the RDF combustion. CaO has also been shown to be useful for in bed HCl capture in coal gasification plants[10] and in steam gasification of PVC waste in fluidized bed[11]. CaO react easily with HCl forming $CaCl_2$ with melting points 772°C[12].

The main reaction is:

$$CaO(s) + 2HCl(g) \leftrightarrow CaCl_2(s) + H_2O(g) \tag{8}$$

$$CaCl_2 + CO_2 + H_2O \leftrightarrow CaCO_3 + 2HCl \tag{9}$$

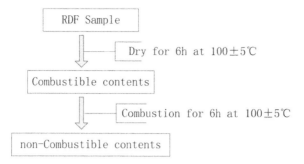

Figure 1. Proximate analysis of RDF.

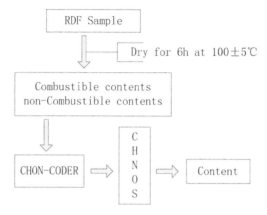

Figure 2. Ultimate analysis of RDF.

3 EXPERIMENTAL METHODS

3.1 *Proximate and ultimate analysis*

RDF samples were analyzed to investigate the properties as well as the characteristics of fuel. The proximate analysis was performed to find out the moisten contents, combustible contents, and non-combustible contents. The proximate analysis process is shown at Fig. 1.

The chemical combustion was also tested using CHDN-CODER. The ultimate analysis process is shown at Fig. 2.

3.2 *Calorific value*

The calorific value of RDF was measured as if bomb calorific water. The calorific value measured by bomb calorific meta was a higher heating value that does not consider latent heat of water. From high heating value measured, lower heating value is calculated by following equation.

$$LHV = HV - 600(W + 9H) \qquad (10)$$

where LHV = Lower heating value, HHV = High heating value, W = Water content, H = Hydrogen content.

3.3 *Samples preparation*

These types of samples were tested to investigate the composition as well as calorific value.

Table 1. RDF samples preparation.

Sample Type	Property	Source
A Type	RDF without organic	W city
B Type	RDF with organic	S landfill
C Type	RDF with CaO addition for combustion experiment	A Type

A TYPE	B TYPE	C TYPE

Table 2. CaO addition in RDF.

Experiment Number	Experiment Number	CaO Addition (CaO-g/RDF-g)	Ca/Cl Ratio (CaCl$_2$ Base)
1	Sample 1 & 2	0% (*0.37%)	0 (0.36)
2	Sample 3 & 4	+1%	0.98
3	Sample 5 & 6	+2%	0.98
4	Sample 7 & 8	+3%	2.94

Note: *0.37% of Ca (CaO-Base) is already contained in the RDF Addition of CaO as free Ca to removal HCl.

Table 3. Combustion air supply.

	Excessive air ratio
Condition #1	1.5
Condition #2	2.5

3.4 *Experimental condition*

CaO was used as the sorbent for binding HCl and the effect of various molar ratios of Ca/Cl on HCl emission was investigated. The CaO addition samples were shown at Table 2.

The amount of combustion air were also changed were shown at Table 3.

4 RESULTS AND DISCUSSION

4.1 *Results of proximate analysis ultimate analysis of RDF*

The proximate analysis and ultimate analysis data are shown at Table 4 and Table 5.

4.2 *Thermal Gravity Analysis of RDF*

The RDF type A and type B was tested for Thermal Gravity Analysis (TGA) and the results are shown in Fig. 3.

Table 4. Approximate and ultimate analysis of W City RDF.

	Sample	RDF
Approximate analysis (wt%)	Moisture	7.4–8.5
	Volatile matters	67.2–72.4
	Fixed carbon and ash	18.7–22.6
Ultimate analysis [wt%]	Carbon	38.9–46.7
	Nitrogen	2.6–2.8
	Hydrogen	4.2–6.5
	Oxygen	20.7–26.1
	Sulfur	0.12–0.21
	Chlorine	0.43–0.97

Table 5. Approximate and ultimate analysis of W City RDF.

	Sample	RDF
Approximate analysis [wt%]	Moisture	4.4–5.5
	Volatile matters	75.2–81.4
	Fixed carbon and ash	16.7–17.6
Ultimate analysis [wt%]	Carbon	59.3–67.1
	Nitrogen	1.6–1.8
	Hydrogen	4.8–7.6
	Oxygen	12.7–18.5
	Sulfur	0.11–0.19
	Chlorine	0.23–0.57

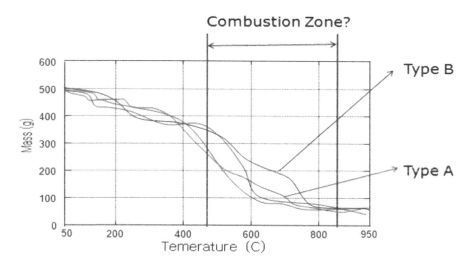

Figure 3. Thermal Gravity Analysis of RDFs.

According to the above TGA results, the type B RDF was pyrolyzed at lower temperature than type A RDF. It may be implied that the type B RDF contents high plastics that has high pyrolytic temperature.

For type A RDF, most of the sample was pyrolyzed temperature higher than 700°C. It is mainly due to the biogenic contents. The type B sample, however, needed to temperature higher than 750 to complete the pyrolysis.

On the basis of TGA test, the testing combustion temperature was set to 950°C.

Table 6.　Approximate and ultimate analysis of CaO addition RDF.

	Sample	RDF
Approximate analysis (wt%)	Moisture	7.4–8.5
	Volatile matters	77.2–80.1
	Fixed carbon and ash	13.6–16.2
Ultimate analysis (wt%)	Carbon	61.4
	Nitrogen	1.6
	Hydrogen	7.1
	Oxygen	18.3
	Sulfur	0.15
	Chlorine	1.2

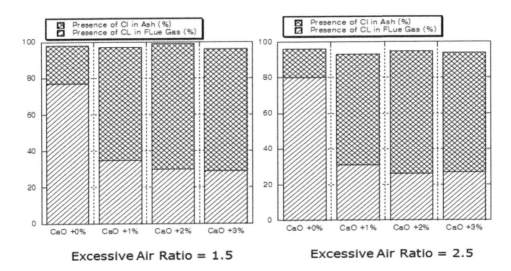

Figure 4.　HCl proportion when excessive air ratio = 1.5 and 2.5.

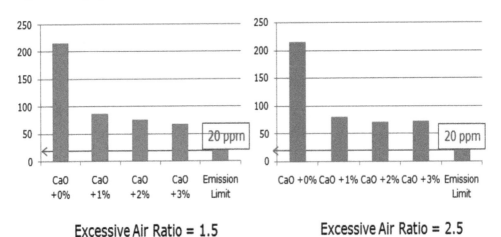

Figure 5.　The influence of CaO addition ratio on HCl emission concentration when excessive air ratio = 1.5 and 2.5.

4.3 Test results of CaO addition RDF

To investigate the influence of CaO addition of RDF, the CaO addition ratio were carried, as well as the excessive air ratio. In Table 6, the composition of tested RDF was illustrated.

Based the results, it was found that the difference between the influence of air supply ratio on HCl captured was almost same. So excessive air ratio did not influence the HCl generation.

It was found that the maximum HCl concentration is 210 ppm when CaO addition ratio is 0%. The HCl concentration, when the CaO addition, is about 60–80 ppm. And the limit emission is 20 ppm. The HCl emission can be reduced by addition of CaO by 65%. It means that CaO addition can be decreased the HCl concentration efficiency.

Based the results, it was found that the HCl concentration is 70–80 ppm when CaO addition ratio is 1%. But the HCl concentration, when CaO addition ratio is 3%, is 60–70 ppm. The difference between 70–80 ppm and 60–70 ppm is not significantly. So increase of CaO addition did not escalate the HCl removal efficiency.

5 CONCLUSION

The results of the experiment about HCl emissions during RDF spiked of CaO-based indicated that CaO-based can be control the HCl emissions. It was not possible to reduce the flue gas HCl concentration less than emission standard of China. The removal efficiency of the HCl using CaO addition was approximately 60%–70%. Based on the results of experiments, it may be told that the chlorine content would be less than 0.4% to keep the emission standard using CaO addition.

REFERENCES

[1] M. Kondoh, R. Yamazaki & S. Mori. (2000) Combustion test of refuse derived fuel in a fluidized bed. *Waste Management*, 20, 443–447.
[2] B. Courtemanche & Y. Levendis. Control of the HCl emission from the combustion of PVC by in-furnace injection of calcium magnesium basedsorbent. *Environ Eng Sci*, 15, 123–135.
[3] G. Piao, S. Aono, S. Mori, et al. (1998) Combustion of RDF in a fluidized. *Waste Management*, 18, 509–512.
[4] H.M. Zhu, X.G. Jiang, J.H. Yan, Y. Chi & K.F. Cen, TG-FTIR analysis of PVC thermal degradation and HCl removal. *Journal of Analytical and Applied Pyrolysis*, 82 (2008), 1–9.
[5] G. Mura & A. Lallai. Analysis of multi-target orthogonal experiment. *Chemical Engineering Science*, 47(9–11), 2407–2411.
[6] A.D. Lawrence & J. Bu. (2000) The reactions between Ca-based solids and gases representative of those found in a fluidized-bed incinerator. *Chemical Engineering Science*, 55, 6129–6137.
[7] B.K. Gullet, W. Jozewicz & L.A. Stefanski. (1992) Reaction kinetics of calcium-based sorbents with hydrogen chloride. *Industrial and Engineering Chemistry Research*, 31, 2437–2446.
[8] J. Abbasian, J.R. Wangerow & A.H. Hill. (1993) Effect of HCl on sulfidation of calcium oxide. *Chemical Engineering Science*, 48, 2689–2695.
[9] W. Jozewicz & B.K. Gullet. (1995) Reaction mechanisms of dry Ca-based sorbents with gaseous HCl. *Industrial and Engineering Chemistry Research*, 34, 607–612.
[10] W. Li, H. Lu, H. Chen & B. Li. (2005) The volatilization behavior of chlorine in coal during its pyrolysis and CO_2-gasification in a fluidized bed reactor. *Fuel*, 85, 1874–1878.
[11] M.J.P. Slapak, J.M.N. van Kasteren. & A.A.H. Drinkenburg. (2000) Design of a process for steam gasification of PVC waste. *Resources Conservation and Recycling*, 30, 81–93.
[12] J. Delgado, M.P. Aznar & J. Corella, Calcined dolomite, magnesite, and calcite for cleaning hot gas from a fluidized bed biomass gasifier with steam: life and usefulness. *Industrial and Engineering Chemistry Research*, 35, 3637–3643.

Engineering Technology and Applications – Shao, Shu & Tian (Eds)
© 2014 Taylor & Francis Group, London, ISBN 978-1-138-02705-3

Temperature control in quick construction of high roller compacted concrete dams

Xubin Du
Institute of Water Conservancy Engineering, Yangling Vocational & Technical College, Xianyang Shaanxi, China

Bo Chen
Xianyang Urban Planning and Design Institute, Xianyang Shaanxi, China

ABSTRACT: Most concrete thermal parameters used for the temperature field simulation of RCC dams are derived from empirical formulae and costly experiments. This paper presents an inverse thermal problem solution based on Genetic Algorithm (GA), which builds an inverse temperature field analysis model according to the characteristics intrinsic to inverse temperature field analysis of RCC dams, defines the thermal parameters to be inverted, and establishes an optimized program based on the optimization of Genetic Algorithm. The temperature field calculated by this feedback calculation is compared with that calculated by variable tolerance inversion. The result indicates that the temperature values calculated with thermal parameters inverted by Genetic Algorithm provides higher agreement with the real measurements; the temperature stress field resulted from these values is able to reflect the real stress conditions of the dam, thereby making it possible to guide temperature control and predict future variation tendency.

Keywords: temperature control; Genetic Algorithm; inverse problem

1 INTRODUCTION

Presently, temperature field simulation for mass concrete is exposed to deviation between the calculated and the actual temperatures. How to select the right temperature characteristic parameters constitutes one of the most critical problems for temperature calculation. The proper selection of parameters directly decides the precision and success of temperature and stress calculation. The complexity of external environmental conditions of a structure during construction and operation makes it very difficult to select the right calculation parameters, most of which are derived from empirical formulae and experiments. As the laboratory is significantly different from the real environment at the locality of the project, parameters so derived often differ widely from the real project[1–6].

Temperature characteristic parameters closely related to temperature include adiabatic temperature rise parameter θ, temperature diffusivity and surface heat transfer coefficient β. These parameters are generally derived by experimenting with a specially designed adiabatic rise measuring device which is costly, highly professional and unavailable for many professional laboratories in China. Furthermore, as the laboratory differs from the real environment at the site, and the adiabatic temperature rise in concrete as well as its regularity is closely related to the variety and proportion of cement used, the variety and proportion of mixed material used in the concrete and the temperature in the concrete itself, concrete prepared with different batches of cement could differ from each other and come of the differences cannot be ignored in calculation, the temperature rise regularity obtained by experiments does not always reflect the reality of the project but rather, leads to errors in the calculation or even misleads the design and construction of the project.

To enable simulation analysis to better reflect the reality of the project, some of the site temperature measurements as well as easily available parameters that are precise enough for the project (e.g. specific heat, density) may be used and inverted to derive thermal parameters of a specific project under different environmental conditions. After more reliable calculation parameters are established by field inverse analysis, feedback analysis may be conducted on the design or construction to determine the operation state of the structure and predict its future variation tendency. This could guide subsequent construction activities, provide more scientific, effective construction measures, serve the real project timely and correctly and form a highly functional, complete "closed-circuit" system comprising planning and design – construction – operation monitoring and inverse feedback. In a word, field inverse analysis of concrete temperature field is not only cost saving, but also derives fairly reliable thermal parameters. It is both economically valuable but also has positive implications on engineering construction.

This paper presents an optimization of Genetic Algorithm based on the essentials of inverse analysis and develops a 3D finite element inverse program for the temperature field of roller-compacted concrete (RCC) dams with a view to deriving more reliable thermal parameters that are more close to the construction reality. It also validates the reliability of this 3D finite element inverse program for the temperature field of RCC dams with a real project.

2 MODELING

2.1 Building a temperature field model

The 3D nonsteady temperature field thermal conductivity is expressed as:

$$\frac{\partial T}{\partial \tau} = \alpha \left(\frac{\partial^2 T}{\partial x^2} + \frac{\partial^2 T}{\partial y^2} + \frac{\partial^2 T}{\partial z^2} \right) + \frac{\partial \theta}{\partial \tau} \tag{1}$$

Initial conditions: The temperature field is a known function $T_0(x, y, z)$ of the known function of coordinates (x, y, z), i.e.:

$$T(x, y, z, t) = T_0(x, y, z)$$

First boundary conditions: $T(\tau) = f_1(\tau)$

Second boundary conditions: $\lambda \frac{\partial T}{\partial n} = f_2(\tau)$

Third boundary conditions: $-\lambda \frac{\partial T}{\partial n} = \beta(T - T_a)$

where: θ – Adiabatic temperature rise of RCC;
α – Thermal diffusy of RCC;
n – Surface external normal direction;
λ – Coefficient of thermal conductivity;
β – Surface heat transfer coefficient;
T_a – Air temperature.

Given the initial conditions and boundary conditions, when all the RCC thermal parameters are known, this expression forms a forward problem of 3D stochastic temperature field. When the material thermal parameters are known, from additional conditions:

$$T\left(x_i, y_i, z_i, \tau_j\right) = T_m\left(x_i, y_i, z_i, \tau_j\right) \qquad (x_i, y_i, z_i, \tau_j) \in \partial\Omega \tag{2}$$

where: T is the calculated temperature at the measuring point; T_m is the measured temperature at the measuring point; M is the number of measuring points; N is the number of observations at the measuring point.

This constitutes an inverse problem for identifying 3D transient heat transfer thermal parameters.

2.2 *Objective function*

In our study, the solution of the inverse problem is defined by nonlinear optimal constraint optimization control. Generalized least square estimation is used. The square of the error between the temperature calculated by finite element and the measured temperature is used as the objective function for parameter inversion, i.e.:

$$J(X) = \sum_{i=1}^{M} \sum_{j=1}^{N} \omega_{ij} \left(T(x_i, y_i, z_i, \tau_j) - T_m \left(x_i, y_i, z_i, \tau_j \right) \right)^2 \tag{3}$$

where: ω_{ij} – Weighted coefficient of measuring point i at time j, taken as 1 herein;
    M – Number of measuring points;
    N – Number of observations at a measuring point;
    X – Design variable, i.e. $[\theta_0, n, \alpha, d]$.

From this, a mathematic model for inverse problem of RCC temperature field can be established: Look for the design variable X, fulfill the constraint condition (2) and makes the value in (3) the smallest.

3 OPTIMIZATION

3.1 *Preprocessing system*

The principal functions of a preprocessing system are to generate the calculation information documents necessary for forward and inverse analysis of temperature field and the mapping information documents necessary for grid display and post-processing. Calculation information documents includes two parts: (1) unit information, base control data (e.g. poured layer thickness, horizontal joint interval), material division and material thermodynamic parameters, construction information, air temperature and water storage information, water pipe cooling and surface heat insulation. Fig. 1 shows the structural chart of the preprocessing module of the inverse temperature field analysis module.

3.2 *Optimization process*

Take the optimization program of Genetic Algorithm as the main program. Read in the calculation information documents generated in the preprocessing system and form the initial population

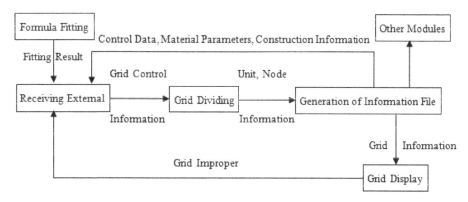

Figure 1.   Structural chart of the preprocessing module.

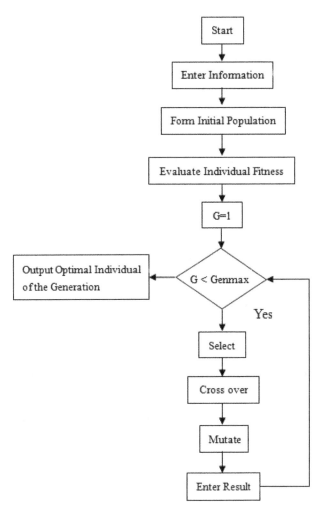

Figure 2. Flowchart of the optimization system.

randomly. By using the temperature field forward operator program, calculate the fitness value of each individual in the population, evaluate these individuals and record the best one. Judge the terminal conditions. If these conditions are met, jump out of the circulation and output the best individual. Otherwise select, cross over and mutate to obtain a new generation of population and start a new round of individual evaluation. Select the best individual of this generation and compare it with the recorded best individual. If it is better than the recorded best individual, take it as the recorded best individual. Replace the worst individual in this generation of population with the recorded best individual. Now we have obtained the second generation of population. Repeat until the terminal conditions are fulfilled. Fig. 2 shows the flowchart of the optimization system.

4   CASE VALIDATION AND ANALYSIS

To validate the correctness and superiority of the Genetic Algorithm program, apply it into the shape optimization design of an RCC dam, conduct optimization using Complex Method, and then compare the optimization results of these two methods.

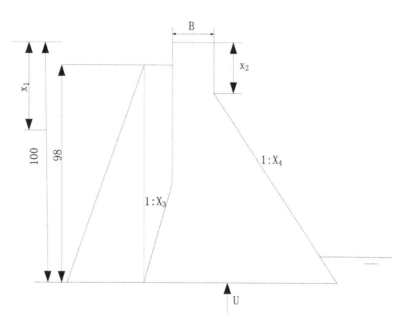

Figure 3.   Shape optimization model for gravity dams.

4.1 *Base data*

In the case of a non-overflow section of a solid concrete gravity dam, the crest height is 100 m; the design water level is 98 m; the downstream tailrace level is 10 m; the crest width is taken as 9 m to allow for traffic and equipment placement; the uplift reduction coefficient is 0.24; the concrete bulk density is 24.5 kN/m³; the water bulk density is 10 kN/m³. The foundation plane concrete to bedrock shear friction coefficient $f = 0.70$. Constraint conditions are as follows:

(1) Constraint of the upstream dam slope ratio: $0 \leq m \leq 0.2$
(2) Constraint of the downstream dam slope ratio: $0.5 \leq n \leq 0.9$
(3) Height of upstream slope-break to dam height ratio: $0 \leq h_1/\text{H} \leq 0.2$
(4) Height of downstream slope-break to dam height ratio: $0 \leq h_1/\text{H} \leq 1$
(5) Heel stress $\sigma_u = (\sum W/B + 6\sum M/B^2) \geq 0$
(6) Toe stress $\sigma_d = (\sum W/B - 6\sum M/B^2) \leq [\sigma_d]$
(7) $k = (f\sum W/\sum P) \geq [K_c]$

where: $[K_c]$ – Permissible slip-proof force; $[\sigma_d]$ – Permissible dam foundation pressure stress.

4.2 *Optimization model*

Fig. 3 shows the optimization model for gravity dams.

Design variables: x_1 – Height from starting point of upstream dam slope to dam crest; x_2 – Height from starting point of downstream dam slope to dam crest; x_3 – Upstream dam slope ratio; x_4 – Downstream dam slope ratio.

As the cost of a gravity dam is primarily decided by the dam concrete cubage, the dam concrete cubage per unit length is taken as the objective function. $V(\{X\})$, then look for design variables that fulfill the constraint conditions of ①–⑦ so that the objective function: $V(\{X\})$ is the smallest.

During the optimization, Limit State Design Method of Partial Factors is used. The slip stability and toe compression strength constraints are calculated for the ultimate limit state (ULS). The heel stress constraint is calculated for normal limit state.

Table 1. Comparison of optimization results.

Solution	Design variable				Stress (kPa)		Slip-proof force	Objective function (m^3)
	x1	x2	x3	x4	Heel	Damsite		
Complex Method	20.8	16.1	0.16	0.61	0.18	208.0	1.2	3522.1
GA	18.2	23.4	0.19	0.63	0.97	201.4	1.1	3445.8

4.3 Optimization results

Shape optimization design is conducted for this gravity dam using Genetic Algorithm at the initial number of populations of pop = 30, crossover probability of $p_c = 0.90$, mutation probability of $p_m = 0.1$ and maximum evolutional algebra of $T = 70$. Table 1 compares the optimization results of the two methods.

From this table, the objective function derived by Genetic Algorithm is: $V(\{X\}) = 3445.977$, which is 75.429 m^3 or 2.14% less than that derived by Complex Method, and is therefore more economic. Compared with Complex Method, the upstream dam slope ratio of the section is closer to the critical value, the downstream dam slope ratio is larger; the toe pressure stress is reduced by about 7 kPa, the heel stress reserve is larger; the slip-proof force is closer to the permissible level. This suggests that the optimization result of Genetic Algorithm is closer to the optimal global solution.

5 CONCLUSIONS

The high non-linearity and unfitness of inverse problem of RCC concrete dam temperature field make it unrealistic and impossible to obtain its gradient information.

In this paper, Genetic Algorithm is used to search for the solution space directly, for which only the value information of the objective function is needed instead of any high-valence information like gradient. Furthermore, Genetic Algorithm also bypasses the unfitness intrinsic to the inverse operation of operators involved in inverse problem solution and opens up a new path for solving the unfitness of inverse problem.

This paper establishes an optimization program for Genetic Algorithm based on the essentials of Genetic Algorithm to optimize the shape of a non-overflow section of a solid concrete dam, and compares its optimization result with that of Complex Method to validate the reliability of the Genetic Algorithm program and the superiority of Genetic Algorithm.

ACKNOWLEDGEMENT

This work is supported by the Research on Temperature Control Method of Roller Compacted Concrete Dams (GN: A2012005); 2012 Scientific Research Fund Program of Yangling Vocational & Technical College (Director: Du Xubin).

REFERENCES

[1] Miguel, C., Javier, O. & Toma's, P. (2000) Simulation of construction of RCC dams. I: Temperature and aging. Asce, *Journal of structural Engineering*, 126(9), 1053–1061.
[2] Mei, F.L. & Zeng, D.S. (2002) New method for calculation of transient temperature field in concrete structure. *Journal of Hydraulic Engineering*, (9), 74–76.
[3] Zhu, B.F. (2010) On pipe cooling of concrete dams. *Journal of Hydraulic Engineering*, 41(5), 505–513.

[4] Zhu, B.F. (2003) The equivalent heat conduction equation of pipe cooling in mass concrete considering influence of external temperature. *Journal of Hydraulic Engineering*, (3), 49–54.

[5] Huang, G.Y., Liu, W.Q. & Liu, X.J. (1993) Inverse problems and computational mechanics. *Computational Structural Mechanics & Application*, (8).

[6] Liu, Y.X., Wang, D.G., & Zhang, J.L., et al. (2000) Identification of material parameters with gradient-regularization method. *Chinese Journal of Computational Mechanics*, 17(1).

Engineering Technology and Applications – Shao, Shu & Tian (Eds)
© 2014 Taylor & Francis Group, London, ISBN 978-1-138-02705-3

Mechanical design of railway bed cleaning based on the virtual prototype technology of engineering machinery

Chen Chen & Chunhua Li
Baoding Vocational and Technical College, Baoding, Hebei, China

ABSTRACT: Railroad is the lifeblood in a nation's transportation and communications, therefore regular maintenance is essential for the railroad departments, among which sand cleaning shall be addressed on a strict basis. Since the accumulative sand in railway is hard to clean up, more appropriate and efficient machines are required to support the cleaning operation. Taking such concern into account, the author redesigned the sand cleaning machinery for railway bed in this paper. Based chiefly upon the analysis over the key principles of virtual prototype technology in engineering machinery, the author designed a railway bed sand cleaning machine, and conceived a set of overall design proposal for sand cleaning machinery as reference.

Keywords: engineering mechanical virtual prototype; mechanical design and manufacturing; railway bed cleaning; mechanical and electrical engineering

1 TECHNICAL ANALYSIS OVER ENGINEERING MECHANICAL VIRTUAL PROTOTYPE

The general technology of virtual prototype system proceeds from the overall situation and addresses the systematic problems, it also takes the various parts of the virtual prototype into consideration, and prescribes and coordinates the operation of each subsystem, which is then formed into an organic entity to realize information and resource sharing and ultimately, the overall goal.

Virtual prototype technology is a brand-new product development method and a digital design approach based on computer simulation model of product. Combined with the development models from different engineering fields, these digital models can "simulate the real products in appearance, functions and behaviors". To some degree, the virtual prototypes of complex products are the additional products generated in the development process of such products; they are the representation of some or the entire appearance, functions, performance and behaviors of products.

1.1 *The structure of engineering mechanical virtual prototype*

The earlier product virtual prototypes were built on the basis of CAD model and VR technology, they attached importance to the simulation from the aspects of appearance, spatial relationship, dynamics and kinematics etc., and emphasized the operation of models by users and the interaction with users from various perspectives so as to make qualitative judgment over the behavior and characteristics of products. The German Furlong Hoff Computer Graphic Institute provided earlier reference architecture of virtual prototype (see Fig. 1).

With the development and application of concurrent engineering, the collaborative design supporting CVP application emerged. CVP is a new IPPD-based design/development norm, and aims to build IPT key enabling technologies between enterprises and the government, it's chiefly constituted by the sub-layer support function, the interactive integration framework as well as the IPT member interaction service layer provided by the upper layer (see Tab. 1).

CVP enables the IPT members to realize concurrent interactions via the digital model and data interchange, and accomplish the test, experiment and appraisal over the function, performance and

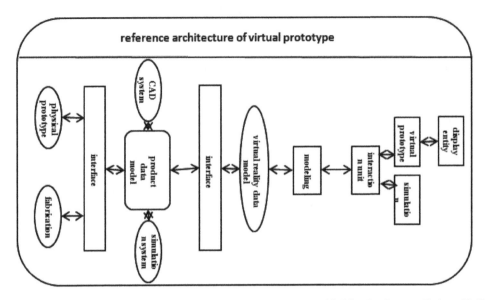

Figure 1. The earlier reference architecture of virtual prototype provided by the German Furlong Hoff Computer Graphic Institute.

Table 1. CVP architecture.

	User interface			
Service	IPT and the tool integration	Product and the process (quasi-real time interaction)	Functional performance & other analysis assessment application	Product and process model and data generation
Integration framework	Product and process data management Object management and information sharing Distributed computing environment and network			

process before the real system being constructed; In addition, it will pay more attention to the economy, supportability and maintainability over the total-life cycle, and accelerate the production process through VM.

1.2 *The co-simulation techniques of engineering mechanical virtual prototype*

The current simulation chiefly focuses on the machinery, control, electron or other individual fields, there are little or no interrelations mutually, and as a result, the need to satisfy the overall design of the engineering machinery system can be hardly met. However, the development process of engineering machinery may involve multiple disciplines and techniques such as electron, machinery, control, hydraulic pressure and computer soft/hard wares, to name a few; and product itself is also a unified entity, composed by many components, parts and subsystems; furthermore, even each component or part would be again made up of a number of components and parts, and every subsystem is, to different degrees, connected with each other, thus it's very difficult to carry on a complete and accurate simulation analysis over the products with complex structure via one single simulation. Therefore, we need to conduct corresponding simulation analysis over the entire system, which is composed by many subsystems from many different disciplines and fields, namely, to carry out multi-disciplinary co-simulation, thus to enable an overall appraisal and analysis on engineering machinery products at the product design phase.

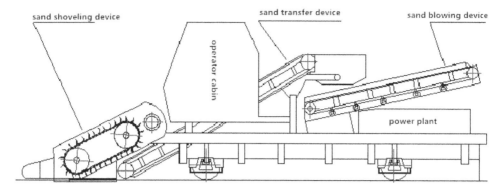

Figure 2.   The sketch map of machinery principle and planning for railway bed cleaning machine.

We need, more often than not, to adopt different simulation software to start modeling for the simulation of complex products. The interaction of the generated models from different software may be very close; the output of one model is likely to be the input of another. During the simulation run process, the different models generated from different software realize information exchange via the step size in simulation discrete time and the cross-process communications, then they use their respective equation solvers to obtain the results so as to realize the simulation of the whole system, that's what people defined "co-simulation".

2   MECHANICAL DESIGN OF RAILWAY BED CLEANING BASED ON THE VIRTUAL PROTOTYPE TECHNOLOGY OF ENGINEERING MACHINERY

2.1   *Principle of design*

The overall principle and conceptual design sketch map for the targeted large-size railway bed cleaning machine is shown in Fig. 2.

The designed sand cleaning machine is made of the vehicle body, and attached power plant (prime engine), sand transfer device, sand blowing device as well as the sand shoveling device fixed at the head of the operator cabin. With a view to ensuring the sand-damaged trunk railway could be activated timely upon emergency, the sand cleaning machines are designed with travelling gear and the working gear. The sand cleaning machine may reach the sand hazard zone at travelling speed or tugged by a locomotive, then lower its sand collecting plate down to the appropriate level on the rail surface through hydraulic control and start to work at working speed.

During the operation, the sand is collected into the plate of the sand shoveling device on a constant basis. The cleaning scope could be changed by adjusting the ambilateral removable blades of the plate. With the operation proceeding, the sand and dust on the track, at both sides of the rails as well as in the rail tracks can be accumulated, when the accumulated sand reaching a height of 200 mm, the scraper on the scraper conveyor of the sand shoveling device will be activated; driven by the chain gearing, the sand will be conveyed to the Lv. 1 sand storage funnel along the sand collecting plate by the scraper on a constant basis, then via the storage funnel, the sand and dust will be again conveyed to a Lv. 2 storage funnel from which they are carried to a pivoted conveyer belt of the sand blowing device; the pivoted conveyer belt will cast them to the outer sides of track selectively with a maximum sand blowing distance ±5000 mm.

2.2   *The primary structure of the sand cleaning machine*

2.2.1   *Framework*
The framework and other devices are welded with the advanced welding technology. The welding steel girder and plate steel can ensure that the framework is equipped with very high stability.

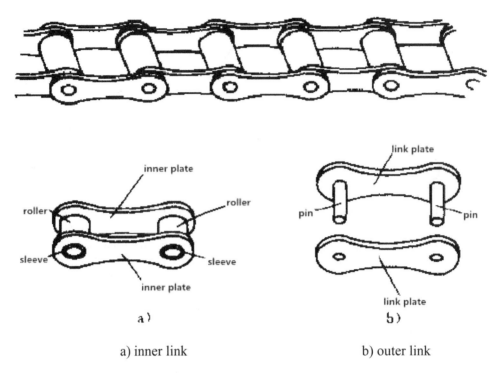

a) inner link   b) outer link

Figure 3.   The conveyor chain structure.

2.2.2   *The sand shoveling device*

The sand shoveling device is located in front of the operation cab. This device consists of the larger and smaller sand-collecting wings as well as a sandbox. The front end of the sand collecting device is added with wings in design so as to increase the sand collecting scope. The lifting obstacle avoidance and crosswise stretching places are adjusted by hydraulic gear. Hydraulic drive is adopted to realize the stepless speed regulation and elevation of the sand collecting device thus to satisfy the main conditions of operation. The scraper facilities of sand collecting device are driven by hydraulic efficiency motor with sprockets and chains to make sand shoveling come true. As to the sand shoveling device, the author chiefly studied the design of chain gearing in this paper. The conveyor chain structure is shown in Fig. 3.

Chain gearing is widely used in practice, for cases with greater center distance, multiaxial transmission, transmission requiring high-precision average transmission ratio, cases requiring exposed driver at hostile environment, low-speed heavy duty transmission as well as well-lubricated and high-speed transmission etc., we may adopt chain gearing successfully. By function, the chains can be divided into drive chain, conveying chain and lifting chain. According to production and application of chains, the conveying short-pitch precision roller chains (or roller chain for short) are the overriding chains to be used in practice. Normally, the transmission power of roller chain is below 100 kW with a speed less than 15 m/s. Yet the modern advanced chain gearing technology has enabled the high-quality roller chains to run with a transmission power up to 5000 kW at a speed of 40 m/s. As to the efficiency of chain gearing, its value is approx. 0.94~0.96 for regular transmission, but the efficiency of high-precision transmission with circulating pressure feeding lubricating oil could reach a value of 0.98.

Sprocket is an important structure of chain gearing with multiple structure formations, the currently popular sprocket shape is the impregnated teeth-linear sprocket, the cogging shape of which is shown in Fig. 4. When selecting this sprocket and the corresponding standard cutting tools to process such sprocket, we needn't draw its shape on the work sheet, instead, we only have

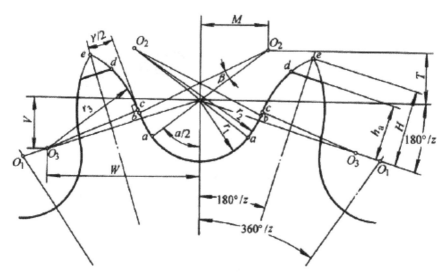

Figure 4. The shape of the impregnated teeth-linear sprocket.

to note the basic parameters and principal dimensions of the sprocket and manufacture it by the standard of 3RGB/T 1243-1997.

2.2.3 The sand transfer device

The sand transfer device is located at the central section of the machine. The sand collected by the shoveling device is conveyed to the Lv.1 sand storage funnel by the scraper and then sent to the lengthways transport tape of the central transporter. The sand transfer device is constituted by a belt conveyer due to its simple structure and mature technology. The scraper facilities adopt chain gearing; the sprocket is driven by hydraulic motor and the sand transfer device uses two hydraulic motors to drive head roll directly so as to drive the transport tape and realize sand transport. The pivoted transport tape of the sand blowing device adopts a rotary motor as the driving force.

The faster the conveyer belt runs, the higher the delivery rate is, yet the more the power absorption will be. In order to conserve energy, we should take all the relevant factors into consideration and choose the optimal tape speed if the prerequisite that the maximum transport capacity can be satisfied. As for the foundry, the manufacturing personnel shall consider whether there being a plough unloader in the delivery system, the dust nuisance in old sand, the material spilling and other factors. The normal delivery speed is 0.8, 1.0 and 1.25 m/s, not exceeding 1.6 m/s in general. Upon feeding, the tape speed shall be slowed down appropriately and set at 0.4~0.8 m/s. When conveying an object with larger volume, the loading and unloading methods as well as the productivity shall be taken as reference. The tape speed is usually set at 0.4~0.8 m/s.

The conveyer belt functions as a tractor and load bearer, usually the upper side is the carrying section and the lower side runs with no load. The conveyer belts are divided into regular rubber belts and plastic belts. The regular rubber belts are the most widely used type in practice, which is made of the belt core and the rubber cover. The rubber belt structure is shown in Fig. 5.

2.2.4 The sand blowing device

The sand and stones conveyed from the sand transfer device will be delivered to a pivoted transport tape of the sand blowing device at the rear of the machine via a sand storage funnel. The pivoted transport tape will cast the collected sand to both sides of the track selectively and its pivoted transport tape is driven by a rotary motor. The sand blowing device is also constituted by a belt conveyer due to its simple structure and mature technology. As for the roller spindle, it's normally made of type-45 steel; the scantlings of the structure and the torque diagram are shown in Fig. 6.

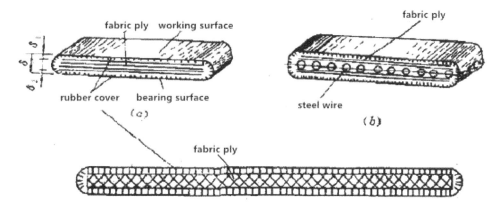

Figure 5.   Common rubber belt.

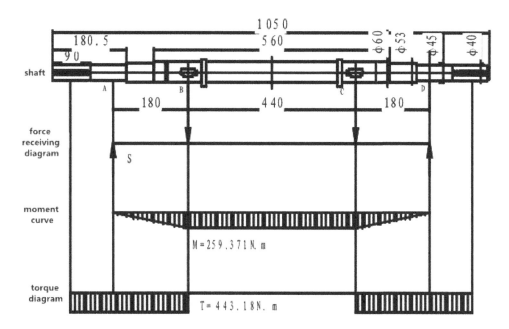

Figure 6.   Load distribution of the spindle and the torque diagram.

The sand blowing device of the railway bed cleaning vehicle consists of the transport facilities and the traversing mechanism; the transport facilities adopt troughing idlers and are driven by belt with dual hydraulic wheel motors as the power source, the 3D solid modeling assembly of which is shown in Fig. 7.

The traversing mechanism is an important component in the sand blowing device of the railway bed cleaning vehicle, the function of which equals to that of several sets of regular bearing combinations, and is of great significance to the sand cleaning vehicle's performance. It's made up of the following parts: the slewing bearing, the driving motor and the slewing platform. The slewing bearing is set between the revolving frame and the under-frame, serves as the connector of the vertical structure, thus is required to possess large bearing capacity, small structure dimension and small steering resistance, furthermore, the platform shall not turn over at slewing. The slewing bearing is designed the same as the regular bearing, consisting of the rolling body and the rolling

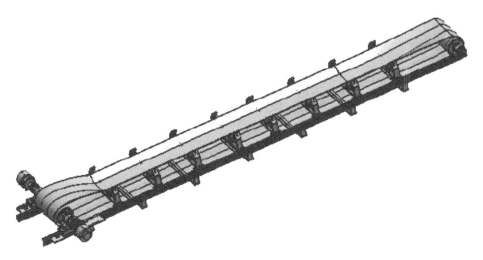

Figure 7.   The transport facility assembly drawing of the sand blowing device.

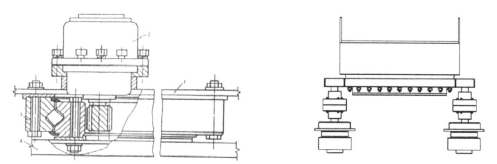

Figure 8.   The layout form of the traversing mechanism.

ring with raceway; but there are many differences between them as well. The slewing bearing is of larger size with diameter at 0.4~1.0 m, and designed with mounting holes, annulus/external gear, oil bores and gland seal, one set of slewing bearing can function as many a set of regular rolling bearings. Under the different requests, the slewing bearing may fit different situations with corresponding structure formations and can satisfy the main unit's function with different manufacturing precision levels. In most cases, the slewing bearing won't slew continuously; they only rotate back and forth within a certain angle. The core part or component of the traversing mechanism is the slewing bearing. Fig. 8 is the layout form of the traversing mechanism and Fig. 9 is its load-carrying diagram

3   CONCLUSION

In this paper, the author summed up the current status of the engineering mechanical virtual prototype technology, studied the integral technology accordingly and summarized some experiences in technical design, upon which an overall design on the railway sand cleaning machine was carried out and discussed. To be specific, the author studied the railway sand cleaning machine mainly from the perspectives of its framework, the design of its sand shoveling, transfer and blowing devices etc. Via the design of the sand cleaning machine, the author discovered that the sand cleaning efficiency and degree could be affected by the size of each devices, the length and running speed of

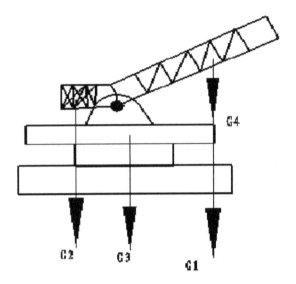

Figure 9. The load-carrying diagram of the slewing bearing.

the conveyer belt as well as the selection of drive etc. Therefore, when designing the sand cleaning machine, we shall analyze the affecting factors on a detailed basis and refer to the actual situation thereof, not all the problems could be settled purely by theoretic knowledge. The low efficiency in China's sand cleaning is exactly caused by such factors mentioned in this paper; therefore, it's the inevitable choice for our nation to design a sand cleaning machine that could overcome such affecting factors. This paper is aimed to address such problems and offers some innovative designs and conceptions for sand cleaning machines with the best hope that such probes could be of some theoretical and technical significance for the railroad departments.

REFERENCES

[1] Ning, Q. (2006) Research on technologies of virtual prototype collaborative modeling and simulation for mechatronic products. Sichuan, Sichuan University.
[2] Lars, G. & Roger, M.C. (2007) Systems engineering, test and evaluation: maximising customer satisfaction. *Innovations in Systems and Software Engineering*, (2).
[3] Hunor, E. & Doru, T. (2009) Virtual prototyping of a car turn-signal switch using haptic feedback. *Engineering with Computers*, (2).
[4] Shao, Y.Y. (2005) Research on Complex Product Developing Process Management Based on Virtual Prototype. Shandong, Shandong University of Science and Technology.
[5] Wang, H.L., Gong, L.H. & Xiao, B.N. (2008) Simulation and application of construction machinery based on virtual prototyping technology. *Machine Tool & Hydraulics*, (07), 145-147+168.
[6] Du, P.A., Yu, D.J. & Yue, P. (2007) Research on technology and methodology architecture of virtual prototyping technology. *Journal of System Simulation*. (15), 89–93.
[7] Wang, Y. (2009) Application and development of virtual prototyping technology used in the combined mechanism design. *Journal of Wuhan University of Technology*, (13), 156–158.
[8] Han, B.J. & Yu, R.B. (2008) Virtual prototype technology and its dynamic simulation analysis. *Mechanical Engineer*. (03), 142–143.
[9] Wang, J.J. & Zhang, H. (2009) The application of virtual prototype technology in mechanical design. *Popular Science & Technology*, (03), 111–119.
[10] Wan, J.J. & Li, Y. (2006) The application of virtual prototype technology in design of engineering machinery. *Jiangxi Building Materials*, (04), 51–52.
[11] Tao, B.D. & Yao, G.L. (2010) The application of virtual technology in r&d of modern mechanical products. *Journal of Hubei University of Education*, (02), 109–11.

Engineering Technology and Applications – Shao, Shu & Tian (Eds)
© 2014 Taylor & Francis Group, London, ISBN 978-1-138-02705-3

Influencing factors and control method of mudstone slaking in drilling coring

Wendong Ji, Yuting Zhang, Yafei Jin & Xiaoyu An
Tianjin Research Institute for Water Transport Engineering, Tianjin, China

ABSTRACT: In underground oil and gas reserved engineering, coring is an important method to acquire the stratum information and study the properties of underground rocks. The integrity of cores is prerequisite for the accuracy of follow-up research. With the analysis of the perniciousness of the fact that drilling cores are more tending to slake and fracture, it can be found that mudstone slaking has adverse effects on such aspects as the collecting of drilling information, recovery of cores and accuracy of experiments on cores, as well as packing, transport and storage of cores. At the same time, mudstone structure, stress modes, water-rock interaction and surrounding environmental factors constitute the main factors influencing mudstone slaking. What's more, methods to prevent mudstone slaking are proposed in view of engineering practical operations, which may provide certain reference to coring in drilling engineering.

Keywords: drilling, oil and gas storage, mudstone, coring, slaking, control method

1 INTRODUCTION

The construction of underground oil and gas reserved has entered into a rapid development period in our country. The gas transmission projects, which is from West to East and Sichuan to East, are both constructing self-contained gas storage. Therefore, it is very important that the research of the properties of bedded salt rock is to be strengthened. In the research of bedded salt rock, the mudstone is the key factor of the stabilities and tightness of this storage, and how to get the right-layer and integrity mudstone for test is an important problem to engineering practice.

Drilling coring is the most direct methods to get the information of rock layer. The samples that obtained by drilling in deep layer can reflect kinds of properties of underground rock in the most direct and accurate ways. Meanwhile, that the sample is integrity or not can decide whether the status and property of underground rock could be reflected accuracy and truly. In actual production, the transportation and preservative of drilling rock samples will be carried out by the special person with the formulary method. However, after drilling, logging, packing, transportation and etc., the samples that drilled from the mudstone layer in bedded salt rock and transported to laboratory are often slaked and fracture. As shown in figure 1, that is the situation of slaking and fracture mudstone, which is drilled out from the well. In the figure, the mudstone cores are split into different thickness sheet and the utilization rate is greatly reduced.

Since that the characteristic of mudstone, which is easy to airslake, would cause a serious impact to a project, some scholars have been studied on it, such as Tan Luorong, Huang Hongwei, Wu Daoxiang, Wu Qinghua, Liang Weibang, Liu Changwu and etc. In their research, the engineering geological characteristics and structural features of clay interlayer in water conservancy and Hydropower Engineering Foundation are analyzed, besides the Physical and chemical factors in the formation and development trend of Siltized Intercalations are studied, and variation of physical and mechanical properties after shale disintegration are analyzed. The result is that the special properties of intercalated clay layer are mainly due to the large number of clay mineral in the rocks, which is high dispersion and high hydrophilic substances.

Figure 1.   Slaking and fracture of coring mudstone.

Most of these results are not directly related to the underground oil and gas reserved engineering, nor analyzed through combining the environment of drilling field or operating procedures or other factors. Therefore, their application in practice is not strong. In this paper, the problems of slaking and fracture of mudstone in drilling coring are discussed, and its slaking and fracture reasons are researched based on its adverse effects and the method of combining operating procedures. Meanwhile, the control method according to the problem is advanced and its result provides a valuable reference.

2   THE ADVERSE EFFECTS OF SLAKING OF MUDSTONE CORES

The slaking and fracture of mudstone in drilling coring would not only affect the rock sample testing results, but also affect the operating convenience of multiple process, which reflect specifically in the following aspects.

2.1   *To affect the collection of the cores information*

Taking out of the earth, the cores' information is collected through the artificial and electronic combination method, which is to identify the character of the rock, record the depth of its layer, analyze its composition and some other information, build a three-dimensional information database by taking photos, scanning and some other methods.

If the coring sample has been slaked, its layer would be not identified correctly and some information of the rock layers may be even lost. Besides this, the number of the connection sections that obtained by scanning would increase significantly, which would make the three-dimensional information database incorrect.

2.2   *To affect the recovery of the cores*

There are relevant standards towards various experiments on the cores in laboratory, including shape and size [8]. The cores' layer is also specified in some professional experiments. For example, in the experiment of the permeability and mechanical properties' assignment between different layers, the cores must locate the connection position of two layers. But if the slaking of the sample is too serious, it is difficult to find the sample that meets the requirements.

When the sample length is less than the standard length of experiment for the reason of slaking and fracture, this sample is nearly unavailable and some remedial methods must be taken. Nevertheless, the remedial materials can't fully consistent with the actual ones and the accuracy of this experiment will descend. For this reason, remedial methods are not recommended in this situation. In a word, the slaking and fracture of mudstone will decrease the utilization rate of cores, make some them losing their values and directly cause huge economic losses because of the huge investment in drilling coring.

2.3 *To affect the accuracy of experiments*

The slaking and fracture of cores will cause a bigger contact area of the rock sample with outside environment. This will increase the conversion probability of the cores' components and characters through the way of the sample components' leaking and the outside inclusions' attaching. And then, because of these conversions, the accuracy of this experiment would be affected.

2.4 *To affect packing, transport and storage of cores*

The cores' packing is carried out in accordance with the principles of block packing. When the cores are relative integrity and their length are uniform, the packing and storage work will be all right and the time will be short. And this will directly reduce the contact time of the samples with outside environment. Thereby, the interference of external factors on cores will be reduced too.

The integrity and uniform cores will make the storage and transport easy. But when it is slaked, small cores will increase and their losing possibility in the process of storage and transport will increase too.

3 THE OPERATIONAL PROCESS OF DRILLING CORING

With the progress of science and technology, the technology of drilling coring also appeared various methods, such as colloid coring, motor coring, shaping coring, totally enclosed coring with hydraulic lifting and etc. However, because of the costs and technical constraints, the widely used method is still the conventional method. Its general process can be described as follows: firstly, drill to the coring depth using the ordinary bit; secondly, replace the coring bit and coring barrel; lastly, lift the drill at the right time according to the drilling depth and time, which can prevent the cores losing too much in the drilling process.

Having lifted the drill, the cores would be taken out of the barrel, washed, cataloged, labeled and put into the cases that labeled in advance. After that, the electronic imaging information and some other first-hand information will be taken at this step. Then, it's the packing step, which is to band up each section with kraft paper, dip it into melted paraffin wax and form a protective shell with a good sealing after taking out and solidifying. These series operations can prevent the air and water to seep into the cores and change their weightness.

After that, the packed cores will be put into cases in order and transport to the storehouse in relevant units. And when it needs to be tested, the right core will be found according to the catalog information and made a further processing according to the experiment standards.

4 THE FACTORS INFLUENCING MUDSTONE

There are kinds of reasons that the cores tend to slake and fracture, including mudstone structure, the transformation of stress mode, the water and temperature influence and etc. The combined effect of these factors determines the characteristic of mudstone, which is tending to slake and fracture.

4.1 *The mudstone structure*

Mudstone, which has an obviously sedimentary structure, is a typical sedimentary rock. Because the process of sedimentation is a long period, the natural climates are different every year during this process. Different rainfall and evaporation will cause different sedimentations. Therefore, each sedimentary section of mudstone has different thickness, compositions and etc. and this is the typical reason that the mudstone has a sedimentary structure, which is shown in figure 2.

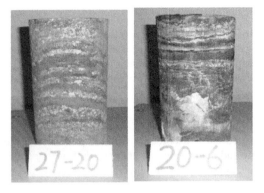

Figure 2. Typical sedimentary of mudstone.

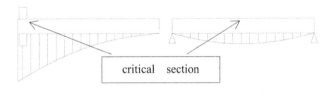

(a) cantilever situation   (b) simply supported situation

Figure 3. Section moment in different situation.

There is a large number of flabby interlayer in mudstone layer, such as the layer full of organic matter and sand. This type of layer has a breakable structure itself. Therefore, the cores will be fractured more easily when receiving an external stress.

4.2 The stress on the core

Different type of rock have a common characteristic that the ensile strength is lower than the compressive strength. According to some relevant research reports, the ensile strength of mudstone is less than 10% of compressive strength. Therefore, mudstone tends to fracture when receiving a pulling force.

Because of the drilling method and the restriction of sampling conditions, the drilling core general appears to be a long cylinder. And the longer the axial length, the stronger the stress on the key component.

According to some relevant research reports, the ensile strength of mudstone may be less than 0.1 MPa when considering the influence of water. Take a section of core as example, which is 80 cm length, has a diameter of 5 cm and a density of 2.65 g/cm^3. When the sample is in cantilever situation, the moment is 6.7 kN·M at the section that has a biggest stress on, which will be 0.14 MPa. When the sample is in simply supported situation, the moment is 1.7 kN·M at the section that has a biggest stress on, which will be 0.035 MPa. If considering the shaking and pounding, the stress on the dangerous section will increase. Therefore, the unreasonable stress mode would cause the core fracture directly.

4.3 The factor of water-rock interaction

The property of mudstone tending to slake and fracture has a very close relationship with its composition. It contains a large number of clay minerals (such as montmorillonite, llite, kaolinite and etc.), which have a characteristic of highly absorbent and dispersion and will be significantly expanded after absorbing water. This problem was discussed in detail by Tan Luorong. The result is that the mudstone will dehydrate when it is taken out of underground. And because of the rapidly dehydrating of surface, the mudstone will have a different contraction inside and outside. Thus, a tensile stress appears and a fracture may be caused. Finally, the crack makes more water losing from the mudstone and it will be bigger and bigger.

However the dehydration will make the space of montmorillonite contracted, it will be back to, even exceed the original space if absorbing enough water. This will create expansion force in the crack and make the mudstone slaking from inside. Therefore, If the mudstone experienced several reciprocating cycles of wetting and drying, its slaking degree will be very high.

4.4 *Environmental factor*

Before the step of packing is accomplished, the mudstone is exposed to air at most of time. During this time, the mudstone will be influenced by the surrounding environment factors obviously. In the environment of hot and humid in South China, the core's surface will appear a layer of condensation water when it is placed outdoor in the night. While at noon, it will become dry quickly under the sun. This forms a cycle between dry and wet. In most areas of northeast and west of China, there is a greatly changes in temperature. When the temperature is below zero, the expansion force which is generated by the conversion of water into ice will cause the mudstone slaking.

5 THE CONTROL METHODS OF THE MUDSTONE SLAKING

Considering the actuality factors that can be controlled and the restriction of the cost and technology, some effective control methods combining with the engineering practice are advanced in this paper. Using these methods, the slaking and fracture of the cores during the process of drilling coring can be prevented and its utilization ratio is increased. The realizations of these methods are mainly through the ways of mechanic and circumstances.

5.1 *Optimizing the stress mode*

According to the contrast in figure 4, the maximum sectional stress is decreased effectively because of the reasonable stress mode is adopted. Then, the possibility of mudstone's slaking and fracture will be lower.

As a result, the suggestion is that when the length of cores is more than 50 cm, the usage of cantilever mode is must be forbidden, the simply supported mode should be used fewer and the multi-points uniform support mode should be more adopted during the processes of transport, and handling.

5.2 *Using the shock absorption measures*

To prevent the cores to receive a bigger shock directly, the case that is used in the processes of and transport should be made of a flexible material and spread a soft mat inside. Besides this, during the transport and handling processes, it's forbidden to use a high frequency shaking machine to contact with the cores directly.

5.3 *Controlling the influence of surrounding environment*

The key factor of the environmental controlling is to reduce the disturbance of water to the cores, which is to avoid the repeatedly variations from wet to dry and dry to wet since some natural or artificial reasons. To stabilize the environment, the effective methods include adjusting the process step at scene, keeping humidity artificially and etc. When finishing the recording step, the cores must be packed as soon as possible to reduce the time of exposing to air.

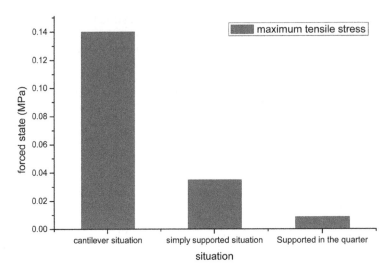

Figure 4. The maximum tensile stress in different situations.

6 CONCLUSIONS

In this paper, the perniciousness of mudstone slaking is analyzed completely. And the main factors influencing mudstone slaking are researched. Then, some practical control methods are advanced and the conclusions are that:

1) The slaking and fracture of mudstone in drilling coring has adverse effects on such aspects as the collecting of drilling information, recovery of cores, accuracy of experiments, as well as packing, transport and storage of cores.
2) The main factors influencing mudstone slaking are mudstone structure, stress modes, water-work interaction and surrounding environmental factors.
3) In view of engineering practical operations, optimizing the stress mode, using the shock absorption measures and controlling the influence of surrounding environment can reduce the possibility of mudstone slaking.

REFERENCES

[1] Zheng, Y.L. & Zhao Y.J. (2010) General situation of salt cavern gas storage worldwide. *Oil & Gas Storage and Transportation*, 29(9), 652–655.
[2] Ji, W., Yang, C.H., Liu, W., et al. (2013) Experimental investigation on meso-pore structure properties of bedded salt rock. *Journal of Rock Mechanics and Engineering*, 32(10), 2036–2044.
[3] Xu, J.L., Song, S.L. & Wei, C. (2000) Foreign new technology in core drilling (1). *China Petroleum Machinery*, 28(09), 53–56.
[4] Tan, L.R. (2001) Discussion on mechanism of disintegration and argillitization of clay-rock. *Rock and Soil Mechanics*, (1), 1–5.
[5] Huang, H.W. & Che, P. (2007) Research on micro-mechanism of softening and argillitization of mudstone. *Journal of Tongji University (Natural Science*, 35(7), 866–870.
[6] Wu, D.X., Liu, H.J. & Wang, G.Q. (2010) Laboratory experimental study of slaking characteristics of red-bed soft rock. *Chinese Journal of Rock Mechanics and Engineering*, 29(A02), 4173–4179.
[7] Wu, Q.H., Feng, W., Li, B., et al. (2008) An approach to mudstone characteristics of Dadingzi Mountain and application of research result. *Port & Waterway Engineering*, (5), 71–73.
[8] Liang, W.B. (2007) Experimental study on weathering of mudstone under different environmental conditions and suggestion on its utilization. *Water Resources and Hydropower Engineering*, 38(12), 86–88.

[9] Liu, C.W. & Lu, S.L. (2000) Research on mechanism of mudstone degradation and softening in water. *Rock and Soil Mechanics*, 21(1), 28–31.

[10] Ji, W.D., Yang, C.H., Yao, Y.F., et al. (2011) effects of loading strain rate on mechanical performances of salt rock. *Chinese Journal of Rock Mechanics and Engineering*, 30(12), 2507–2513.

[11] Yang, C.H., Li, Y.P., Qu, D.N., et al. (2008) Institute of rock and soil, Chinese Academy Of Sciences, et al. advances in researches of the mechanical behaviors of bedded salt rocks. *Advances in Mechanics*, 38(4), 484–494.

[12] Zhou, C.Y., Deng, Y.M., Tan, X.S., et al. (2005) Experimental research on the softening of mechanical properties of saturated soft rocks and application. *Chinese Journal of Rock Mechanics and Engineering*, 24(1), 33–38.

Engineering Technology and Applications – Shao, Shu & Tian (Eds)
© 2014 Taylor & Francis Group, London, ISBN 978-1-138-02705-3

Research on the cause and control of cracks in reinforced concrete based on stress analysis

Chuan Yu

Capital Construction Division, Shandong Women's University, Jinan, Shandong, China

ABSTRACT: Concrete is widely used in building construction and has made considerable contribution to modern architecture with its superb quality. In this paper, the cause of concrete cracks during construction was examined. First, the cracking in concrete was outlined from the properties of the material used and from the mechanical mechanism underlying this phenomenon. Next, the effect of the internal composition and external environment of concrete on the generation plastic shrinkage cracks, autogeneous shrinkage cracks, dry shrinkage cracks and temperature shrinkage cracks in concrete was investigated using literature review, experiment and graphical methods, and crack control measures were derived from data graphs. Lastly, comprehensive control measures for concrete cracks were summarized according to theoretical analysis and experimental results throughout the paper and the outlook on the research topic was noted with a view to making contribution to the prevention and control of concrete cracks.

Keywords: concrete cracks; concrete hardening; tensile stress; shrinkage cracks; crack control measures

1 INTRODUCTION

There is no need to discuss how widely concrete is used across the globe. However, against its development goal of high intensity and high performance, cracks in concrete during and after construction have proved to be an inevitable chemical-physical phenomenon. In order to control concrete cracks to a reasonably low level, many specialists and researchers have given recommendations some of which include highly incisive propositions. Chen Z.Y. et al (2006) discussed cracks in reinforced concrete and their control from four aspects: the permissible degree, generation mechanism, typical shrinkage cracks and possible control measures[1]. Lu L. et al (2012) outlined the design, construction, measurement and crack repair of concrete according to the cause of temperature shrinkage with an aim to characterizing cracks in concrete as a result of temperature reduction, and summarized control methods used currently for mass concrete cracks. They also analyzed existing problems and their practical implications and predicted the development trend and application potential of crack control researches[2]. Lu Y.F. (2013) summarized the type, cause and general identification methods in relation to mass concrete cracks, and discussed and concluded on possible control methods for plastic shrinkage cracks, plastic settlement cracks, dry shrinkage cracks and temperature difference cracks[3]. Wang B. et al (2014) investigated the cause of cracks in underground wallboard structures with their many years' experience in engineering structural design, discussed the generation and expansion of cracks from both microscopic and macroscopic perspectives, proposed BIM-based crack control design methods and made selective crack control design, which provided empirical result for the quick development of crack control[4].

In this paper, the cause and stress characteristics of concrete shrinkage cracks were examined on the basis of previous findings with a view to understanding the direction of experiment data collection in connection with the concrete cracking process, laying foundation for data validation and data exploration and providing theoretical basis for scientific control measures against concrete cracks.

Table 1. Average bulk density of concrete and its components.

Material	Avg. bulk density
Concrete	1.400039 g/mL
Water	1.550043 g/mL
Stone	1.600045 g/mL
Sand	1.000028 g/mL
Cement	2.260063 g/mL

2 MECHANISM OF CONCRETE SHRINKAGE AND STRESS ANALYSIS

Concrete is a building material composed of cement, sand, stone and water that are organically mixed together. Its extensive application almost covers all building areas. Concrete has a higher volume than any of its components during concrete hardening. In other words, concrete has larger weight per unit volume when it is hardened. The causes behind this phenomenon are the higher compaction caused by pore filling on the one side, and shrinkage during the hardening process on the other side. Table 1 shows the bulk densities of concrete and its components.

2.1 *Theoretical analysis on concrete shrinkage in relation to temperature difference*

From this table, shrinkage of concrete is actually a hardening process that contributes to the larger volume, higher compaction and higher strength of concrete. By time period, mechanism and conditions of occurrence, concrete shrinkage can be divided into three setting, hardening and shrinkage processes: plastic shrinkage, autogeneous shrinkage and dry shrinkage, during which there are generally restraints in the space that right explain why concrete cracks when its shrinkage has come to a certain extreme. To describe this cracking process, an ultimate tensile-strain variable may be incorporated, which keeps changing during concrete hardening and does not stabilize until the 28th day. Shrinkage is also attributable to the hydration heat released during the hydration process. The chemical equation of this thermal reaction can be expressed as (1) below.

Portland cement + water →hydrated calcium silicate (CSH) + calcium hydrate (CHO)

+ hydration heat (500 kJ/kg) (1)

From expression (1), a kilogram of cement releases 500 kJ heat. It is this heat that causes higher temperature in concrete which, in mass concrete, can be as high as 80°C. This gives rise to internal/external temperature difference of the concrete which in turn leads to temperature stress. The larger the temperature difference is, the larger temperature tensile stress is produced. Once it comes beyond a certain limit, cracks will appear in the concrete. This is also the mechanism underlying temperature shrinkage.

2.2 *Analysis on shrinkage stress inside concrete*

Given that concrete is ideally a homogeneous dielectric elastomer, the tensile stress of a concrete member can be characterized under restraint at both sides, continuous restraint at the bottom and restraint inside the concrete member.

In the first scenario, i.e. the tensile stress under restraint at both ends, the shrinkage of concrete caused by temperature reduction of hydration heat during temperature shrinkage and the dry shrinkage in setting and hardening can be expressed by ε_{ct}; the shrinkage caused by drying can be expressed by ε_{cd}; the total shrinkage ε_c is the summation of the two; and the temperature shrinkage is equal to the expansion coefficient multiplied by α the temperature difference ΔT. However, as it is hardly impossible to characterize the effect of dry shrinkage separately, it is often converted

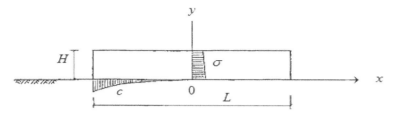

Figure 1. Sketch showing the reaction stress at the bottom of concrete on semi-infinite foundation.

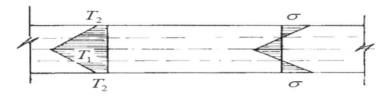

Figure 2. Reaction stress induced by the internal/external temperature difference of a concrete member.

through temperature stress analysis before the total shrinkage is calculated. The total shrinkage is expressed as (2) below:

$$\varepsilon_c = \varepsilon_{ct} + \varepsilon_{cd} = \alpha(\Delta T + \Delta T') = \alpha \Delta T_e \qquad (2)$$

In expression (2), $\Delta T'$ is the equivalent temperature difference; ΔT_e is the synthetic temperature difference. Given that the deformation in the member is homogeneous, the tensile stress σ can be expressed as (3) below:

$$\sigma = E_c \varepsilon_c = \alpha \Delta T_e E_c \qquad (3)$$

In expression (3), E_c is the elastic modulus of concrete.

As for the wall on the semi-infinite foundation shown in Fig. 1, when homogeneous shrinkage takes place in the wall, the restraint at each point will be different across the concrete member. The restraint at the bottom will be larger than that at the top. The horizontal tensile stress induced will be as expressed in (4) below:

$$\sigma = R E_c \varepsilon_c = R E_c (\alpha \Delta T_e) \qquad (4)$$

In expression (4), R is the restraint coefficient smaller than 1.

As shown in Fig. 1, when $x = y = 0$, the restraint degree at the bottom of the central cross section is the smallest, i.e. $R = 1$; when the slenderness ratio of the member is larger, the tensile stress on the point of the central cross section is nearly homogeneous and the horizontal tensile stress on the central cross section where $x = 0$ is also fairly homogeneous. The larger stress at the bottom is caused by the restraint of the underlying foundation where $y = 0$ on the horizontal shear stress on the member. When the slenderness ratio is smaller, the upper part of the central cross section may not only be free from any tension but also exposed to pressure.

The restraint inside concrete members comes from many sources, the most principle of which is the internal restraint induced by internal/external temperature difference. As shown in Fig. 2, the internal temperature of the member is T_1 and the external temperature is T_2, and the former is larger than the latter. Given that the temperature field is in linear distribution as shown in Fig. 2, the member could be exposed to pressure inside it and tension at the upper and lower edges. As the fiber above the average temperature line is neutral, it is not exposed to either tension or pressure.

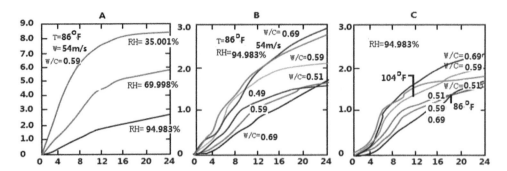

Figure 3.   Data graph showing the effect of air temperature, relative humidity and air velocity on plastic shrinkage as the pouring time gains.

The internal reaction stress shown in Fig. 2 is expressed as (5) below, in which the temperature above the average temperature line is a half the internal/external temperature difference.

$$\sigma = E_c \varepsilon_c = E_c \left[\alpha \left(\frac{1}{2} \Delta T \right) \right]$$
(5)

3   ANALYSIS AND CONTROL OF CONCRETE SHRINKAGE CRACKS

3.1   *Analysis and control of plastic shrinkage cracks*

Plastic shrinkage can be subdivided into plastic dry shrinkage and plastic settlement. The former is caused by water evaporation on the surface of concrete during setting before it is hardened, which generally occurs on the exposed surface of poured concrete members, while the latter is the bleeding of water in fresh concrete when it moves upwards.

The shrinking process of fresh concrete includes a period when it bleeds faster than it evaporates, a period when it evaporates and dries faster than it bleeds, a period when it becomes sticky because of setting and a period when it hardens after it is finally set. In the first period, shrinkage does not take place. In the second, concrete starts to shrink. In the third, plastic cracking may occur. In the fourth, the hardening concrete dries and shrinks. When the capillary pores amid the solid particles reduce in response to evaporation, the meniscus formed could produce tension and cause the concrete that just plasticizes to shrink.

Increase in the amount of water used for concrete mixing will enlarge the capillary pores, thereby reducing the capillary water tension. In practical operation, however, increase in the amount of water used will also increase the plastic shrinkability, resulting in plastic cracking. The reason why this happens is that the rate of plastic shrinkage is also associated with the ability of the slurry to resist shrinkage. If the slurry is harder, its ability to resist shrinkage is larger. A large amount of water used will right reduce its consistency, thereby increasing its eventual plastic shrinkage.

Major contributors to plastic shrinkage include the air temperature, relative humidity and air velocity, which may finally be reduced to the variation in the rate of water evaporation. By experiment, a data graph showing the effect of these three contributors on plastic shrinkage as the pouring time gains was derived as shown in Fig. 3.

In Fig. 3A, RH is the relative humidity. B indicates that the plastic shrinkage of concrete under different water/cements ratios at the air velocity of 15 km/h is much larger than when there is no air at all. C indicates the relationship between the water loss and plastic shrinkage of the same cement mortar under different environmental exposures.

The sinking of aggregate and rise of moisture will result in gaps that gather water in the bottom of horizontal reinforcement. The rising water will also stay at the bottom of coarse aggregate, resulting in gaps over the interface between slurry and aggregate, thereby affecting the permeation and freezing resistance of concrete. When vertically sinking solid particles deposit on a horizontally placed reinforcement, settlement difference will be produced with the surrounding concrete, resulting in plastic settlement cracks in the top surface of the concrete. If concrete for beams, slabs or columns is poured at the same time, plastic settlement cracks will most likely appear over the interface between the different members. When the protection is too thin, the plastic settlement cracks will run into the surface and along the length of the reinforcement. These longitudinal cracks are essentially different from horizontal ones that develop across the reinforcement in their degree of hazard to reinforcement erosion. The former should be totally avoided while the latter are not so hazardous.

Control of plastic shrinkage cracks may include cooling, damping the air over the concrete by spraying, providing air barriers to minimize the local air velocity, installing a shelter to prevent direct sun exposure, limiting the mixing time to the minimum necessary, shortening the time between mixing and pouring, damping the formwork before pouring, covering the concrete with plastic membrane immediately after pouring, and eliminating any crack by plastering and smoothing.

Prevention of plastic settlement may include, to the extent that the workability of the concrete is guaranteed, minimizing the slump of concrete to maintain satisfactory consistency and water retention; pouring the concrete layer by layer when implementing columns, beams and slabs that are associated with each other but have different depths; increasing the protection thickness for the surface reinforcement; proper vibration; adding air entraining admixture; ensuring the formwork rigidity and foundation stability; surface plastering and smoothing.

3.2 *Analysis and control of autogeneous and dry shrinkage cracks*

Autogeneous shrinkage is a shrinkage induced by the hydration of cement. The continued hydration of unhydrated cement in hardened cement slurry is the main contributor to autogeneous shrinkage. As hydration causes the pores to reduce in size, if they are not replenished with external water, capillary water negative pressure will occur and cause pressure on the hardened hydration product leading to autogeneous shrinkage.

Autogeneous shrinkage takes place mainly in the early days of concrete hardening. As it is generally accepted that concrete shrinks autogeneously within a few or some ten plus days after it sets, autogeneous shrinkage should be measured immediately after the concrete sets. When the water/cement ratio is lower, the autogeneous shrinkage will be larger. Silicon powder is a material that accelerates the autogeneous shrinkage of concrete. On a construction site, however, the best way of improving autogeneous shrinkage without conflicting with other parameters of concrete is timely and adequate curing with water.

Dry shrinkage is an ongoing process when plastic flow ends, the concrete is initially set and starts to harden. In some projects, dry shrinkage in concrete can last a few and even tens of years. From this respect, dry shrinkage is an inherent property of cement-based concrete. Hardened cement slurry contains many pores inside it. Theoretically, when the water/cement ratio is smaller than 0.4 and proper curing conditions are present, the porosity can be reduced to the minimum of 28%. Main contributors to the dry shrinkage of concrete include cement, type and amount of aggregate used, amount of water used, water/cement ratio, chemical admixture, fly ash or other pozzolanic admixtures, environmental conditions and the size of the concrete member.

Of these contributors, the type and amount of the aggregate used decide the degree of dry shrinkage of the concrete. When the shrinkage reduction caused by aggregate is larger, this indicates that the elastic modulus of the aggregate is higher. When the amount of aggregate is larger and the maximum grain size has a larger proportion, the degree of shrinkage will be reduced. The effect of the water/cement ratio, the amount of aggregate and the amount of water and cement used on the dry shrinkage of concrete and cement mortar is presented in Fig. 4. The shrinkage measurements of the same cement under different parameters are given in Table 2.

In Fig. 4, A: the horizontal axis is the amount of aggregate used, and the longitudinal axis is the ratio of concrete shrinkage to cement slurry shrinkage; B: the horizontal axis is the amount of

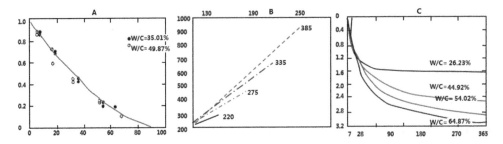

Figure 4.   Data graph showing the effect of the amount of main concrete components on dry shrinkage.

Table 2.   Shrinkage measurements of the same cement under different parameters.

Aggregate type	Specific density	Water absorption	Shrinkage
Sandstone	2.470069 g/mL	5.002%	0.116815%
Slate	2.750077 g/mL	1.293%	0.068355%
Granite	2.670075 g/mL	0.801%	0.047073%
Limestone	2.740077 g/mL	0.198%	0.041282%
Quartzite	2.660074 g/mL	0.296%	0.072894%

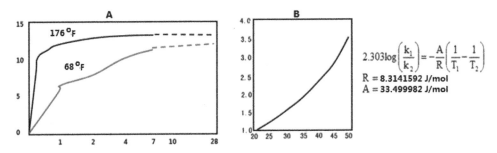

Figure 5.   Data graph showing temperature-based hydration degree vs hydration rate.

water used, and the longitudinal axis is the degree of dry shrinkage; C: the horizontal axis is the age (days) of concrete, and the longitudinal axis is the degree of dry shrinkage, leading us to a conclusion that, when the amount of aggregate is larger and more aggregate of the maximum grain size is used, the shrinkage can be reduced; the dry shrinkage increases as the amount of water and cement increases; and a larger water/cement ratio can increase the shrinkage.

Control of shrinkage cracks in set and hardened concrete could include preparing low-shrinkage concrete, delaying surface water loss, using shrinkage-compensating concrete and late-poured band construction, and using late-poured band expansion joints.

3.3   Analysis and control of temperature shrinkage cracks

Temperature plays an important role in the hydration rate. From the data graph showing the effect of temperature on the hydration degree and hydration rate shown in Fig. 5, the higher the temperature, the higher the hydration degree; and the velocity ratio under 20°C, 30°C, 40°C and 50°C is 1:1.57:2.41:3.59, leading us to a conclusion that a higher temperature will shorten the setting and hardening and speed up the strengthening of concrete.

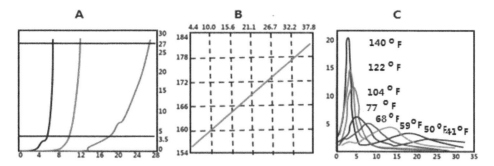

Figure 6.   Data graph showing temperature-based concrete condensation, water demand vs hydration heat release.

In Fig. 5, A: the horizontal axis is the age (days), and the longitudinal axis is the degree of hydration; B: the horizontal axis is the temperature variation, and the longitudinal axis is the ratio of hydration rate.

With experiment data, the effect of concrete setting, the effect of temperature on the amount of water for fresh concrete and the hydration heat release of C_3S shown in can be derived as shown in Fig. 6.

In Fig. 6, A: the horizontal axis is the time (hours), and the longitudinal axis is the penetration resistance (MPa); B: the horizontal axis is the humidity of fresh concrete, and the longitudinal axis is the water content (kg/m^3); C: the horizontal axis is the time (hours), and the longitudinal axis is the heat released (cal/g×h). From the data in this graph, a higher temperature will speed up the setting of fresh concrete. When the pouring temperature of concrete increases, so does the amount of water used. The heat released by the main component of cement, tricalcium silicate (C_3S), when hydrating under different temperatures demonstrates that concrete is more exposed to cracking when the internal/external temperature difference in a member is more than 20°C.

4   COMPREHENSIVE CONTROL MEASURES AND RESEARCH OUTLOOK

4.1   *Comprehensive control measures for concrete cracks*

1) Select the right components for concrete and the right mixing ratio of these components. This includes the use of cement of lower hydration heat, well-graded medium or coarse sand, well-graded rubble or pebble, the use of slow-condensing high-efficient water reducer, adding shrinkage-compensating swelling agents having the ability to increase crack resistance, and minimizing the amount of water used to the extent that the workability of the mixture is ensured;
2) Controlling the temperature to a reasonable level. This requires that the pouring temperature should be lower than 30°C; the maximum internal temperature should not be lower than 65°C; the internal/external temperature of concrete should be smaller than 20°C; and the surface/curing water temperature difference of concrete should be smaller than 15°C;
3) Crack control methods like control joints, late-poured bands, sliding layers and structural reinforcement should be used properly used;
4) The construction technical criteria should be met. In controlling component contents, the water content should be emphatically measured. In mixing concrete, forced mixers should be used. The free fall height of concrete should be higher than 2 m and the concrete should be vibrated to compaction by inserting the vibrator directly into the concrete. The formwork should be installed securely and the chute and form should be damped with water before concrete is poured. The top and bottom slab should be cured with water.

4.2 *Research outlook*

1) The mechanical model of concrete cracks and the stress variation during the cracking process was theoretically analyzed, which offers direction for experiment design and effectively data extraction;

2) The generation of shrinkage cracks, autogenous and dry shrinkage cracks and temperature shrinkage cracks and their contributors were analyzed based on measurement results, and control measures for these cracks were summarized according to their respective data characteristics, which provides theoretical basis for crack control during concrete construction;

3) Subject to the limited length of this paper and the experiment facilities available, we were not able to present the effect data of the control measures proposed after they are implemented. If experiment data before and after these measures were compared with controlled variables, more scientific control measures could have been provided;

4) Future microscopic studies on concrete cracks could include finite element analysis and the use of mechanical analysis software to get a better understanding of the degree of hazard in relation to concrete cracks.

REFERENCES

[1] Chen, Z.Y., et al. (2006) Analysis and control of cracks in reinforced concrete. *Engineering Mechanics*, 23(1), 86–105.

[2] Lu, L., et al. (2012) Research and progress of control for mass concrete temperature crack. *Journal of Water Resources and Architectural Engineering*, 10(1), 146–150.

[3] Lu, Y.F. (2013) Mass concrete plastic shrinkage crack criterion and management methods. *Friend of Science Amateurs*.

[4] Wang, B., et al. (2014) Research on crack mechanism and crack controlling design of underground wallboard structure engineering. *Shanxi Architecture*, 40(3), 46–48.

[5] Ji, L.J. (2013) Optimization of disease treatment schemes for reinforced concrete bridges. *Science & Technology Information*, 25, 62–63.

[6] Zhou, X.L. (2013) Research on the cause and countermeasures of cracks in reinforced concrete. *Construction Management*, (1), 113–114.

[7] Liu, X.M. (2013) Cause and prevention of cracks in concrete. *Engineering Technology*, (8), 169.

[8] Chen, J.P. (2013) Application of mass concrete crack prevention in construction engineering. *Science-Technology & Management*, (8), 179–180.

[9] Li, S.D. (2013) Prevention and processing of concrete cracks in building construction. *Applied Technology*.

[10] Wang, Y.C. (2013) Analysis on the control technology of concrete cracks in construction. *Sci-tech Entrepreneurs*, (12), 40.

[11] Zeng, X.P. (2012) A look at the mechanism and countermeasures of cracks in reinforced concrete structures. *Guangdong Science & Technology*, 3(5), 153–155.

[12] Zhao, F. (2013) The concrete cracks caused by temperature stress and the control measures. *Engineering Technology*.

[13] Zhang, T.C. (2013) Discussion on concrete cracks and control measures. *Building Construction*.

[14] Yuan, H.F., et al. (2013) A discussion on the cause and prevention of concrete cracks. *Engineering Technology*.

[15] Guo, W.Q. (2013) Cracks in reinforced concrete structures and their control. *Shanxi Architecture*, 39(19), 33–34.

Engineering Technology and Applications – Shao, Shu & Tian (Eds)
© 2014 Taylor & Francis Group, London, ISBN 978-1-138-02705-3

Study on coupling model for three-dimensional seepage and land subsidence of deep foundation pit dewatering

Xiaoyu An, Dianjun Zuo, Mingming Li, Wenbin Pei & Jiandong Li
Tianjin Research Institute for Water Transport Engineering, Tianjin, China

ABSTRACT: The foundation pit dewatering has an important position in the construction process of deep foundation pit; and the selection of diaphragm wall of foundation pit and dewatering scheme and their rationality of the design have a great impact on the project quality, construction period and project cost. On account of some problems in the existence of diaphragm wall of deep foundation pit and dewatering scheme design currently, the deep foundation pit dewatering scheme in the complex formation condition is optimized and designed with multi-index evaluation method by combining the deep foundation pit dewatering project of Puzhu Road station of Nanjing Metro Line 3, and the conclusion is drawn. The calculation result shows that: Scheme I is optimum dewatering scheme, the buried depth for bottom of diaphragm wall of foundation pit is 41.5 m, and the buried depth of 15 filters arranged inside the foundation pit is 22–37 m of pumping well. This scheme can let the groundwater level meet the excavation requirement during rainfall, and the number of land subsidence outside the foundation pit is within the controllable scope. After comparison with the actual observed value, the built groundwater seepage model is reasonable, and the calculation result is correct and reliable, which has a certain reference value to the deep foundation pit dewatering project design.

Keywords: loose sediments, deep foundation pit dewatering, three-dimensional seepage, numerical simulation, multi-scheme comparison

1 INTRODUCTION

The deep foundation pit dewatering project is a system project related to many factors[1–4], so it needs to unify and coordinate the safety of foundation pit and surrounding environment, cost reducing, construction period shortening and others; and perform many optimization designs, to form an ideal scheme.

While performing the deep foundation pit supporting design, the buried depth of diaphragm wall will have a great influence to the subsequent dewatering project. If the buried depth is too big, it will increase the construction difficulty, delay the construction period, and then improve the project cost greatly; otherwise, it will cause the big deformation of supporting structure of foundation pit, low strength and insufficient stability. The dewatering design should also pay attention to select reasonable well group arrangement and dewatering control scheme, of which includes the pumping flow, well distance and setting of all parameters of well group. For the pumping well, if the design flow is too large and the arrangement is too depth, it may excessively damage the water environment around, cause the differential settlement of buildings around and even collapse, and occur the serious project safety accident; and if the designed pumping flow is too small and the arrangement is too light, it is difficult to play the role of reducing the groundwater and may have impact on the construction progress. It is also very important to select reasonable well distance, if the distance is too small and the well is too many, it will cause inconvenience of construction due to the limitation of field, and then increase the cost; otherwise, if the yield of single well is big, it is easy to exceed its water capacity, and the dewatering effect is poor[5–7].

Therefore, it shall consider a series of technical and economic indexes while optimizing the scheme for design of diaphragm wall of foundation pit and well arrangement, and each scheme has different degrees of difference on these indexes, if part of influencing factors is compared only, the proportion accounted by the human factor influence of scheme selected is big, and it is often easy to ignore other factors, which will cause a certain mistake or waste. It can make more reasonable and correct judgment only by comprehensively considering the influence of many factors. So, this article performs the scheme optimization with the numerical simulation method and statistical method by relying on the deep foundation pit dewatering project of Puzhu Road station of Nanjing Metro Line 3 on account of main goals involved by the problem, in order to improve the scientificity of foundation pit supporting and dewatering scheme selection.

2 PROJECT OVERVIEW

Nanjing Puzhu Road station, located at the intersection of Pukou District, Liuzhou Road and Puzhu Road and roughly arranged from east to west, is the first station of Nanjing Metro Line 3 in Jiangbei. The length of main foundation point of station is 172.3 m, the width of standard section is 24.0 m, and its depth is 19.8 m. The class of foundation pit of station is class I, and the retaining structure is underground diaphragm wall and inner support and constructed with open cut method.

As shown in Figure 1, the stratum of field located by the station is as follows respectively from top to bottom: ①-2 plain fill, ②-3c-3 silt, ②-3d-4-3 silty sand, ②-4d-3-2 silty sand, ②-5d-2-1 silty sand and fine sand, ④-4e pebble, macadam and round gravel, K2p-1 strong weathered siltstone and mudstone. The phreatic aquifer consists of silt and silty sand, where the buried depth of groundwater level is 1.0–1.5 m, and the water inflow of single well is less than 10 m³/d. The micro-confined aquifer consists of intermediate fine sand, medium-coarse sand with gravel, gravel cobble and the like, where the buried depth of groundwater level is 1.0–3.0 m, and the water inflow of single well is more than 1000 m³/d. During construction, it shall decompress and dewater the micro-confined aquifer, so that the groundwater level is at least 1.0 lower than the excavation face of foundation pit, and it shall strive to control the settlement of surrounding environment,

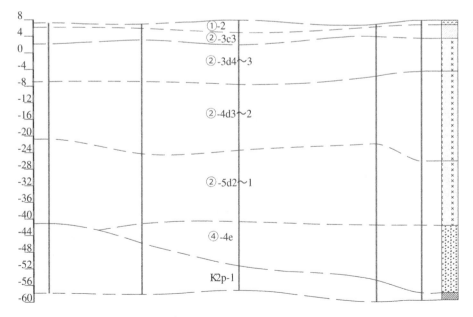

Figure 1.   Map of geological cross section.

to ensure the safety for natural gas station at the south of foundation pit and Venice Water City at its northeast.

3 NUMERICAL SIMULATION FOR THREE-DIMENSIONAL UNSTEADY SEEPAGE AND LAND SUBSIDENCE OF DEEP FOUNDATION PIT DEWATERING

3.1 Mathematical model of groundwater movement

3.1.1 Control equation

The three-dimensional unsteady seepage equation[8–14] of groundwater movement is as follows according to the Darcy law and principle of continuity:

$$\frac{\partial}{\partial x}\left(k_{xx}\frac{\partial h}{\partial x}\right)+\frac{\partial}{\partial y}\left(k_{yy}\frac{\partial h}{\partial y}\right)+\frac{\partial}{\partial z}\left(k_{zz}\frac{\partial h}{\partial z}\right)+W=\mu_{s}\frac{\partial h}{\partial t} \qquad (x,y,z)\in\Omega, t>0 \qquad (1)$$

where k_{xx}, k_{yy} and k_{zz} are the permeability coefficient of anisotropic main direction respectively; h is the water head value of points (x,y,z) at t time; W is source sink term; μ_{s} is water storage rate; t is time; and Ω is calculation zone.

3.1.2 Initial condition

$$h(x,y,z,t)\big|_{t=t_{0}}=h_{0}(x,y,z,t_{0}) \qquad (x,y,z)\in\Omega \qquad (2)$$

where $h_{0}(x,y,z,t_{0})$ is initial water head value at point (x,y,z).

3.1.3 Boundary condition

The boundary condition includes the first boundary condition (Dirichlet condition), the second boundary condition (Neumam condition) and free surface boundary condition:

$$\begin{cases} h(x,y,z,t)\big|_{\Gamma_{1}}=h_{1}(x,y,z,t) & (x,y,z)\in\Gamma_{1}, t>0 \\ k_{xx}\frac{\partial h}{\partial x}\cos(n,x)+k_{yy}\frac{\partial h}{\partial y}\cos(n,y)+k_{zz}\frac{\partial h}{\partial z}\cos(n,z)\bigg|_{\Gamma_{2}}=q(x,y,z,t) & (x,y,z)\in\Gamma_{2}, t>0 \\ h(x,y,z,t)=z(x,y,t) & (x,y,z)\in\Gamma_{3}, t>0 \\ k\frac{\partial h}{\partial n}\bigg|_{\Gamma_{3}}=-\mu\frac{\partial h}{\partial t}\cos\theta & (x,y,z)\in\Gamma_{3}, t>0 \end{cases} \qquad (3)$$

where Γ_{1}, Γ_{2} and Γ_{3} are the first boundary, the second boundary and free surface boundary respectively; $h_{1}(x,y,z,t)$ is water heat value of the first boundary; k is the permeability coefficient of free surface boundary; μ is saturation deficit while the free surface rises or specific yield while the free surface falls (dimensionless); θ is the angle of outer normal of infiltration curve of free surface boundary with vertical direction; $q(x,y,z,t)$ is replenishment quantity of unit area on the second boundary; $\cos(n,x)$, $\cos(n,y)$ and $\cos(n,z)$ are cosine of angles in the coordinate axis direction and flow boundary outer normal direction respectively.

3.2 Mathematical model of land subsidence

The aquifer compression is caused by groundwater recession, and its calculation model[15] is:

The amount of elastic deformation Δb and amount of inelastic deformation Δb^{*} of phreatic aquifer are as follows respectively:

$$\begin{cases} \Delta b=-\Delta h\left(1-n+n_{w}\right)\mu_{ske}b_{0}=-\Delta h\mu_{fe} \\ \Delta b^{*}=-\Delta h\left(1-n+n_{w}\right)\mu_{skv}b_{0}=-\Delta h\mu_{fv} \end{cases} \qquad (4)$$

73

Figure 2.   3-D subdivision graph for the study area.

The amount of elastic deformation Δb and amount of inelastic deformation Δb^* of confined aquifer are as follows respectively:

$$\begin{cases} \Delta b = -\Delta h \mu_{\mathrm{ske}} b_0 = -\Delta h \mu_{\mathrm{fe}} \\ \Delta b^* = -\Delta h \mu_{\mathrm{skv}} b_0 = -\Delta h \mu_{\mathrm{fv}} \end{cases} \qquad (5)$$

where: Δh is changed water head value (m); n is porosity; n_w is the moisture capacity as a part of total volume of porous medium above groundwater level; μ_{fe} is elastic storage factor of soil skeleton[16–17]; μ_{fv} is inelastic storage factor of soil skeleton; μ_{ske} is the elastic water storage rate of soil skeleton; and μ_{skv} is inelastic water storage rate of soil skeleton.

3.3   Model generalization and discretization

The author generalizes the whole model as heterogeneous anisotropy on the three-dimensional space, taking boundary around the foundation pit in the calculation zone on the horizontal direction as starting point, and expanding 800 m respectively towards east, south, west and north, where the plane dimension of model is 1774×1626 m. According to the groundwater systematic characteristics of study area, the boundary around is generalized as the boundary of fixed water level, the phreatic aquifer and micro-confined aquifer are arranged from top to bottom respectively in the vertical direction, the bottom is taken from the bottom plate of strong weathered siltstone layer, where the hydraulic connection is occurred among the layers. In the process of deep foundation pit dewatering, the groundwater outside the foundation pit flows around to enter into the interior through the bottom of underground diaphragm wall of foundation pit, the flow state of groundwater is three-dimensional unsteady flow, and the unwatering decompression well in the foundation pit is the only source sink term.

Figure 2 is the plane, three-dimensional and vertical subdivision grid of study area. As shown in the figure, the whole calculation zone is rectangular, with total area of about 2.88 km². The three-dimensional subdivision shall be performed according to the structural feature of actual hydrogeology of calculation zone. The whole calculation area is subdivided into 43×75 rectangular grid cells on the plane. Vertically, considering the influence of pumping well, observation well and layer thickness, it is totally subdivided into 6 layers, where the amount of valid cell is 19350, and the amount of valid node is 23408.

3.4   Identification and verification of model

The observed value of groundwater level and observed value of number of land subsidence required by inverse analysis shall be obtained by the deep foundation pit dewatering information and observational data of land subsidence. This numerical simulation arranges 15 dewatering wells in the third floor and fourth floor inside the foundation pit, and arranges 10 observation wells in the second

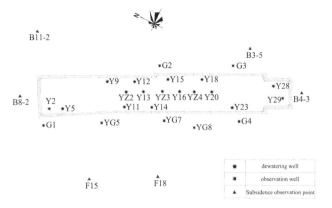

Figure 3. The schematic diagram for well point.

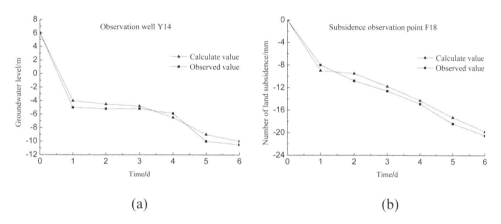

(a)   (b)

Figure 4. Groundwater level and land subsidence fittings chart of observation points.

floor and third floor of model, where Y2, Y14 and 29 observation wells are arranged in the pit; and G1, G2, G3, G4, YG5, YG7 and YG8 observation wells are arranged outside the pit. According to the different distance away from the foundation pit, six representative observation points of land subsidence F15, F18, B3-5, B4-3, B8-2 and B11-2 are used for fitting for number of land subsidence, seeing Figure 3 in detail.

Select six days from August 19, 2011 to August 25, 2011 as the identification and verification period of model, and totally divide into 6 stress periods, where each stress period has one time step. The identified model is totally divided into 120 parameter partitions. Figure 4(a) is the fitting chart of calculated value and observed value of groundwater level of Y14 observation well, and Figure 4(b) is the fitting chart of calculated value and observed value for settlement of fitting chart of calculated value and observed value of observation point of land subsidence.

It can be seen from the figure above that the fitting precision for calculated curve and measured curve of groundwater level is relatively good, the total variation trend is consistent, the calculated curve and measured curve of land subsidence is also basically coincided, and the error is small, which can verify that the model built is correct and reliable. The various parameters of model are reasonably, which can accurately reflect the substantive characteristics of groundwater system in the area, and can be used for simulation and prediction for deep foundation pit dewatering and land subsidence of Puzhou Road station in Metro Line 3.

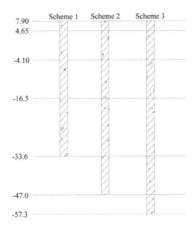

Figure 5.   The depth of continuous diaphragm wall of three schemes.

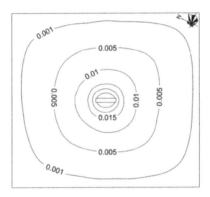

Figure 6.   Land subsidence contour map after foundation pit dewatering 66 days.

4   MULTI-SCHEME COMPARATIVE ANALYSIS

With the identified and verified numerical model of three-dimensional unsteady seepage and land subsidence of groundwater, it plans to use three different containments (of which the insertion depth is different) and dewatering well layout scheme (of which the position and pump output are different) for simulation. The depth of continuous diaphragm wall of three schemes is shown in Figure 5.

The water head value inside and outside the foundation pit in three schemes and number of land subsidence caused by dewatering around the foundation pit are predicted and analyzed. According to the deep foundation pit dewatering and land subsidence control requirement, the groundwater level needs to be reduced to 1 m below the bottom plate of foundation pit, namely the groundwater level is reduced to -12.9 m (of which the buried depth is 20.8 m).

(1) Scheme I

Scheme I totally selects 15 wells, i.e. Y5, Y9, Y11, Y12, Y13, Y14, Y15, Y16, Y18, Y20, Y23, Y28, YZ2, YZ3 and YZ4, as the pumping well, where the flow of pumping well is reduced from 2160 m³/d initially to 1440 m³/d finally. The diaphragm wall of foundation pit is inserted into the bottom plate of silt layer in the fourth floor of model, the depth of bottom of diaphragm wall is about 41.5 m and meet the excavation requirement of foundation pit after dewatering 10 days, namely, the groundwater level inside the foundation pit is reduced to -12.9 m (of which the buried depth is 20.8 m). Figure 6 is land subsidence contour map of the third plane after dewatering 66 days, and

Figure 7. Vertical section land subsidence contour map after foundation pit dewatering 66 days, x = 900 m (unit: m).

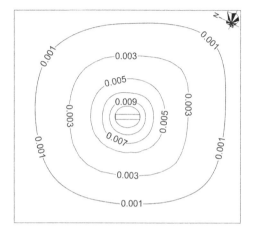

Figure 8. Land subsidence contour map after foundation pit dewatering 66 days.

Figure 9. Vertical section land subsidence contour map after foundation pit dewatering 66 days, x = 900 m (unit: m).

Figure 7 is vertical section land subsidence contour map after foundation pit dewatering 66 days, x = 900 m.

(2) Scheme II

Here, select totally 7 wells, i.e. Y5, Y11, Y15, Y16, Y18, Y23 and Y28, as the pumping well, where the flow of pumping well is reduced from 1200 m³/d initially to 300 m³/d finally. The diaphragm wall of foundation pit is inserted into the bottom plate of gravel layer in the fifth floor of model, the depth of bottom of diaphragm wall is about 54.9 m and meet the excavation requirement of foundation pit after dewatering 10 days, namely, the groundwater level inside the foundation pit is reduced to −12.9 m (of which the buried depth is 20.8 m). Figure 8 is land subsidence contour map of the third plane after dewatering 66 days, and Figure 9 is vertical section land subsidence contour map after foundation pit dewatering 66 days, x = 900 m.

(3) Scheme III

Scheme II selects Y11 and Y20 as the pumping well, where the flow of pumping well is reduced from 500 m³/d initially to 50 m³/d finally. The diaphragm wall of foundation pit is inserted into the bottom plate of gravel layer in the sixth floor of model (to bedrock), and the depth of bottom of diaphragm wall is about 65.2 m. Theoretically, there is no hydraulic connection inside and outside the model pit and seepage phenomenon of diaphragm wall around the foundation pit,

Table 1. Project cost statistics.

	Diaphragm Wall [yuan]	Pumping [yuan]	Total [yuan]
Scheme I	41514400	1800000	43314400
Scheme II	47988600	840000	48828600
Scheme II	52965000	240000	53205000

and the seepage field outside the foundation pit is not changed, and does not also have land subsidence.

5 OPTIMIZATION DESIGN OF DEEP FOUNDATION PIT DEWATERING

At present, there are mainly the following two problems in the excavation process of deep foundation pit:

(1) Some deep foundation pit project accidents with severe economic losses are often caused by unreasonable design and construction, which will not only have influence to the foundation pit engineering itself, but also damage the building, municipal engineering and other facilities around.
(2) The design of foundation pit support is very conservative, causing the great waste of resources. The second is ignored more easily in the actual project. From the economic point of view, half of total project cost is often occupied by the cost of foundation pit project. The construction period of foundation pit project also decides the construction period of whole project to some extent, and the safety and stability of foundation pit decides the safety and stability of large buildings in one sense.

The selection of optimal scheme of deep foundation pit dewatering involves many factors, which can be divided into the following four as a whole: technical reliability and advancement, feasibility of construction and others; economic benefit evaluation; evaluation of environmental influence; and comparison of construction period. The author obtains the following conclusions through systematic analysis and demonstration:

a) For the technical reliability and feasibility of construction, the diaphragm wall in the Scheme II and Scheme II shall be passed through the gravel-cobble layer, and due to high gravel hardness, the construction difficulty is increased greatly, and the feasibility of construction is poor.
b) For economic benefit, Table 1 is the project cost statistics of three different schemes. It can be seen from the table that the cost required by Scheme I is least, and the economic benefit is significant.
c) For environmental influence, it is easy to discover that the number of land subsidence outside the pit in Scheme I and Scheme II meets the requirement, and has no threat to the buildings around; and in Scheme III, since the diaphragm wall passes through the whole model, there is no hydraulic connection inside and outside the pit, therefore, theoretically, the land subsidence will not happen around the foundation pit, and have no influence to geological environment.
d) For the construction period, without double, the depth of Scheme I is relatively light, and the time required by the foundation pit project is shorter than that in Scheme II and Scheme III, which will also directly impact on the economic benefit.

Figure 10 enumerates the curve chart for groundwater level and duration of land subsidence of part of observation points in Scheme I. The Figure shows that the groundwater level of groundwater level and land subsidence in Scheme I have little difference with the actual value, which proves that the diaphragm wall structure and pumping scheme of Scheme I designed by the author meet the actual construction requirement.

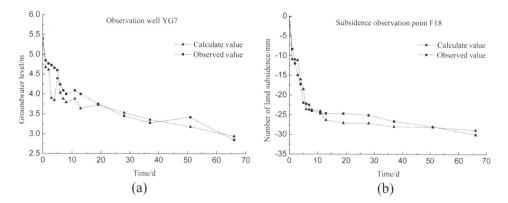

Figure 10.   Fittings of groundwater level and subsidence with data from the simulation and observation.

In conclusion, Scheme I is relatively good scheme, which can be safe and reliable, be convenient for construction, shorten the construction period, and can bright considerable economic benefit and social benefit.

6 CONCLUSION

On the basis of summarizing the design theory and calculation method of diaphragm wall and dewatering scheme of deep foundation pit and, the author stimulates and calculates the three diaphragm wall supporting and dewatering schemes of deep foundation pit dewatering of Puzhu Road station in Nanjing Metro Line 3 with coupling mathematical model of three-dimensional unsteady seepage and land subsidence of groundwater by taking the deep foundation pit dewatering project of Puzhu Road station in Nanjing Metro Line 3 as the example, and obtains the pumping well of which 15 filters are arranged in the foundation pit and the buried depth is 22.0–37.0 m. The pit will be dewatered when the flow is reduced from 2160 m^3/d initially to 1440 m^3/d finally, so as to obtain the optimum dewatering and diaphragm wall supporting design scheme of which the depth for the bottom of diaphragm wall of foundation pit is about 41.5 m. While meeting the dewatering requirement in the pit, the number of land subsidence outside the pit meets the control requirement, and the number of land subsidence at 100 m around the foundation pit is less than 25 mm, which will not have influence on the safe and stable operation of buildings around, and can bright considerable economic benefit and social benefit.

REFERENCES

[1] Hu, C.L. & Yang, H.J. (2002) Study on deep well dewatering optimization design in deep foundation pit and engineering application. *Journal of China University of Geosciences*, 13, 78–82.
[2] Li, L. (2007) Study on effect of engineering dewatering on the behavior of deep foundation pit and surrounding environment. Tongji University.
[3] Wang, J.X., Feng, B., Yu, H.P., Guo, T.P., Yang, G.Y. & Tang, J.W. (2013) Numerical study of dewatering in a large deep foundation pit. *Environmental Earth Sciences*, 69, 863–872.
[4] Liu, C.A. & Zhang, Y. (2011) *Foundation Pit Dewatering Design*. Beijing, Geological Publishing House.
[5] Hua-Yong Chen, Chang-Yu Wu, Jin-Hua Ding, et al, Muti-objective Fuzzy Decision Making in Dewatering Plans during Excavation of Deep Foundation Pit, J. Journal of Yangtze River Scientific Rescearch Institute, 6 (2008) 86–89.
[6] Lin-Gao Wu, et al, Design and Executin of Dewatering & Theory of Seepage in Deep excavation, M. Beijing, China Communications Press, 2009.
[7] Lin-Gao Wu, Li Guo, Zhao-Chang Fang, et al, Dewatering Case History For Excavation, M. Beijing, China Communications Press, 2009.

[8] Zhi-Fang Zhou, Hong-Gao Zhu, Jing Chen, et al, Nonlinear Coupling Calculation Between Dewatering and Settlement of Deep Foundation Pits, J. Rock and Soil Mechanics, 25 (2004) 1984–1988.

[9] Yang Ping, Shi-Wei Bai, Yan-Ping Xu, Numerical Simulation of Seepage and Stress Coupling Analysis in Deep Foundation Pit, J. Rock and Soil Mechanics, 22 (2001) 37–41.

[10] N Q Zhou, P A Vermeer, R X Lou, et al, Numerical Simulation of Deep Foundation Pit Dewatering and Optimization of Controlling Land Subsidence, J. Engineering Geology, 114 (2010) 251–260.

[11] K J Bathe, M R Khoshgoftan, Finite Element Free Surface Seepage Analysis without Mesh Iteration, J. International Journal for Numerical and Analytical Methods in Geomechanics, 3 (1979): 3–22.

[12] Wei Xiong, Gen-Lin Tian, Li-Xin Huang, et al, Solid-Fluid Coupling Phenomenon in Deformable Porous Media, J. Journal of Hydrodynamics, 17 (2002) 770–776.

[13] Xiao-la Feng, Wen-Lin Xiong, Tao Hu, et al, Application of 3D Coupling Model of Seepage and Stress for Simulation Deep Foundation Fits Dewatering, J. Chinese Journal of Rock Mechanics and Engineering, 24 (2005) 1196–1201.

[14] Jian-Jun Luo, Cheng-Song Qu, Tian-Qiang Yao, Dewatering Project at the Tower Building of Shanghai Global Finance Center, J. Chinese Journal of Underground Space and Engineering, 1 (2005) 645–649.

[15] Zu-Jiang Luo, Lang Li, Tian-Qiang Yao, et al, Coupling model of three dimensional seepage and land-subsidence for dewatering of deep foundation pit in loose confined aquifers, J. Chinese Journal of Geotechnical Engineering, 28 (2006) 1947–1951.

[16] H C Wen, 3D-Groundwater Modeling with PMWIN: A Simulation System for Modeling Groundwater Flow and Transport Processes, M. USA Springer, 2001.

[17] T J Burbey, D C Helm, Modeling Three-Dimensional Deformation in Response to Pumping of Unconsolidated Aquifers, J. Environmental and Engineering Geoscience, 5 (2004) 199–212.

Electrical and computer engineering

Engineering Technology and Applications – Shao, Shu & Tian (Eds)
© 2014 Taylor & Francis Group, London, ISBN 978-1-138-02705-3

Research and application of consumer power utilization monitoring in intelligent power distribution and utilization system

Huijuan Guo & Liang Shi
Jiaozuo Power Supply Company, State Grid Henan Electric Power Company, Jiaozuo, Henan, China

ABSTRACT: This paper is intended to build a system structure based on the business integration requirements between marketing and distribution network. Through the analysis over the correlated techniques and the application system for the smart operations of distribution network and according to the features and functions of the distribution system, the author proposed, based on a distribution system reliability evaluation method used in non-sequential simulation, a corresponding reliability evaluation system for power supply, thus to facilitate monitoring and management of power consumption. By the research, a user power monitoring and management platform for the intelligent power distribution & utilization system was set up, so were the data integration bus and enterprise service bus, and their respective functions realized. In implementation of the integration, various business systems were incorporated on an orderly basis according to the current business system status and expansion needs of the distribution network, thus providing a systematic basis for maintenance efficiency improvement and outage period reduction; furthermore, through and development and installation of reactive compensation device or other corresponding equipment, the quality of power supply is improved, line losses reduced, and the business objectives of improving customer satisfaction and quality of power consumption are thus realized.

Keywords: intelligent power distribution and utilization, power supply & operation syncretism, power monitoring and management

With the rapid development of intelligent power distribution network, the reliability evaluation and application of distribution network have evolved into an important branch of reliability technology as well as the top issue in distribution automation field. Distribution network is at the terminal of the electric power system and an important link in power distribution and supply to terminal users; it's designed to connect the power system or transmission & substation system with the user facilities, including the entire distribution network and facilities like the distribution substations, distribution circuit, service lines etc. According to the statistics collected by power companies, 80% of the power failure can be rooted in distribution network fault[1], thus the improvement of distribution network reliability is deemed as one of the major and vital means to ensure the reliability of power supply, so it has become a top concern in national electrical network planning and reconstruction programs of many countries.

1   INTRODUCTION

With the development of electric power industry and the gradual introduction of theories in reliability engineering since 1960s, the power system reliability evaluation has started to enter into a utility stage[17,18]. In 1964, the Markov process mathematical model was introduced into the power system reliability evaluation by Desineo and Stine for the first time. Later Billinton and Stanton, by computing the results of the linear algebraic equation constituted by the transfer-rate matrix from the Markov process model, they obtained the mean time to failure and repair for the system under a long-term probability distribution. Under the impetus of rapid sci-tech updating, it's of high necessity to enhance the comprehensive automation level of distribution network, and improve the quality and reliability of power supply on a constant basis [11].

Statistics show that the measures to improve the reliability of its distribution network in Western developed countries are not always the same[12]. Canada takes the power supply reliability and the user's reliability level into integral consideration, customer lines with different importance degree will be set with different power supply reliability standards. Japan gives priority to the level management of network structure and switching capacities and improves the automation level of its entire distribution network, thus ensuring the swift load transfer upon power failure to reduce the outage time; so far, ideal running effect has been achieved. By comparison, the UK and the USA have developed their own detailed guidelines on reliability, and issued the concrete reliability evaluation methods, equipment reliability data, basic theories on reliability and efficiency relation analysis as well as power failure loss data etc., so as to provide reference to their power supply companies, the guidelines are revised on an annual basis.

The electric system reliability study is relatively delayed, only started since 1970s. In 1985, the Ministry of Water Conservancy and Electric Power funded the establishment of the Electric Power Reliability Management Center, since then the work on power generation, power transmission and transformation, reliability statistics of distribution system as well as the research and development of relevant standards were under way, in addition, the electric system reliability engineering management was carried out in an in-depth way, many universities and research institutes have started to carry out theoretical investigation and training work on electric power system reliability [14, 15, 16].

With regard to the content of distribution network reliability evaluation research, it's typically divided into the following aspects [13]:

(1) Defining the indicators of distribution network reliability evaluation;
(2) Statistics and analysis over the reliability evaluation indicators of the distribution network, and the application of such results to serve as a guide in each production processes of the prevailing system, including the design, manufacturing, installation, debugging, running, maintenance and overhaul;
(3) Various effective measures and countermeasures to realize the guidance function for distribution network reliability analysis and prediction;
(4) The coordinative relation between the cost and economy in distribution network reliability evaluation as well as the actual application of Power System Reliability Economics etc.

Based on the all-rounded analysis over the distribution system reliability researches and according to the features and functions of the distribution system, the author proposed, with the help of a distribution system reliability evaluation method used in non-sequential simulation, a corresponding reliability evaluation system for power supply, thus to facilitate monitoring and management of power consumption.

2 DESCRIPTION OF THE SYSTEM

The Project mainly studied: 1) the business integration requirements between marketing and the distribution network and the system structure to satisfy this requirement; 2) key technology and its application modes like the relevant fault detection and location technologies for intelligent operation of distribution network, communications group system techniques of distribution network[2], front-end collecting device power supply technology, distribution and transformer monitoring technology, distribution network reactive power compensation technique etc.; 3) the development of distribution network data measurement devices, collecting device and compensation equipment according to research results, and the establishment of a smart operation all-in-one platform for the distribution network based on the "power supply & operation syncretism" concept, as well as the verification over the feasibility and practicability of such technology and programs.

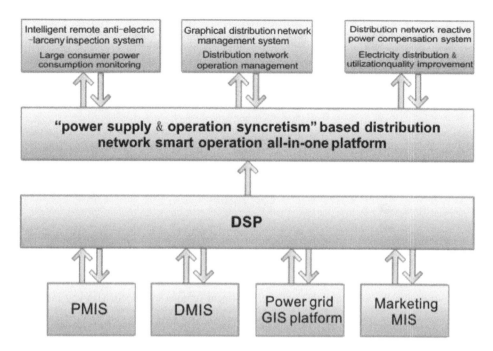

Figure 1. System logic structure diagram.

The focus in this Project are to develop a data acquisition unit for high voltage side[3], thus to realize the data collection of the high voltage side and secondary sides of distribution transformer; to conduct in-depth analysis over the collected data, and calculate the loss of distribution transformer; to describe the running status according to the calculating results; to analyze the service modes of transformer and realize the status-based forewarning and analysis; to develop a corresponding dynamic reactive power compensation device thus realizing user three-phase equilibrium and reactive dynamic compensation; and eventually realizing the purpose of transformer condition monitoring and overhaul, line loss minimization as well as improved quality consumption.

3 FUNCTIONAL MODULES OF THE SUBSYSTEM

Through related analytical investigation and according to the "Three Intensive Managements and Five Huge Systems" requirements, a user power monitoring and management supporting platform for the "power supply & operation syncretism" smart distribution network is required to build up, this Project is mainly intended to realize the function of power monitoring and management over the large consumers for the marketing inspection monitoring platform, the production and operation management platform for power distribution, as is mentioned in the distribution network outage management requirements, thus to improve the power supply reliability analysis; meanwhile, corresponding distribution network reactive-load compensation equipment shall be developed, thus realizing the reactive power optimization of distribution network lines and improving the consumption quality for users.

3.1 Intelligent remote anti-electric-larceny inspection system

This independently operable system[4] is constituted by the high voltage current collecting device, data receiving terminal and an intelligent anti-electric-larceny inspection master station. It can not only monitor whether there are electricity stealing activities but identify likely high-tech electricity

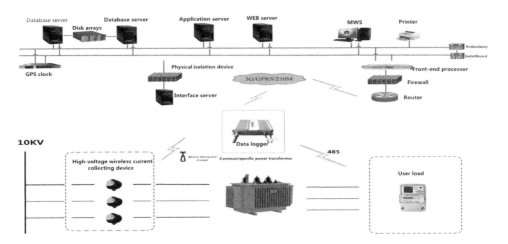

Figure 2.   System network topological structure.

stealing ways as well. Structure chart of the intelligent remote anti-electric-larceny inspection system.

3.2   *Graphical distribution network management system*

The graphical distribution network management system consists of two subsystems: the drawing subsystem and operation management subsystem. The functions are as follows: 1) realizing workflow-based shift change, the dispatcher may from time to time modify the distribution network graph according to the power grid status; 2) realizing auto retrieve of equipment parameters in SCADA, coloring the switch liaison diagram of the distribution network according to the changes in power network status, thus reflecting the status quo of power grid; 3) giving early warning when faulty operation or malfunction of grid occur. Operation structure chart of the system.

3.3   *Distribution network reactive power compensation system*

The distribution network reactive power compensation system offers a solution to the usual switching oscillation problem under light load yet giving inadequate compensation[8] under heavy load, as is normally occurred under former switching criteria. Compared with the traditional simple reactive power compensation devices, this device can not only make automatic compensation to the reactive power of the load of each phase, but also make automatic adjustment against the asymmetry in three-phase load active power, thus adjusting it to a symmetrical or quasi-symmetrical (so much as to meet the national specified standards) situation.

4   RESEARCH ON THE DISTRIBUTION RELIABILITY

The distribution network is at the terminal of the electric power system and deemed as an important link in power distribution & supply to terminal users, it's designed to connect the power system or transmission & substation system with the user facilities, including the entire distribution network and facilities like the high and low voltage electricity distribution lines and distribution substations etc. Based on the whole distribution network, this paper made an attempt to adopt the non-sequential simulation method in complex distribution system so as to explore the reliability of power distribution.

Monte Carlo stochastic modeling method is intended to obtain a certain amount of sample data according to some given probability distribution (probability density function or probability

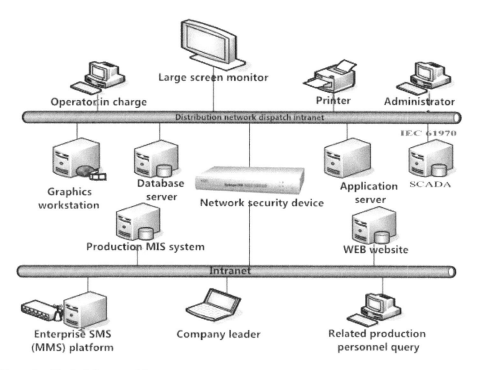

Figure 3. Physical diagram of the system.

distribution function), thus to ensure the distribution could have totally identical statistics features. The basic principle of this method is rooted in the Bernoulli Law of Great Numbers. Provided the occurrence of random event Xi is $P\{xi\}$, and in the N independent sampling experiments, the occurrence frequency is m/N (m refers to the event, Xi refers to the occurring times within these N experiments), then for any given tiny positive number $\epsilon > 0$, we can get:

$$\lim_{N \to \infty} P\left\{ \left| \frac{m}{N} - P\{x_i\} \right| < \varepsilon \right\} = 1 \tag{1}$$

i.e., when the independent experiment times N is greater enough, its frequency m/N will converge to $P\{xi\}$ from probability 1. Thus the probabilistic convergence of the simulation method can be guaranteed.

Non-sequential analog simulation method is often deemed as state sampling method, which has been widely used in reliability evaluation projects. The judgment criteria are as follows: taking system state as a combination of state of all components, and the state of every component can be ascertained by its occurrence probability in such state as is calculated in sampling. Each component is simulated by a uniform random number at [0,1] interval [19]. Each component has working or failed state, and the failed state is caused by independent factors. Si is used to indicate the state of component i, Qi is its failure probability, then a random number Ri is generated for component i, when $Ri > Qi$, the state of Si is 0, indicating the working state; when $0 < Ri < Qi$, the state Si is 1, indicating a failed state.

When all the state probabilities of each main feeder network (herein it refers to the radial distribution network simplified in the principle of equivalence which does not contain branch feeders) have passed the sample assessment, the malfunction probability, failure frequency, average duration of failure and other reliability indexes of system can be calculated with loading points and the system reliability indexes. In this paper, the author assumes the interconnection switch and sectionalizing switch are absolutely reliable, i.e. the error rate being 0. As to whether the fuse protector is disconnected or not, it's designed as: when a random number is given in simulation, if this random number is greater than the probability of the fuse protector's reliable disconnection, it indicates failure in disconnection and vise versa.

As for the simple main feeder system, the key steps for its reliability simulation evaluation are as follows:

(1) Set the analogue simulation number of years $k = 0$;
(2) $k = k + 1$; if it reaches the k year, the random number group $\{r1, r2, r3, \ldots rn\}$ is generated to simulate the running state each zone node of the entire network;
(3) Find out the fault components of the entire distribution network, and investigate the types of fault elements;
(4) Count the power failure times and outage hours of each loading point by year and sum the results up on an accumulated and yearly basis;
(5) If the simulation convergence conditions are met, stop the sampling analog computation, otherwise repeat step (2);
(6) Work out the mean failure rate, average interruption hours and the annual average interruption hours of each loading point within the overall simulation years, and calculate the reliability index of the system.

The above non-sequential analog simulation process may be described with the following flowchart.

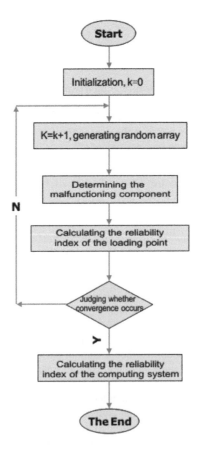

Figure 4. Arithmetic flow chart.

5 SYSTEM APPLICATION

This Project chiefly explores how to make the daily routine in distribution network done on the basis of workflow, and how to realize unified management as well as the reactive dynamic compensation for distribution network, and enables the unified management over the switch liaison system diagram of the distribution network. All the equipments deployed or managed in urban districts can correlate with each other in the system diagram and single line diagram, thus providing an intuitive and high-efficient management tool to the dispatchers. It takes on the features of being convenient to use, easy to maintain, practical, advanced in technology as well as other advantages, and turns out to be a high-efficient supplementary means for dispatchers to fulfill their daily operational and maintaining work, thus helping them raise work efficiency and ensure safety of the distribution network. The realization of reactive power compensation system can meet the requirements of dynamic reactive power compensation standards in electric system, thus making distribution network transformer reactive real-time compensation come true.

6 CONCLUSION

With the development of the modern information technology and according to the "Three Intensive Managements and Five Huge Systems" requirements, a user power monitoring and management platform for the "power supply & operation syncretism" based intelligent power distribution and utilization system is built up, which mainly realizes the function of power monitoring and management over the large consumers, and the production and operation management platform for power distribution, as is mentioned in the distribution network outage management requirements[10], thus to improve the power supply reliability analysis and provide a systematic basis for maintenance efficiency improvement and outage period reduction; furthermore, through the development and installation of reactive compensation device or other corresponding equipment, the quality of power supply is improved, line losses reduced, and the business objectives of improving customer satisfaction and quality of power consumption are thus realized.

REFERENCES

[1] Anonymous. Smart Grid Enabling the Development of Low-carbon Economy [J]. Power System and Clean Energy, 2010.

[2] Zhang Gang. Study on Policies to Promote Development of China's Smart Grid System [J]. State Grid, 2011(5).

[3] Wang Chengshan, Wang Shouxiang and Guo Li. Prospect over the Techniques of Smart Distribution Network in China [J]. SOUTHERN POWER SYSTEM TECHNOLOGY, 2010, 4.

[4] Chen Yanhui. Research and Implementation on the Field Intelligent Communications Devices for Distribution Network Terminal [D]. Master Dissertation of Central South University, 2003.

[5] Li Juanjuan and Chen Feng. Discussion on Integration Reform of Power Dispatching and Control in Distribution Networks [J]. ELECTRIC POWER IT, 2011.

[6] Gong Jing. Integrated Automation Technology in Distribution Network [M]. China Machine Press, 2008.

[7] Chen Fuming. Common Information Model Extension and Case Simulation of Distribution System [D]. Master Dissertation of Tianjin University, 2007.

[8] Xie Kai et al. The Vision of Future Smart Grid [J]. Electric Power, 2008.

[9] Wang Jun. Optimal Design on the 20kv Medium-voltage Distribution Network [D]. Shenyang: Shenyang Agricultural University, 2007.

[10] Sun Yimin et al. A Grading Solution for Building Digital Station [J]. Automation of Electric Power Systems, 2007.

[11] Billinton. R, Kmuar. S, Chowdhury. N, et al. A Reliability Test System for Educational Purposes-basic Data. IEEE Transactions on Volumes, 1989, 4(3): 1238–1244.

[12] Guo Yongji. The Theory and Application of Power System Reliability [M]. Beijing: Tsinghua University Press, 1986, 3–20.

[13] R. N. Allan, R. Billinton. Bibliography on the Application of Probability Methods in Power Systems Reliability Evaluation. IEEE Transactions on PAS, 1994, 9(1): 41–48.

[14] Ding Ming, Zhang Jing and Li Shenghu. A Sequential Monte-carlo Simulation Based Reliability Evaluation model For Distribution Network [J]. Power System Technology, 2004, 28(3): 38–42.

[15] Xie Kaigui, Zhou Ping and Zhou Jiaqi et al. Reliability Evaluation Algorithm for Complex Medium Voltage Radial Distribution Networks Based on Fault-spreading-method [J]. Automation of Electric Power Systems, 2001, 25(4): 45–48.

[16] Xu Yingqi, Li Rilong and Chen Shuting. A New Reliability Evaluation Method for Distribution System Implemented by Visual Basic [J]. Power System Technology, 2004, 28(3): 48–54.

[17] G. Tollefson, R. Billinton. Comprehensive Bibliography on Reliability Worth and Electrical Service Consumer in Terraption Costs. IEEE Transactions of PAS, 1991, 6(4): 1508–1513.

[18] R.N. Allan, R. Billinton. Bibliography on the Application of Probability Methods in Power Delivery, 1989, 4(1): 561–568.

[19] Goel L. Monte Carlo Simulation-based Reliability Studies of a Distribution Test System. Electric Power Systems Research, 2000, 54(1): 55–56.

Engineering Technology and Applications – Shao, Shu & Tian (Eds)
© 2014 Taylor & Francis Group, London, ISBN 978-1-138-02705-3

The design and realization of an HV breaker monitoring system

Hongbing Li, Yuxiang Liao & Zhiyong Li
Jiangbei Power Supply Company, State Grid Chongqing Electric Power Company, Chongqing, China

ABSTRACT: The significance of studying and designing condition monitoring techniques for high-voltage circuit breakers is discussed by discussing what important role high-voltage circuit breakers play in power systems. After summarizing the current achievements and challenges relating to the condition monitoring and evaluation systems for high-voltage circuit breaker, a new monitoring system is designed which further refines the monitoring contents and system operation flow and allows the monitoring of coil current, vibration signals, contacts and breaking times. The new system is a reform from conventional scheduled maintenance to real-time monitoring and offers a thorough solution for the safe, efficient operation of HV breakers.

Keywords: high-voltage circuit breaker; condition monitoring; fault rate; functional diagnosis

1 SIGNIFICANCE OF HV BREAKER CONDITION MONITORING TECHNOLOGY

The HV circuit breaker in a power grid is designed as a control and protection device [1] which, to guarantee normal service of the electric equipment, connects or disconnects the equipment or circuit as necessary to change its switching mode so that once a failure occurs, the equipment quickly switched over and the grid is still able to run sound and safe. In this way, HV circuit breakers play a very important role in the safe operation of the power grid.

Statistics of Jianbei Power Supply Bureau show that as much as 60% of the substation maintenance cost is spent on the inspection and repair of HV circuit breakers. The colossal quantity of HV circuit breakers in service forms a challenge as maintenance could involve large volume and high cost. Scheduled maintenance, which is currently the prevailing way of maintenance, is often aimless and unselective. Each maintenance event is a routine blanket check that not only takes a lot of labor and material, but also affects the normal service of other systems and equipment. If major service or disassembly is involved for an HV circuit breaker, a lot of money and time would be spent. The normal service of other equipment would also be affected leading to unnecessary depletion, shorter service life or even other failures that could result in immeasurable losses to the normal service of the power grid. In this context, there is pressing need to develop a well-targeted real-time HV breaker condition monitoring and evaluation system.

2 CURRENT RESEARCH ON HV BREAKER CONDITION MONITORING AND EVALUATION SYSTEMS AND CHALLENGES

2.1 *Current HV breaker condition monitoring technology*

Research on HV circuit breakers did not start as early as that on other devices contained in a power grid like transformers or generators, but has experienced rapid growth [3].

At present, means of monitoring the conditions and evaluating the failures of HV circuit breakers mainly includes:

1) Receiving and analyzing acoustic signals [4] and/or vibration signals to identify and solve early failures.

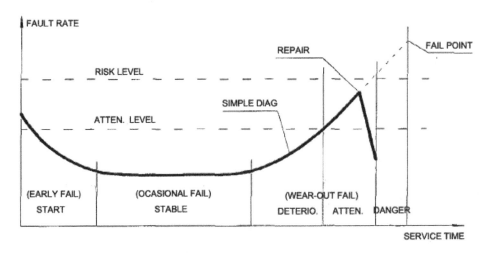

Figure 1. Fault rate to service life relation curve.

2) Summarizing the action frequency, 3-phase opening/closing and historic data of the circuit breaker as basis for diagnosing its operation conditions.

3) Fully recording and analyzing the current waveform of the tripped coil [5], and comparing it with that observed under normal operation to ensure accurate monitoring of the breaker conditions.

4) Measuring the voltage at the ends of the contact and the current of the opening/closing coil when the breaker is actuated, and extracting the opening/closing time, arcing time [6], arc voltage and arc energy from these signals to determine the conditions and life expectancy of the breaker.

5) At large, breaker condition monitoring products available so far are still under exploration and immature whether in or out of China. They are not completely suited to application yet.

2.2 One of the greatest technical challenges facing the online monitoring of HV circuit breakers

1) Their diverse working conditions. HV circuit breakers are extensively used and operated very differently, ranging from frequent to occasional. This difference makes the online monitoring of HV circuit breakers even more complex and difficult.

2) Difference in product structure. At present, many manufacturers are making HV breakers. Their products are largely different from each other in terms of the working principles of the operation mechanism and mechanical elements. The different manufacture years and parameter standards also add to the influence on, and difficulty in, the monitoring of HV breakers.

3) Available HV breaker monitoring systems so far incorporate merely data acquisition, which separates condition monitoring from data analysis and fault diagnosis so that the latter have to be performed by hand later on. This is not only time and labor-consuming, the manual diagnosis result is also liable to errors. Furthermore, as a breaker works from normal conditions to failure [7], showing one kind of signature or another during its operation, the identification of which according to existing information is beyond the ability of the monitoring and diagnosis means we have today.

3 SYSTEM DESIGN AND APPLICATION

By function module, an HV breaker monitoring system is designed to monitor:

(1) Accumulated switching times and frequency of the breaker;
(2) Switching time, tripping times and waveform of the breaker;

(3) Opening/closing coil current, voltage and path of the breaker;
(4) Conducting part temperature the HV breaker;
(5) Moving contact speed of the HV breaker;
(6) Insulation condition of the HV breaker;
(7) Mechanical vibration of the HV breaker [8].

3.1 *System design*

The system is designed according to the condition monitoring contents of the breaker. Its workflow is shown in figure 2 [9].

Before starting test, the host and the lower PC perform self-test and initialization, then establish communication connection between IPC and monitoring host PC. Select test option, add test parameter and click "START TEST". The lower PC receives test order and get ready for test; acquires test data, pre-processes them and send them to the host PC for diagnosis. Then it receives test data. The page refreshes and the test data are calculated. Diagnosis is performed on real-time basis according to historic data or curve of the HV breaker. If a fault occurs, a fault alarm is prompted so that the fault is located and warned against.

3.2 *Functional design*

1) Coil current monitoring and diagnosis: This is designed to monitor the coil current of the HV breaker with settings beyond which the system will automatically generate alarm signals.

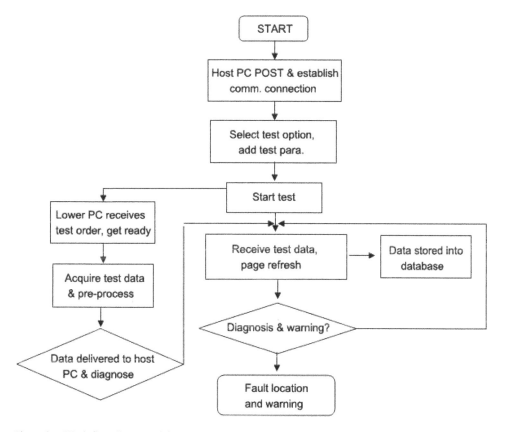

Figure 2.   Work flow diagram of the system.

2) Vibration signal monitoring and diagnosis: This is designed for the time domain analysis, energy analysis and identification calculation of breaker vibration signals to develop the eventual waveform of the HV breaker vibration signals with which to determine the operation conditions of the HV breaker [10].

3) Contact travel monitoring and diagnosis: This system is designed to monitor the travel, overtravel, opening/closing time, acceleration and speed of the breaker by measuring the moving contact of the breaker and incorporates thresholds [11] beyond which the system will automatically generate alarm signals.

4) Current breaking and breaking times monitoring and diagnosis: This calculates the wear level of the breaker contact according to the breaking current and breaking times to monitor the electric life of the breaker.

5) Information display and printout, which allows the inquiry of monitoring parameters of the breaker in question, including the coil current, vibration signals and breaker contact signals, any abnormality of which will trigger automatic alarm signals. The information can also be printed out.

6) Statistics, which counts the operation times of all breakers covered by the system within a period of time as basis for related operations.

7) Data sharing, which allows all the monitoring data in the system to be displayed and downloaded at the terminal so that the monitoring data are shared across the system.

4 CONCLUSIONS

The system proposed provides for real-time online monitoring of HV breakers. It eliminates the need of scheduled maintenance and allows early discovery, early diagnosis, early treatment and early prevention of breaker failure. It is a great improvement in minimizing the excess labor and material normally spent by conventional blanket service and allowing immediate updating of the operation conditions of all breakers involved.

REFERENCES

[1] Huang, Y.Z. (2000) Integrated substation automation technologies. *China Electric Power Press*, 3.
[2] Chen, H.G., Pan J.K. & Wu Y.H. (2000) *Fault Diagnosis and Treatment of High/low-voltage Switching Devices*. Beijing, China Water Conservancy & Electric Power Press.
[3] Liu, L.S. (2000) *Microcomputer Interface Technology and Application*. Wuhan, Huazhong University of Science and Technology Press.
[4] Li, G., Lin L. & Ye W.Y. (2002) *TMS32OF206DSP: Structure, Essentials and Application*. Beijing, Press of Beijing University of Aeronautics and Astronautics.
[5] Zhang, X.W., Chen, L. & Xu, G.L. (2003) *Principle, Development and Application of DSP Chips*. 3rd edition. Beijing, Publishing House of Electronics Industry.
[6] Yuan, S. (2001) *Condition Monitoring and Diagnosis Technology for High-voltage Switchgear*. Beijing, China Machine Press.
[7] Wang, X.M. (2002) *SCM Control of Electric Motors*. Beijing, Press of Beijing University of Aeronautics and Astronautics.
[8] Zhang, A.H. (1995) *Condition Monitoring and Fault Diagnosis Technology for Electric-Mechanical Equipment*. Xi'an, Northwestern Polytechnical University Press.
[9] Zeng, F.T. (2001) *VHDL Program Design*. Beijing, Tsinghua University Press.
[10] Weng, G.Q. (2003) *The Study on a Novel Substation Automation System*. Postgraduate thesis. Southwest Jiaotong University.
[11] Xu, J.F., Huang, Y.L., & Xu, G.Z. (2001) Study of travel-monitoring for high voltage circuit breaker. *High Voltage Apparatus*, 37(1), 13.

Engineering Technology and Applications – Shao, Shu & Tian (Eds)
© 2014 Taylor & Francis Group, London, ISBN 978-1-138-02705-3

An automatic white balance algorithm based on an improved method of specular reflection

Wuqi Chen
College of Information Science and Engineering, Ningbo University, Ningbo, Zhejiang, China

Ming Yang
College of Information Science and Engineering, Ningbo University, Ningbo, Zhejiang, China
High-tech Medical Optoelectronic Equipment Research and Development Center, Ningbo Zhejiang, China

Jianbo Duan & Yulong Fan
College of Information Science and Engineering, Ningbo University, Ningbo, Zhejiang, China

ABSTRACT: In response to the rigid application conditions and limited correction effect intrinsic to classic white balance algorithms, an improved specular reflection algorithm is introduced which divides an entire image into a number of regional blocks, looks for white dots in each of these blocks, derives the estimated values of unknown light sources by calculating the average color information of the dots and eventually corrects the white balance of the image. Our algorithm makes full use of the abundant color information over the image and results in highly reliable light source information. Experiment demonstrates that our algorithm provides an ideal effect of correction.

Keywords: white balance; gray world; specular reflection; white balance correction

1 INTRODUCTION

Different light sources vary in their spectral composition and distribution. This is what we call "color temperature". The term color temperature is often used to describe the color tone of different lights in Kelvin (K). For example, the color temperature under the sunlight of a fine day is approximately 6000K, 2000K at sunset and 1000K under candlelight. As the color temperature varies, so does the color of the target. This is particularly true for white objects: the scenery will turn out bluer than reality under a high color temperature and redder than reality under a low color temperature. To minimize the effect of external light sources on the color of the target object and restore the original color of the target scenery, color correction is necessary to achieve color balance. That is why white balance adjustment is used.

White balancing normally includes three steps: color estimation, which calculates the color temperature across the image using a certain algorithm; channel gain calculation, which derives the gain of red and blue channels according to the color temperature measurements; and color cast correction, which enforces the calculated channel gain on related components so as to correct the original image.

Among the many sophisticated White Balancing algorithms available so far, the most classic are Specular Reflection[1] and Gray World[2]. Auto White Balance algorithms introduced more recently include Artificial Neural Network (ANN)[3], Color Correlation[4], Gamut Mapping Model[5], Fuzzy Rule and White Balancing based on edge detection. These algorithms, however, are limited in one way or another though they do provide for ideal processing results for different images. For example, an algorithm based on Gray World could be deviation-prone if the image processed does not provide rich scene color and has a lot of solid-color objects in it. An algorithm based on Specular

Reflection could also fail if there is a very bright light spot in the picture processed. White Balancing has been studied for three decades outside China compared to the merely 10-plus years in China. All major research findings are from overseas researchers, making it imperative to take a deeper step into the research of White Balancing algorithms.

2 RELATED RESEARCHES

2.1 Gray World (GW)

Gray World algorithm is based on the assumption that given an image with sufficient amount of color variations, the average value of the RGB components of the image should equal over the entire image and the average reflection of this scene should be able to offset the color difference. The average of the three channels should average to a common gray value, i.e. $R = G = B = 128$. Gray World corrects the white balance of each image by calculating the gain of the three channels. For an M*N image, it is expressed as:

$$R'_{ij} = R_{ij} * 128 / \bar{R}, \quad G'_{ij} = G_{ij} * 128 / \bar{G}, \quad B'_{ij} = B_{ij} * 128 / \bar{B}$$

(1)

where: i and j are the row and column pixel numbers of the image; R_{ij}, G_{ij} and B_{ij} are the original pixel values; $\bar{R}$, $\bar{G}$ and $\bar{B}$ are the corresponding means; R'_{ij}, G'_{ij} and B'_{ij} are the corrected pixel values. Gray World has the advantage of simple calculation and easy implementation.

2.2 Specular reflection

This method estimates the color of the light source with the reflected light of an image and white balances the image with the estimated color of the light source. The maximum value at the brightest point of the image is defined as $R = G = B = 255$, believed by Specular Reflection to be the point of the color cast white dot under the ambient color temperature. Using the corrected coefficient of this point as the one for processing the entire white balance gain, an M*N image is processed as follows:

$$R'_{ij} = R_{ij} * 255 / R_m, \quad G'_{ij} = G_{ij} * 255 / G_m, \quad B'_{ij} = B_{ij} * 255 / B_m$$

(2)

where: i and j are the row and column pixel numbers of the image; R_{ij}, G_{ij} and B_{ij} are the original pixel values; R_m and $B_m G_m$ are the maximum values; R'_{ij}, B'_{ij} and G'_{ij} are the corrected pixel values.

This algorithm is very convenient and, in the presence of a mirror surface in the image, allows for very good restoration. However, in the absence of a mirror surface in the image or if the overall brightness of the image is low, its performance is not satisfactory.

3 IMPROVED SPECULAR REFLECTION

Among the many White Balancing algorithms, the most classic should be Gray World and Specular Reflection which, however, does not guarantee good performance unless particular assumptions are met. Even the Auto White Balancing that combines the merits of the two does not guarantee ideal effect. The assumptions for Gray World will not be met if an image is obviously color cast with mass color blocks or if the scene color is not abundant. Direct average weighting of the pixel RGB components will result in significant deviation and poor reliability. Specular Reflection assumes that that there is a pure white "mirror surface" in the image and extracts light source information directly from this surface. However, this method is deviation-prone if there is no such pure dot or if there is a very bright light dot in the image[7]. When Auto White Balancing based on color temperature estimation is used, the image has to be converted from the RGB color space to the

YCBCr color space during which the precision can be an issue, since if the conversion coefficient matrix is not precise enough, RGB crosstalk will be involved[8].

To allow for application and achieve better adjustment, we made some improvements on Specular Reflection. As in the RGB color space, our eyes are most sensitive to the information of G channel, and the CFA in bayer format provides for the maximum aperture range for the information volume of G channel[9], we tried to look for white dots through G channel. Besides, to fully reflect the information over the entire image and ensure that the resultant white dot information is as reliable as possible, we divided the entire image into blocks and looked for white dots in each of these blocks. We also derived the estimated value of unknown light sources by calculating the average color information of the dots and eventually corrected the white balance of the image. Our algorithm was implemented in the following steps:

Step 1: Divide an image with M*N pixels into P*Q blocks and marks the pixels of each block as m*n, in which m and n should meet:

$$m = \begin{cases} M/P, & M\%P = 0 \\ int(M/(P-1)), & Others \end{cases} \tag{3}$$

$$n = \begin{cases} N/Q, & N\%Q = 0 \\ int(N/(Q-1)), & Others \end{cases} \tag{4}$$

where: int() is the integer part taken. To enable data processing, when $M\%P \neq 0$ or $N\%Q \neq 0$, the image blocks corresponding to the right and lower edges of the image were zero-filled so that they also have m*n pixels.

Step 2: Measure the maximum value of the G channel in each block and the corresponding R and B values:

$$G^k_{max}(itemp, jtemp) = max(G^k_{ij})(i = 1,2\cdots m, j = 1,2\cdots n; k = 1,2,\cdots P*Q) \tag{5}$$

where: $(itemp, jtemp)$ is the row and column numbers at the maximum value of G^k_{ij}.

$$R^k_{max} = R^k(itemp, jtemp) \tag{6}$$

$$B^k_{max} = B^k(itemp, jtemp) \tag{7}$$

Step 3: Calculate the maximum G value of each block and the corresponding maximum R and B values R_{max}, G_{max} and B_{max}:

$$R_{max} = max(R^k_{max}) \tag{8}$$

$$G_{max} = max(G^k_{max}) \tag{9}$$

$$B_{max} = max(B^k_{max}) \tag{10}$$

Step 4: Calculate the gain of each channel:

$$R_{gain} = \frac{aver(R_{max}, G_{max}, B_{max})}{R_{max}} \tag{11}$$

$$G_{gain} = \frac{aver(R_{max}, G_{max}, B_{max})}{G_{max}} \tag{12}$$

$$B_{gain} = \frac{aver(R_{max}, G_{max}, B_{max})}{B_{max}} \tag{13}$$

where: aver means average.

Step 5: Correct the pixel values.

$$R'_{ij} = \begin{cases} R_{ij} * R_{gain}, & R_{ij} * R_{gain} < 255 \\ 255, & R_{ij} * R_{gain} \geq 255 \end{cases} \tag{14}$$

$$G'_{ij} = \begin{cases} G_{ij} * G_{gain}, & G_{ij} * G_{gain} < 255 \\ 255, & G_{ij} * G_{gain} \geq 255 \end{cases} \tag{15}$$

$$B'_{ij} = \begin{cases} B_{ij} * B_{gain}, & B_{ij} * B_{gain} < 255 \\ 255, & B_{ij} * B_{gain} \geq 255 \end{cases} \tag{16}$$

where: R_{ij}, G_{ij} and B_{ij} are the RGB pixel values of the image before it is processed; i and j are the row and column pixel numbers of the image; R'_{ij}, G'_{ij} and B'_{ij} are the RGB pixel values after auto white balancing, for which the upper limit is defined to be 255 to avoid overflow.

4 EXPERIMENT RESULTS AND ANALYSIS

To verify the effect of our improved algorithm of Specular Reflection, we compared it with the classic White Balancing algorithms and examined its White Balancing effect. In Experiment 1, the same blue photograph taken under high color temperature was processed by using our algorithm, Gray World, Specular Reflection, the Auto White Balance described in Reference 11 ("Auto White Balance" as a means of comparison. In Experiment 2, however, a whiteboard photo under the color temperature of 6500 K was processed using our algorithm and the built-in white balance feature of the camera and the processing results were analyzed.

4.1 *Comparison of different white balancing algorithms*

In the analysis above, Specular Reflection failed and the RGB gain remained a constant 1 throughout the process (the processed image is not presented here). This should have been caused by the presence of particular bright spots in the original picture. From the four images above, our improved algorithm of Specular Reflection has the best processing result. The correction made by Gray World shows severe color cast. The correction made by Auto White Balance does not exhibit obvious effect and tends to be color casting. By algorithm corresponding to these processing results, Table 1 compares the merits of each of the algorithms used in the experiments.

4.2 *White balance correction based on whiteboard image*

To quantify the effect of our improved algorithm of Specular Reflection, this section combines whiteboard and color temperature adjustment together. Fig. 2(a) shows a photo taken by the MT9P031 5 mpx camera system under the color temperature of 6500 K; Fig. 2(b) is a photo taken under the same color temperature but with the built-in auto white balance feature in the camera turned on, representing the white balance processing result of Fig. 2(a) by the built-in auto balance feature in MT9P031. Fig. 2(c) shows the processing result of Fig. 2(a) by our algorithm.

From Fig. 2, both our algorithm and the Camera Auto White Balance are highly effective and it is hard to tell which is better. To enable quantitative comparison between these two methods, we calculated the maximum, minimum, mean and standard deviation of the RGB of the image before it is processed, and after it is processed by Auto White Balancing and by our algorithm, as presented in Table 2. From this table, the maximum and minimum values of the image before and after correction remain unchanged, but its mean and standard deviation are significantly different.

For a whiteboard image, the RGB means are theoretically equal. For a color-temperature image, however, the RGB values are never equal but rather, significantly different. For an image processed

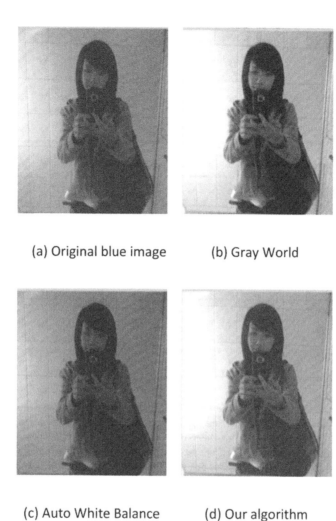

(a) Original blue image   (b) Gray World

(c) Auto White Balance   (d) Our algorithm

Figure 1. Processing result of different White Balancing algorithms.

Table 1. Comparison of different White Balancing algorithms.

	Gray World	Auto White Balance	Specular Reflection	Our algorithm
Calculation amount	Simple & quick	Slightly larger	Small calculation amount	Modestly increased
Correction effect	Possible color cast	Relative difference	Failed here	Failry good

by an algorithm, the closer and less different the R, G, B values are, the closer to white are the processed image, and the better performance is the processing algorithm used. Before the color-temperature image is processed, its means are 87.94, 131.33 and 110.87, and the standard deviations are 43.39, 63.03, 52.57. After the image is processed with the White Balance feature of the camera, its means are 87.45, 113.52, 118.32, and the standard deviations are 68.09, 68.44 and 71.57. After the image is processed by our algorithm, its RGB means are 99.65, 118.98 and

(a) Color-temperature picture   (b) Processed by Camera Auto White Balance   (c) Processed by our algorithm

Figure 2.   Comparison of corrected images.

Table 2.   Comparison of RGB values before and after correction.

	Before processing			Camera White Balancing algorithm			Our algorithm		
	R	G	B	R	G	B	R	G	B
Max.	255	255	255	255	255	255	255	255	255
Min.	0	0	0	0	0	0	0	0	0
Mean	87.94	131.33	110.87	87.45	113.52	118.32	99.65	118.98	109.68
Standard deviation (Std)	43.39	63.03	52.57	68.09	68.44	71.57	47.60	55.92	54.68

109.68, and the standard deviations are 47.60, 55.92 and 54.68. Before the image is processed, the maximum difference of its RGB means is a significant 22.93. When the image is processed by the White Balance feature of the camera, the maximum difference of its RGB means is 30.87, which is slightly smaller. When the image is processed by our algorithm, the maximum difference of its RGB means is 19.33 and the three means are very close to each other. Furthermore, the standard deviations of all the RGB values of the image corrected by our algorithm are smaller than their counterparts of the image corrected by the Camera White Balancing Algorithm. This demonstrates that our algorithm is substantially advantageous in correcting color-temperature images.

5   CONCLUSIONS

This paper presents an improved algorithm of classic Specular Reflection and demonstrates through concrete experiments that this improved algorithm is superior over Gray World, Specular Reflection and Auto White Balance. A MT9P031 5 mpx camera system was used to examine a color-temperature image before it was processed, and after it was processed by Auto White Balance integrated into the camera and by the improved algorithm and compare the RGB means and standard deviations. The results indicate that the improved algorithm is effective and practically usable. The improved algorithm of Specular Reflection presented herein has an implication on further studies of the white balancing of images.

ACKNOWLEDGEMENT

This paper is supported by Ningbo Natural Science Foundation (GN: 2012A610042); Disciplinary Project of Ningbo University (GN: xk1088); Scientific & Technological Innovation Activity Plan of College Students in Zhejiang Province 2012 (GN: 2012R405069).

REFERENCES

[1] Varsha, Chikane & Chiou-Shann Fuh. (2006) Automatic white balance for digital still cameras. *Jouornal of Information Science & Engineering*, 22, 497–509.

[2] Han, Q., Rong, M. & Liu, W.J. (2009) Algorithm research of auto white balance in hardware-based ISP. *Information Technology*, (11), 55–59.

[3] Yin, J.F. (2004) Color correction methods with applications to digital projection environments. In: *WSCG'2004. Plzen: CzechRepublic.*

[4] Graham, D.F., Steven, D.H. & Paul, M.H. (2001) Color by correlation: a simple, unifying framework for color constancy. *IEEE Transactions on Pattern Analysis and Machine Intelligence.*

[5] Kobus, B., Vlad, C. & Brian, F. (2002) A comparison of computational color constancy algorithms. I: Methodology and experiments with synthesized data. *IEEE Transactions on Image Processing.*

[6] Li, Y. (2002) Improvement and implementation of automatic white balance algorithm. Xidian University.

[7] Yan, G. (2007) Design of white balance digital process circuit for CMOS image sensor. Tianjin: Tianjin University.

[8] Zhang, Y., Liu, X. & Li, H.F. (2007) The precision of RGB color space convert to YCbCr color space. *Journal of Jiannan University*, 6(2), 200–202.

[9] Wang, M. (2011) Auto white balance algorithm based on color-temperature estimation. Tianjin: Tianjing University.

[10] Tian, L.K., Liu, X.H., Li, J., et al. (2013) Improvement and implementation of self-adaptive automatic white balance algorithm. *Electronics Optics & Controls*, 12, 37–41.

[11] Gan, B., Wei, Y.C. & Zheng, R. (2011) Auto white balance algorithm for CMOS image sensor chip. *Liquid Crystals & Displays*, 26 (2), 224–228.

Engineering Technology and Applications – Shao, Shu & Tian (Eds)
© 2014 Taylor & Francis Group, London, ISBN 978-1-138-02705-3

Application of BP neural network to power prediction of wind power generation unit in microgrid

Dongxun Wu & Haiming Wang
Taiyuan University of Technology, Taiyuan, Shanxi, China

ABSTRACT: With the intensifying world energy crisis and environmental concerns, new energy power generation is favored by all countries. Wind power generation, in particular, has become a focus of research for the academic community. However, the intermittence and stochasticity inherent in wind energy itself also challenge the security, stable operation and energy quality of the microgrid system. Accurate short-term wind power prediction can effectively resolve this problem. Conventional wind power prediction methods mostly involve the use of a single neural network tool, which frequently leads to problems like complex model structure and excess sensitivity of the prediction results to the sample data used. Therefore, this paper proposes an idea of using fuzzy clustering technique to select similar days before establishing the BP neural network. The simulation experiment in this paper is based on the August wind speed data of an area, selecting August 27 as the forecasting day. The relative error of the forecasting results is within 5%, with a couple of singular points though. The experimental results show that the method of combining fuzzy clustering with BP neural network provides better prediction accuracy, and can be applied to wind power prediction.

Keywords: wind power generation; fuzzy clustering technique; BP neural network; wind power prediction

1 INTRODUCTION

With the deepening of the world's energy crisis and environmental issues, the research of new energy is prevailing in the world[1]. Microgrid, which integrates a variety of distributed energy, has become the new direction of the world's electric power sector.

Among the many distributed power sources, wind power generation is the focus of development and research for the ubiquity of wind energy. However, the intermittence and stochasticity inherent in wind itself also challenge the security, stable operation and energy quality of the microgrid system. Accurate short-term wind power prediction can effectively resolve this problem. This both helps with the operation management of the microgrid system, and allows for the secure, stable and economical operation of distributed power supply under high permeability. From this perspective, wind power prediction is very significant.

2 FUZZY CLUSTERING TECHNIQUE

Clustering is to investigate and deal with the classification of a given item mathematically according to the similarity level of things. Conventional clustering is a hard process that allocates items in an "all-or-nothing" manner with clear boundaries[3]. In practical application, however, as most objects are intermediary in their behavior and generic, the limitations of rigid classification are increasingly evidenced. In 1965, Zedah offered power theoretical support for flexible classification with his fuzzy set theory. Fuzzy clustering, which establishes uncertainty description of classes

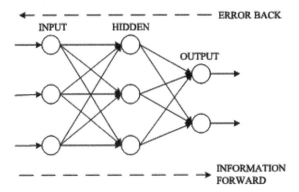

Figure 1.   BP neural network.

and thereby better reflects the objective reality, has become a hot spot for clustering research. Ever since Ruspini introduced the idea of fuzzy partition, many clustering methods have been developed on this basis. Some of the typical methods include method based on similarity relations and fuzzy relations, clustering method based on transitive closure of fuzzy relation, the largest tree method based on fuzzy graph theory, and convex decomposition, dynamic programming and indiscernibility relation based on datasets[4].

3   BP NEURAL NETWORK

BP neural network algorithm is a nonlinear fitting process in which the trained neural network is also able to provide proper outputs for inputs near their sample set. As BP neural network is highly capable of nonlinear fitting, particularly for the air velocity and temperature factors in weather forecasting, plus its learning rules are simple for computer realization, the BP learning algorithm is often used to train the neural network[5]. The general structure of the BP neural network is as shown in Fig. 1.

Here each node represents a neuron. The BP neural network is actually a feedforward network of a multilayer perceptron consisting of an input layer, a hidden layer or layers and an output layer. The purpose of using learning algorithm to optimize weights is to find the weight w that limits the error to the minimum. Learning algorithm can be defined as modifying the weights and thresholds of a given sample through forward propagation of information and back propagation of errors until the correct outputs are obtained[6].

4   SHORT-TERM WIND POWER PREDICTION BASED ON FUZZY CLUSTERING TECHNIQUE COMBINED WITH BP NEURAL NETWORK

In practical application, wind power prediction is performed in the following steps: Classify the historic data using fuzzy clustering technique to form several sample data and train different networks according to their classes. After selecting the forecasting day, allocate the group of the forecasting day according to the meteorological information of the forecasting day (e.g. temperature, relative humidity, weather) by mode identification, select the corresponding network and perform the prediction. Raw data used herein are the air velocity data in a certain area in August involving 31 days, each having 24 air velocities at one hour internals.

4.1   *Classifying samples with fuzzy clustering technique*[7]

The combined action of factors affecting the air velocity was included for by fuzzy clustering analysis. Here we incorporate the term "similarity", which describes the degree of similarity between two samples in their characteristic indices. A "sample set of similar days" refer to a data set similar to the air velocity characteristics of the forecasting day, i.e. a set of sample data whose similarity to the forecasting day is smaller than a given threshold. The fuzzy clustering was performed in the following steps:

1. Standardizing data

Historic air velocity data: The air velocity data were normalized into [0, 1] using the expression below.

$$x_i' = (x_i - x_{min})/(x_{max} - x_{min}) \tag{1}$$

Here x_{max} and x_{min} are the maximum and minimum input variables in the training sample set; x_i and x_i' are the unnormalized and normalized values of the input sample.

The relative humidity is expressed as per cent. The temperature (maximum, minimum, average) and weather (fine, cloudy, rainy) were normalized to certain rules.

2. Calibrating

Among the many calibration methods like Euclidean distance, absolute exponent, absolute reciprocal, exponent similarity coefficient and maxminimum method, we selected Euclidean distance method.

$$d_{ij} = \sqrt{\frac{1}{n}\sum_{k=1}^{n}(x_{ik} - x_{jk})^2} \tag{2}$$

3. Clustering

A proper threshold $\lambda \in [0, 1]$ was selected. As λ decreased from 1 to 0, the clustering result became thinner and thinner. In general, the exact value of λ is determined empirically.

4. Pattern recognition

Pattern recognition of the forecasting day, i.e. the selection of sample set of similar days, was also performed using Euclidean distance method.

After clustering analysis, the historic air velocity data were allocated to different classes. Then, we calculated the Euclidean of the forecasting day from each of these classes, and established a BP neural network for prediction using the group with the shortest Euclidean distance as the sample set of similar days. The calculation formula is shown in expression (2).

A 0.05 threshold was used for fuzzy clustering. The air velocity data were divided into several groups using Euclidean distance method. Here only one group is selected for predictive analysis. After fuzzy clustering the August days of the area, the 3rd, 5th, 9th, 12th, 13th, 15th, 21st, 25th and 27th days were allocated into one group. The 27th day was selected as the forecasting day. A BP neural network model was built for prediction using the air velocity data of the rest of these days as the training samples and that of the 27th day as the test sample.

4.2 *Establishing a prediction model based on BP neural network*

As has been theoretically confirmed that, to the extent that the number of nodes in the hidden layer is not limited, a three-layer (containing only 1 hidden layer) BP neural network is strong enough to perform any nonlinear mapping, the BP neural network we used is a three-layer structure. The number of nodes in the input and output layers can be determined according to the nature of the subject studied.

As 24 input variables are involved in an air velocity predicting BP neural network model, there are 24 nodes in the input layer. As the output of the network is the air velocity of the forecasting day, there are 24 nodes in the output layer. The number of nodes in the hidden layer was selected according to the empirical formula $n = \sqrt{n_i + n_0} + a$, in which n is the number of nodes in the hidden layer; n_i is the number of nodes in the input layer; n_0 is the number of nodes in the output layer; a is a constant between 1~10[9]. Thus the number of nodes in the hidden layer is in the

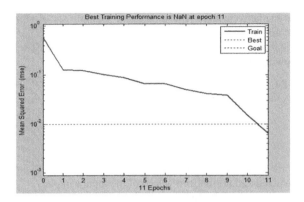

Figure 2.   Network training curve.

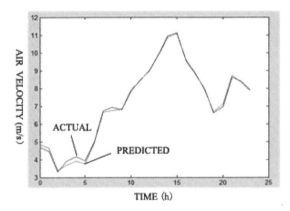

Figure 3.   Air velocity prediction curve.

range of 7~17. Comparison of each of these numbers with the result of experiment showed that satisfactory training time and prediction results can be achieved when the number is 16, as shown in Fig. 2.

4.3   *Predicting air velocity*

The neuron transfer function of the hidden layer was selected to be an S-tan function tansig, and that of the output layer is a trainlm function. After training the network many times, we obtained the predicted air velocities as listed in Table 1.

Analysis shows a mean relative error of the prediction result is 8.49% and a mean square error of 0.014. Of the 24 points in the prediction result, 21 points, or 79.2% of the total forecasting points, have their relative error smaller than 10%, suggesting that our BP neural network model well conforms to the accuracy requirements for prediction results.

4.4   *Wind power prediction*

As we have already model and predicted the air velocity and obtained the optimal air velocity predictions above, now we are going to predict the output of a wind generator based on the typical power characteristic curve. A 1 MW unit capacity variable pitch wind generating unit of Dewind is used as an example. Its cut-in, rate and out-out air velocities are 2.5 m/s, 11.5 m/s and 23 m/s, respectively. Its approximate typical power characteristic curve is as shown in Fig. 4[10].

Table 1.　Data used for air velocity prediction.

Time series	Actual (m/s)	Predicted (m/s)	Relative error	Time series	Actual (m/s)	Predicted (m/s)	Relative error
0	4.83	4.67	9.80%	12	8.97	9.02	0.84%
1	4.64	4.47	11.86%	13	9.85	9.87	0.33%
2	3.38	3.31	41.47%	14	10.88	10.95	0.85%
3	3.71	3.94	45.98%	15	11.07	11.13	0.70%
4	3.92	4.15	31.91%	16	9.58	9.54	0.56%
5	3.76	3.94	32.69%	17	8.84	8.78	1.07%
6	4.97	5.02	2.62%	18	7.92	7.96	0.88%
7	6.66	6.76	3.00%	19	6.66	6.61	1.53%
8	6.75	6.94	5.22%	20	7.13	6.96	4.25%
9	6.82	6.79	0.71%	21	8.73	8.61	2.13%
10	7.79	7.91	2.66%	22	8.36	8.41	0.89%
11	8.47	8.41	1.06%	23	7.86	7.89	0.71%

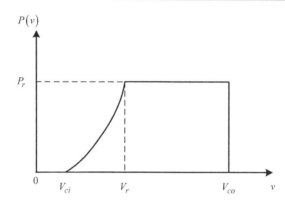

Figure 4.　Typical power characteristic curve.

The output of the wind generator can be approximated with a piecewise function expression as:

$$P = \begin{cases} 0 & 0 \leq V \leq V_{ci} \\ (A + BV + CV^2)P_r & V_{ci} \leq V \leq V_r \\ P_r & V_r \leq V \leq V_{co} \\ 0 & V_{co} \leq V \end{cases} \tag{3}$$

Here P is the output power of the wind generator; V is the air velocity at this time; P_r is the rated power of the wind generator; V_{ci}, V_r and V_{co} are the cut-in, rated and cut-out air velocities of the wind generator; A, B and C are their parameters.

As the wind resources vary, so do the calculated A, B and C. The calculation expression is as follows.

$$\begin{cases} A = \dfrac{1}{(V_{ci} - V_r)^2} \left[V_{ci}(V_{ci} + V_r) - 4V_{ci}V_r \left(\dfrac{V_{ci} + V_r}{2V_r} \right)^3 \right] \\ B = \dfrac{1}{(V_{ci} - V_r)^2} \left[4(V_{ci} + V_r) \left(\dfrac{V_{ci} + V_r}{2V_r} \right)^3 - (3V_{ci} + V_r) \right] \\ C = \dfrac{1}{(V_{ci} - V_r)^2} \left[2 - 4 \left(\dfrac{V_{ci} + V_r}{2V_r} \right)^3 \right] \end{cases} \tag{4}$$

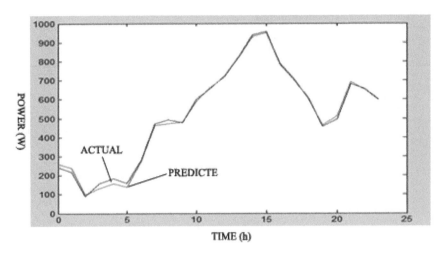

Figure 5. Wind power prediction curve.

Table 2. Wind power prediction result.

Time series (h)	Actual (kW)	Predicted (kW)	Relative error	Time series (h)	Actual (kW)	Predicted (kW)	Relative error
0	258.89	241.11	6.87%	12	718.89	724.44	0.77%
1	237.78	218.89	7.94%	13	816.67	818.89	0.27%
2	97.78	90.00	7.95%	14	931.11	938.89	0.84%
3	134.44	160.00	19.01%	15	952.22	958.89	0.70%
4	157.78	183.33	16.20%	16	786.67	782.22	0.56%
5	140.00	160.00	14.29%	17	704.44	696.78	0.95%
6	274.44	280.00	2.02%	18	602.22	606.67	0.74%
7	462.22	473.33	2.40%	19	462.22	456.67	1.20%
8	472.22	493.33	4.47%	20	514.44	495.56	3.67%
9	480.00	476.67	0.69%	21	692.22	678.89	1.93%
10	587.78	601.11	2.27%	22	651.11	656.67	0.85%
11	663.33	656.67	1.01%	23	595.56	598.89	0.56%

According to the predicted air velocities, the wind power prediction curve as calculated from formulae (3) and (4) is shown in Fig. 5.

Table 2 shows the wind power prediction result.

Analysis shows that, as the air velocities of the hours on the selected training sample days and the forecasting days are all in the range of 2.5~11.5 m/s, the mean relative error of the power prediction is smaller than that of the air velocity prediction. The relative error of the power prediction is 4.09% and the mean square error is 171.04, both meeting the minimum requirements for wind power prediction.

5 CONCLUSIONS

Studies have revealed that the prediction accuracy of the BP neural network is largely dependent on the training sample set selected. In our study, fuzzy clustering technique is used in combination with the BP neural network, and the air velocity and wind power of the forecasting day are predicted using the air velocities of similar days, resulting in much higher prediction accuracy.

The experimental results indicate that our BP neural network model is correct with satisfactory prediction errors and logic theoretical research.

Despite the satisfactory prediction results presented, we are aware that there are still some shortfalls like the sensitivity of the fuzzy clustering analysis to the weather forecast accuracy, the singular data used for the input layer of the network, the failure to reflect all factors that affect the air velocity, and the existence of singular points in our prediction results. Hence, there is still room to improve the prediction accuracy and practical applicability. Further efforts may include the improvement of the BP neural network model with all necessary weather forecasting data and more intensive characterization of air velocity.

REFERENCES

[1] Ye F. New energy power generation – The realization of sustainability of the human being [J]. Energy and Environment, 2008, 03: 55–57+62.

[2] Yin M., Ge X.B., Wang C.S., Zhang Y.B., Liu G.P., Jin X.L. Analysis of issues about China's large-scale wind power development [J]. China Electric Power, 2010, 03: 59–62.

[3] Wang J., Wang S.T., Deng Z.H. Survey on challenges in clustering analysis research [J]. Control and Decision, 2012, 03: 321–328.

[4] Rong H., Zhang J.S. The fuzzy PID algorithm for feedforward neural network and its application in electric power load prediction [J]. Transactions of China Electrotechnical Society, 1998.

[5] Yuan C.R. Artificial neural network and its application [M]. Beijing. Tsinghua University Press, 1999.

[6] Shi H.T., Yang J.L., Ding M.S., Wang J.M. A short-term wind power prediction method based on wavelet decomposition and BP neural network [J]. Automation of Electric Power Systems, 2011, 16: 44–48.

[7] Li Z., Zhou B.X., Lin N. Classification of daily load characteristics curve and forecasting of short-term load based on fuzzy clustering and improved BP algorithm [J]. Power System Protection and Control, 2012, 03: 56–60.

[8] Zhao Y.F., Li H., Gao H.Y. Design mode classification research based on weighted Euclidean distance [J]. Technology Wind, 2009, 12: 19.

[9] Xiao Y.S., Wang W.Q., Huo X.P. Study on the time-series wind speed forecasting of the wind farm based on neural networks. Energy-Saving Technology, 2007.

[10] Fang J.X. The research on short-term air velocity and wind power prediction modeling. Beijing Jiaotong University.

Engineering Technology and Applications – Shao, Shu & Tian (Eds)
© 2014 Taylor & Francis Group, London, ISBN 978-1-138-02705-3

Internet consumption analysis based on impact factors: Model and experiment

Yujuan Guan

School of Management, Xi'an University of Architecture and Technology, Xi'an, Shaanxi, China
College of Economics and Management, Xi'an University of Post & Telecommunications, Xi'an, China

ABSTRACT: This paper presents an impact factors model for the study on brand communication in consumers of Internet virtual community. This study based on Social Cognitive Theory, from brand-related environmental factors and individual consumer motivation factors, sorts out the impact factors on brand knowledge sharing, and thus makes hypothesis. This study applies questionnaire method for data collection, and using AMOS for data analysis and hypothesis testing.

Keywords: virtual community; knowledge sharing; social motivation

1 INTRODUCTION

In the background of Internet, consumers become promoters of innovation, meanwhile knowledge from consumers become important resources and core competencies. Scholars have conducted in-depth studies on user behavior in the virtual community from user perspective or community perspective[1]. However, brand communication is in an increasingly competence due to needs of Internet marketing, brands need to win the consumer's attention and discussion. Therefore, this paper based on Social Cognitive Theory, from the perspective of brand competition for consumer's attention, studies the impact factors on environmental factors and individual motivation factors to brand knowledge sharing in virtual community.

2 RELATED RESEARCH AND HYPOTHESIS

Social Cognitive Theory pointed out that people's behavior is determined by the cognitive of the environment and their own ability (Bandura, 1982), people judge behavior results and form the outcome expectation, and thus determine the behavioral intentions (Bandura, 2001)[2,3].

Chiu[4] (2006) divided outcome expectation into "expectations associated with individual" and "expectations associated with community" in their study on knowledge sharing behavior in virtual community. The purpose of consumer participate in virtual communities are seeking solutions to the problem, getting information, and communicating with community members to establish mutually beneficial relations, seeking respect and recognized (Bock, 2005; Chiu, 2006)[5]. Meanwhile, community members contribute knowledge to enrich the stock of knowledge for the community, make community topics higher quality and maintain community competitiveness, and enable communities to be continuation and development (Bock & Kim, 2002; Kolekofski & Heminger, 2003)[6].

H1: Expectations associated with individual have a positive impact on willingness of brand knowledge sharing in virtual community.

H2: Expectations associated with community have a positive impact on willingness of brand knowledge sharing in virtual community.

Brand-consumer recognition is a cognitive basis of brands topics circle, this common cognitive basis which means quality of consumer-brand relationship (Aaker, 2004)[7]. Social capital theory shows that members obtained more resources from the relationship network, when there is a common cognitive basis in a relationship network. When community members are of a identical brand feelings, communication between them is more smooth and efficient.

H3: Brand-consumer recognition has a positive impact on willingness of brand knowledge sharing in virtual community.

Network effect is related to the size of brand topic circle, that means the brand topic's popularity. The number of participants discussed topics related brand, the greater the network effect of the brand in the virtual community has. The number and frequency of this data will convert to the perception of the consumer, it has a positive effect on the brand.

H4: The size of brand topic circle has a positive impact on expectations associated with individual in the virtual community.

H5: The size of brand topic circle has a positive impact on expectations associated with community in the virtual community.

H6: The size of brand topic circle has a positive impact on willingness of brand knowledge sharing in virtual community.

Relationship quality of virtual community brand-related is interpersonal quality of members in virtual community that means mutual trust among community members. Trust is caused by frequent interaction among members, understand and respect each other, and thus all members make a positive friendly behavior for others. Trust is to believe that someone can act in an appropriate manner (Bhattacherjee, 2002; Gefen, 2003; Eastlick & Lotz, 2011)[8-10].

H7: Trust among the members of brand topic circle has a positive impact on expectations associated with individual.

H8: Trust among the members of brand topic circle has a positive impact on expectations associated with community.

Motivation is the intrinsic driving factor in people's behavior. Motivation is on behalf of the need for a particular outcome. When people are driven by a motivation, they will have the behavior with corresponding demand. Social motivation is an important factor affecting people's social behavior, mainly includes achievement motivation, association motivation, prestige motivation. Interaction of community members are within the scope of interpersonal relationships, which affected by social motivation.

H9: Achievement motivation has a positive impact on expectations associated with individual.

H10: Achievement motivation has a positive impact on expectations associated with community.

H11: Association motivation has a positive impact on expectations associated with individual.

H12: Prestige motivation has a positive impact on expectations associated with community.

3 CONCEPT MODEL

This study proposed research concept model based on the above analysis and hypothesis (see Figure 1).

4 MEASUREMENT AND ANALYSIS

4.1 *Data collection*

This study adopts convenience sampling method, using online (e-mail) and the offline (on-site in classroom) in two forms to collect data. This questionnaire is based on younger age groups student-centered, who are the main groups involved in the network interaction, thus it is consistent with surveys aim of this study. In this questionnaire, respondents are asked to fill in the name of the virtual community they participated, as well as the most interesting sub-topic in the virtual community.

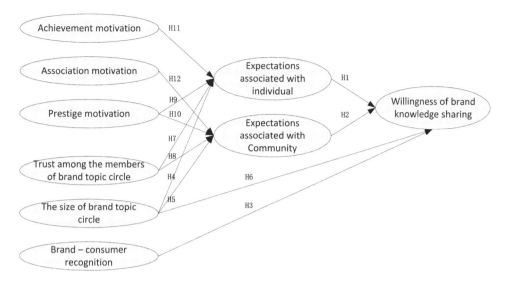

Figure 1. Concept Model.

Table 1. Measurement model of argument, load, composite reliability and AVE.

Variable	Index	Load	Composite reliability	AVE
Achievement motivation	ACM1	0.709***	0.793	0.660
	ACM3	0.904***		
Association motivation	AFM1	0.717***	0.791	0.658
	AFM2	0.895***		
Prestige motivation	PWM1	0.930***	0.828	0.709
	PWM2	0.744***		
The size of brand topic circle	SCS1	0.824***	0.894	0.740
	SCS2	0.978***		
	SCS3	0.765***		
Trust among the members of brand topic circle	SCC1	0.786***	0.868	0.688
	SCC2	0.773***		
	SCC3	0.922***		
Brand – consumer recognition	SCR1	0.774***	0.847	0.649
	SCR2	0.853***		
	SCR3	0.787***		

Respondents can be screened by this problem, some samples do not meet the requirements, or obviously do not understand the intent of the survey was removed after the study.

4.2 *Measurement verification*

This study conducted confirmatory factor analysis for structural equation model by using AMOS, to test the reasonableness of the measurement model. First, implemented testing to the measurement model of argument (see Table 1).

The data (see Table 1) shows that the load of each index is between 0.75 and 0.95, and are under the significant level of $p = 0.001$, that indicating all indexes can reflect the characteristics of

Table 2. Correlation coefficient matrix.

	Trust	Recognition	Size	Prestige	Association	Achievement
Trust	0.805*					
Recognition	0.551	0.830*				
Size	0.354	0.239	0.860*			
Prestige	0.428	0.409	0.413	0.842*		
Association	0.476	0.406	0.296	0.841	0.811*	
Achievement	0.422	0.553	0.428	0.625	0.617	0.812*

Note: *Diagonal has been replaced with the square root of AVE

Table 3. Measurement models of the dependent variable, load, composite reliability and AVE.

Variable	Index	Load	Composite reliability	AVE
Expectations associated with individual	POC1	0.814***	0.899	0.642
	POC2	0.812***		
	POC3	0.810***		
	POC4	0.856***		
	POC5	0.708***		
Expectations associated with Community	OOC1	0.780***	0.912	0.676
	OOC2	0.804***		
	OOC3	0.881***		
	OOC4	0.776***		
	OOC5	0.865***		
Willingness of brand knowledge sharing	KSI1	0.856***	0.885	0.719
	KSI2	0.855***		
	KSI3	0.833***		

the latent variables better. Composite reliability of each variable is greater than 0.5, the minimum is 0.791, indicating a high degree of combination reliability. The AVE of each variable is greater than 0.5, the minimum is 0.649, indicating a high degree of convergent validity.

As can be seen from the table 2, the AVE of each variable is greater than the corresponding correlation coefficients, indicating the discriminant validity of measurement model meet the requirements.

Next, this study conducted testing to measurement models of the dependent variable by using AMOS, implemented confirmatory factor analysis to three endogenous latent variable (see Table 3).

From the data (see Table 3) shows that the load of each index is between 0.75 and 0.95 are, and are under the significant level of $p = 0.001$, that indicating all indexes can reflect the characteristics of the latent variables better. Composite reliability of each variable is greater than 0.5, the minimum is 0.885, indicating a high degree of combination reliability. The AVE of each variable is greater than 0.5, the minimum is 0.642, indicating a high degree of convergent validity.

As can be seen from the table (Table 4), the AVE of each variable are greater than the corresponding correlation coefficient, the dependent variable has a high discriminant validity. This shows that measurement model of the dependent variable defined clear and precise, which can reflect the different characteristics of each variable.

4.3 Model validation

This study conducted hypothesis testing based on the fitting of the data. Among the results in the following table 5, four hypotheses have not significant variable relationship, H2, H4, H5, and H9.

Table 4. Correlation coefficient matrix.

	Willingness of sharing	Expectations Community	Expectations individual
Willingness of sharing	**0.848***		
Expectations Community	0.515	**0.822***	
Expectations individual	0.536	0.763	**0.802***

Note: *Diagonal has been replaced with the square root of AVE

Table 5. Hypothesis testing results.

Hypothesis	Content	Test results
H1	Expectations associated with individual have a positive impact on willingness of brand knowledge sharing	True
H2	Expectations associated with community have a positive impact on willingness of brand knowledge sharing	False
H3	Brand – consumer recognition has a positive impact on willingness of brand knowledge sharing	True
H4	The size of brand topic circle has a positive impact on expectations associated with individual	False
H5	The size of brand topic circle has a positive impact on expectations associated with community	False
H6	The size of brand topic circle has a positive impact on willingness of brand knowledge sharing	True
H7	Trust among the members of brand topic circle has a positive impact on expectations associated with individual	True
H8	Trust among the members of brand topic circle has a positive impact on expectations associated with community	True
H9	Achievement motivation has a positive impact on expectations associated with individual	False
H10	Achievement motivation has a positive impact on expectations associated with community	True
H11	Association motivation has a positive impact on expectations associated with individual	True
H12	Prestige motivation has a positive impact on expectations associated with community	True

The inconsistent conclusions and assumptions are described as follows.

First, H2 is not significant, that means the outcome expectation has not significant impact on the willingness of knowledge sharing. In this regard, previous reports in the literature, the performance of outcome expectation related individuals and community is also unstable in different studies. This study suggests that this result may be caused by the different virtual community selected. Respondent come from multiple different virtual community, be interested in different types of topics, including mechanical, electronic, political and economic, entertainment, business, health and so on. In this case, there are only relatively few respondents in each community, the core members are less likely to appear, reflecting a met mentality that in order to solve their own problems. So it shows that outcome expectation associated with individual has significant effect on the willingness of knowledge sharing, while the role of outcome expectation community-related is not significant.

Second, H4 and H5 are not significant, that means the size of brand topic circles have not significant impact on outcome expectation related individual and community. Previous studies have shown that the larger the size of the network is, the more they may get others to help, resources obtained from members of the network will be more (Nahapiet & Ghoshal, 1998)[11]. The effect about the size of the topic circle to the outcome expectation is not significant, may be related to the purpose of the consumer. The consumer will pay attention to some topics because of its heat degree, but the consumer does not definitely think this topic can bring them the exact benefits.

Third, H9 is not significant, that means achievement motivation has not significant impact on outcome expectations related individual. Achievement motivation refers to the needs that people get a sense of achievement through the completion of a task (Wu & Sukoco, 2010)[12]. In this study, the members of virtual community discussed more interpersonal topics greater than the problem-solving topics. Thus, achievement motivation cannot be met through participation in community discussions, which may be the reason that the path is not significant.

5 CONCLUSIONS

First, to enhance the quality of the relationship between consumers and brands can promote willingness of brand-related knowledge sharing. When brand-consumers recognition is stronger, it means consumer and brand have better relationship quality, thus consumer has a greater willingness to participate in the brand-related discussion in virtual communities. Consumers are more willing to discuss their favorite and familiar brands in virtual community. Conversely they have not interested in participating the stranger brand discussion.

Second, the size of the brand-related topics circle has positive effect on consumer willingness to participate in the discussion. Consumers are more willing to participate in the more popular and mainstream circles to discuss. The more the number of brand-related topics discussed in the circle, the greater the impact on consumers. Third, the trust in virtual community can make consumers believe that participate in the discussion can bring them more benefits, which encourage consumers to participate in brand knowledge sharing. Trust in virtual community has a positive effect on willingness of consumers knowledge sharing, and outcome expectations associated with individual of consumers play an intermediary role.

Fourth, consumers will participate in the discussion of virtual community, in order to meet the needs of their interaction with others, and thus consumers will contribute their knowledge to virtual community. The empirical results of this study show that achievement motivation of consumers can affect consumers' outcome expectation associated with individual, thereby affecting the willingness of consumers to participate in knowledge sharing activities.

REFERENCES

[1] Zhou Tao, Lu Yaobin. A Research on Knowledge Sharing Behavior of Virtual Community Users Based on Social Influence Theory [J]. Research and Development Management, 2009, 21 (4): 79–83.

[2] Bandura, A. (1982). Self-efficacy Mechanism in Human Agency [J]. American Psychologist 37 (2): 122–147.

[3] Bandura, A. (2001). Social Cognitive Theory: An Agentic Perspective [J]. Annual Review of Psychology 52 (1): 1–26.

[4] Chiu, C. and M. Hsu, et al. (2006). Understanding Knowledge Sharing in Virtual Communities: An Integration of Social Capital and Social Cognitive Theories [J]. Decision Support Systems 42 (3): 1872–1888.

[5] Bock, G. and R. W. Zmud, et al. (2005). Behavioral Intention Formation in Knowledge Sharing: Examining the Roles of Extrinsic Motivators, Social-Psychological Forces, and Organizational Climate [J]. MIS Quarterly 29 (1): 87–111.

[6] Kolekofski, K. E. and A. R. Heminger (2003). Beliefs and Attitudes Affecting Intentions to Share Information in an Organizational Setting [J]. Information & Management 40 (6): 521–532.

[7] Aaker, J. and S. Fournier, et al. (2004). When Good Brands Do Bad [J]. Journal of Consumer Research 31: 1–16.

[8] Bhattacherjee, A. (2002). Individual Trust in Online Firms: Scale Development and Initial Test [J]. Journal of Management Information Systems 19 (1): 211–241.

[9] Gefen, D. and E. Karahanna, et al. (2003). Trust and Tam in Online Shopping: An Integrated Model [J]. MIS Quarterly 27 (1): 51–90.

[10] Eastlick, M. A. and S. Lotz (2011). Cognitive and Institutional Predictors of Initial Trust Toward an Online Retailer [J]. International Journal of Retail & Distribution Management 39 (4): 234–255.

[11] Nahapiet, J. and S. Ghoshal (1998). Social Capital, Intellectual Capital, and the Organizational Advantage [J]. Academy of Management Review 23 (2): 242–266.

[12] Wu, W. and B. M. Sukoco (2010). Why Should I Share? Examining Consumers' Motives and Trust On Knowledge Sharing [J]. Journal of Computer Information Systems 50 (4): 11–19.

Engineering Technology and Applications – Shao, Shu & Tian (Eds)
© 2014 Taylor & Francis Group, London, ISBN 978-1-138-02705-3

An intelligent monitoring system based on big data mining

Chen Cheng & Danning Wang
The Second Monitor Centre of China Earthquake Administration, Xi'an, Shaanxi, China

Wenbo Shi
Xi'an Ruiyun Keji Co. Ltd., Xi'an, Shaanxi, China

ABSTRACT: Intelligent power grid monitoring includes diagnosing, eliminating and resolving power grid faults. The quick and accurate diagnosis and restoration of power grid faults is extremely important for keeping the power grid operational and minimizing losses caused by power failure. In this paper, big data mining processes are applied to intelligent power grid monitoring systems. An improved RBF-ANN (Radial Basic Function-Artificial Neural Network) based on big data mining for fault diagnosis was introduced, and the least square method is extended to optimize this RBF neural network. An intelligent monitoring system based on this improved RBF algorithm big data mining and TCP/IP protocol communication. A four-bus transmission network is simulated as the model system. The results show that the monitoring system based on mining big data of past operation is highly realizable, fault tolerant and robust.

Keywords: intelligent monitoring; big data mining; neural network; industrial personal computer

1 INTRODUCTION

The increasing number of large-capacity nonlinear power loads and overloading of power systems as a result of the growing modern electric and electronics industry are posing increasing impact on power systems[1–2]. When a power grid fault occurs, emergent status adjustment will be required. Measures will be taken to shed some of the loads or disconnect the system and restore it to normal operation within the shortest possible time[3–4]. For a power grid early warning system, evaluating, warning against, diagnosing and automatically controlling faults to avoid hidden troubles or limit fault-related losses to the lowest level is critical to ensure the healthy and safe service of the power grid.

So far, fault diagnosis techniques for power grids mainly include Optimization Techniques, Petri Net, Bayesian Network and Neural Network. Neural Network has been widely used in power grid fault diagnosis for its parallel process, associative memory and nonlinear mapping that allow quick computation and comprehensive information processing. Of the various Neural Network models, the most extensively used is the Basic Propagation (BP) Neural Network[5–7]. The BP model is generally trained by the Gradient Descent Algorithm. This learning algorithm converges slowly and possibly to a local minimum. These negative aspects of the BP model have heavily restricted its application to fault diagnosis.

In this paper, we look at an intelligent power grid monitoring system based on artificial neural network (ANN) big data mining, in which RBF-ANN (Radial Basic Function-Artificial Neural Network) is used for data information derived by mining, summarizing and analyzing past operation parameters of power grids. The RBF Neural Network is a Feedforward Neural Network Model (FNN) that is widely used for its global approximation properties and freedom of local minimums. Here, we propose the application of an improved k-mean clustering algorithm in combination of the least square method to the fault diagnosis of intelligent power grid monitoring system.

2 RESEARCH AND APPLICATION OF BIG DATA MINING TECHNOLOGY

2.1 *Concept and methodology of big data mining*

Data mining is a new branch of science that helps us analyze, understand and use information data. Mass data mining is a process that distills undiscovered useful information and knowledge concealed in countless incomplete, vague, random, errored, natural information that is actually acquired. This information and knowledge obtained by data mining is both understandable and easy to store, apply and propagate.

Big data mining provides a full set of processes to resolve practical problems and convert mass irrelevant information to visualized understandable conclusions through the six steps of classification and estimation, prediction, affinity grouping, abstracting and clustering, modeling, description and visualization, and mining complex types of data. By applying these processes to mass data analysis for power grid systems in service, we can expect much higher accuracy and efficiency of an intelligent power grid monitoring system.

2.2 *Application of the analysis methods to grid intelligent monitoring systems*

2.2.1 *Classification and estimation*

This is to classify the data acquired when the power grid is in normal operation according to their types, model the required information according to their types so classified and store them in a model database. Then, correlate the information stored in the database to the time information and structure the discrete data points into simple, continuous datalinks according to their time relationship.

2.2.2 *Prediction*

Prediction is to deep-process the data following the classification and estimation process identify the numerical logics between the data and find out the regularities by correlating the previously continuous data links so as to allow certain prediction of unknown new values. Then, validate the logics so identified with newly acquired data to verify that the analysis result is correct.

2.2.3 *Affinity grouping*

As can be seen from the prediction part of work, each type of data has its internal logics. Affinity grouping is to compare the logics of the different types of data and assign the types whose similar data logics can be associated into the same affinity group. A newly classified affinity group does not need to be predefined as its quantity may change anytime and one type of information may be involved into more than one affinity group.

2.2.4 *Abstracting and clustering*

The abstracting, grouping and prediction processes performed are all based on numerical characteristics and therefore does not have any physical implications. The relationship of the affinity groups created is explained with the physical implications represented by these values. After this abstracting, the practical relationship between different datasets is structured to correlate the physical implications represented by each type of data and abstract the data in different data affinity groups into a complete data relationship network.

2.2.5 *Modeling, description and visualization*

After the abstracting and clustering process, the physical relationship among data becomes a complete relationship network. This relationship network is model and analyzed by incorporating useful data to derive concrete visualized indicators in connection with the normal operation of the power grid. From the available mass data acquired at the time of common power grid failures, the data anomaly characteristics at the time of such anomalies before a fault occurs are extracted through big data mining, centralized, systemized and visualized as useful data basis for the intelligent power grid monitoring system. In general, these five steps will enable us to complete the big data mining and modeling process, which will generate understandable visually described data in a quick and efficient manner.

2.2.6 *Mining complex types of data*

Practice has discovered that, apart from most of the data information in a big data model that can be easily classified and modeled, there is generally some discrete data information in it. This part of information data does not appear to be obviously correlated to other data and cannot be inputted into the model using the method described above. For these data, there are two possibilities.

(1) Noise data. These data do not make sense in themselves and are not associated with the existing data either logically, nor will they be theoretically associated with any other group of data for a period of time. These data are not numerically continuous themselves. Therefore it is impossible to analyze them with normal mathematic theories or interpret their association with available physical theories or phenomena.

These data are not involved in the data mining and classification processes except that the noise data acquired at the same time are retained temporarily during the modeling process and deleted after the model data and analysis results are stored in the system. It has been considered not transmitting data sources that are identified as noise data after the modeling process. This would save both the transmission cost and part of the summarization cost. However, as data transmission and data computation are not very costly nowadays and, under the rapid technology development, these noise data that do not seem useful at present may probably be helpful in future, they are still included for transmission and summarization in our design.

(2) Implicit data. These data do not appear to be associated with the existing model, but can be associated with other existing data by deep processing and analysis. This association is usually invisible when there are not sufficient data samples, but will become more obvious as the volume of data increases when mass data are summarized. Numerically, these data are not significantly associated, but merely tend to be similar to, other data. Physically, they cannot be directly associated with other data. Only after some theoretical processes can they be associated with existing data.

These data will receive a complex process to make them highly associated with other data before they are inputted into the model. These data are sometimes understood as the centralization of variations of many other data such as animal anomalies in the natural world. After they are processed, these data can be associated with an existing intelligent power grid monitoring system though they are a more direct monitor themselves. By mining and processing these implicit data, the intelligent power grid monitoring system will become much more accurate, tolerant and robust.

3 BIG DATA MINING FOR MONITORING SYSTEM DIAGNOSIS

The neural network learning process is to cluster the data inputs mined using the k-means clustering method. That is, to identify the data centers of the hidden nodes in the RBF neural network by unsupervised learning, establish the extension constants of these nodes according to the distance between the data centers, and then train the output weights of each hidden node by supervised learning.

The RBF neural network is an n-h-m, i.e. a structuring having n inputs, h hidden nodes and m outputs. The input vector of the neural network is $\mathbf{x} = (x_1, x_2, \ldots, x_n)^T \in \mathbf{R}^n$, in which $\omega \in \mathbf{R}^{h \times m}$ is the output weight matrix; $\mathbf{b} = (b_1, b_2, \ldots, b_n)^T$ is the input unit offset. The network output is

$$W(x) = \sum_{i=1}^{h} \omega_i \varphi_i(x) \tag{1}$$

Here $\phi_i(x)$ is the activation function of hidden node i. The RBF network uses a variety of activation functions for hidden nodes, of which Gaussian function is a very frequently used one, i.e.

$$\varphi_i(\mathbf{x}) = e^{\frac{\|\mathbf{x}\ \mathbf{c}_i\|}{\delta_i^2}} , \quad (i = 1,2,...h) \tag{2}$$

Here $\mathbf{c}_i = (x_1, x_2, \ldots, x_n)^T$ is the center of hidden node i; δ_i is the extension constant of hidden node i.

The k-mean algorithm is an indirect clustering approach based on inter-sample similarity measurement. It is an unsupervised learning method. The k-means algorithm select k samples out of n samples as the initial cluster centers and assign the other objects to clusters represented by cluster centers that are most similar to them according to the distances to the initial cluster centers. Then, it calculates the cluster centers of the new clusters, which are generally the means of all objects in this cluster. This process is repeated until the standard measuring function starts to converge. Presently, most researchers use the mean square error as the standard measuring function.

h initial cluster centers are created from the samples, and the first h of them are selected by default. $\mathbf{c}_i$ is the center of cluster i; its corresponding mean square error is σ_i. The distance norms from all sample inputs to the initial cluster centers are defined.

$$D_i(\mathbf{x}) = \sigma_i \|\mathbf{x}\ \mathbf{c}_i\|^2, \quad i = 1,2,...h \tag{3}$$

Classify sample input $\mathbf{x}$ according to the minimum distance principle. Then recalculate the new cluster center of each cluster. When the first $D_i(\mathbf{x}) = \min D(\mathbf{x})$ appears,

$$\mathbf{c}_i(k+1) = \mathbf{c}_i(k) + v[\mathbf{x}(k)\ \mathbf{c}_i(k)]; \tag{4}$$

When $D_i(\mathbf{x}) = \min D(\mathbf{x})$ following the first one appears,

$$\mathbf{c}_i(k+1) = \mathbf{c}_i(k)\ \xi v[\mathbf{x}(k)\ \mathbf{c}_i(k)]; \tag{5}$$

For the rest

$$\mathbf{c}_i(k+1) = \mathbf{c}_i(k) \tag{6}$$

Here v is the learning rate of the winner cluster center; ξ is the ratio of the penalty rate of the cluster center ρ to v. Then, the mean square error of the winner cluster center is

$$\sigma_i(k+1) = \mu\sigma_i(k) + (1\ \mu)\|\mathbf{x}\ \mathbf{c}_i\|^2 \tag{7}$$

Here μ is a constant close to 1 but smaller than 1, generally taken as 0.999. Learning rate of a further cluster center

$$v = 1\ \frac{\sum_{i=1}^{h} \overline{v}_i \ln \overline{v}_i}{\ln h} \tag{8}$$

Here $\overline{v}_i = \frac{v_i}{\sum_{j=1}^{h} v_j}$.

If this equation converges, the iteration ends. If it doesn't converge, the distance from the sample to the cluster center has to be circulated. Order $k = k + 1$, then re-cluster and calculate a new cluster center. At the end of the iteration, obtain the optimal cluster center by removing empty ones. Remove a cluster center if it is out of the data collection.

By locating the initial cluster centers outside of the dataset, the improved algorithm from k-mean eliminates surplus competing nodes so that new cluster centers are moved into the dataset and surplus nodes are farther away from the dataset. According to the final position of the centers relative to the dataset[12−14].

The extension constant of a hidden node is often taken as $\delta_i = \kappa d_i$, in which d_i is the distance from cluster center i to other cluster centers, i.e. $d_i = \min_i \|\mathbf{c}_j - \mathbf{c}_i(k)\|$, in which κ is the overlap coefficient.

After determining the cluster center and extension constant of each hidden node, we can obtain the output weight vector $\boldsymbol{\omega} = (\omega_1, \omega_2, \ldots, \omega_h)^T$ using the least square method. The output of hidden node i is

$$O_{ji} = \phi_i \left(\left\| \mathbf{x}_j - \mathbf{c}_i \right\| \right) \tag{9}$$

Then the output matrix of the invisible layer is

$$0 = \begin{bmatrix} O_{11} & O_{12} & \cdots & O_{1h} \\ O_{21} & O_{22} & \cdots & O_{2h} \\ \cdots & \cdots & \cdots & \cdots \\ O_{n1} & O_{n1} & \cdots & O_{nh} \end{bmatrix} \tag{10}$$

With the weight to be determined, the network output vector is

$$\mathbf{y}_o = \mathbf{O}\boldsymbol{\omega} \tag{11}$$

Given the final teacher signal $\mathbf{y}$; assuming the approximation error is

$$e_o = \left\| \mathbf{y} \quad \mathbf{y}_o \right\| = \left\| \mathbf{y} \quad \mathbf{O}\boldsymbol{\omega} \right\| \tag{12}$$

By incorporating the least square method

$$\boldsymbol{\omega} = \mathbf{O}^+ \mathbf{y} = \left(\mathbf{O}^T \mathbf{O} \right)^{-1} \mathbf{O}^T \mathbf{y} \tag{13}$$

In this algorithm, the winner learning rate self-adaptive to the original dataset information volume contained in each cluster center. The penalty rate of the competition center is limited and in proportion to the learning rate. That is, when the center cluster begins, the penalty rate for the competition center is fairly high. This way, redundant competition centers will be away from the dataset become empty nodes that will not be learned anymore. As the calculation goes on, the competition rate among cluster centers diminish too, so should the penalty rate for the competitions. It is obvious that the maximum number of iterations of the algorithm will not be larger than the number of the training samples. Its converging speed will also be quick and can theoretically achieve zero error for the training samples. In this sense, the improved RBF neural network is very effective for a real-time fault diagnosis system.

4 A MONITORING SYSTEM BASED ON RBF NEURAL NETWORK BIG DATA MINING

The monitoring system based on RBF neural network big data mining introduced herein incorporates an industrial personal computer (IPC) as the detection and control core of the system

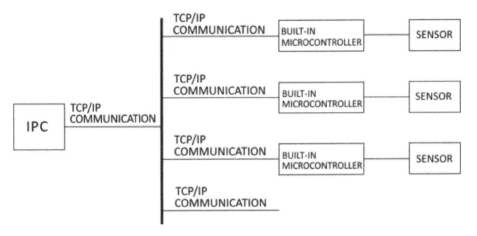

Figure 1. Structure of the monitoring system based on big data mining.

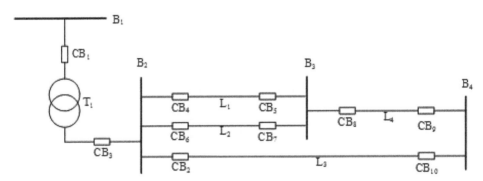

Figure 2. Power transmission test system.

in consideration of the complexity of the algorithm. This system is a distributed control system of a distributed architecture with an integrated panel. Compared with a system of a distributed architecture with distributed panels, this design is more flexible for expansion and maintenance.

Information for the monitoring system is acquired by sensors. A built-in microprocessor transforms and processes the data so acquired, and communicate monitoring information and control signals with the host computer via TCP/IP protocol. Fig. 1 shows the structure of this monitoring system. The far transmission distance, low data cost and good response mechanism of the TCP/IP network communication provide support for the real-time high-speed communication and programmable remote monitoring required for this monitoring system. The pre-processed data of each module make up the input vector of the neural network. The fault units are identified using the improved k-mean clustering algorithm described above and the least square method. The identified faults are then removed and restored.

5 SIMULATION AND ANALYSIS

5.1 *Simulation system and its sample composition*

The basic four-bus power transmission system is used as the test system as shown in Fig. 2. Nine elements are contained in the system, including four buses B_1, B_2, B_3 and B_4, one transformer T_1 and four transmission lines L_1, L_2, L_3 and L_4. In our simulation, we consider a simplified protection

Table 1. Samples selected for failure test.

No.	Protection and breaker with action	Faulty element
1	MBP1 BLP3 CB1	B1
2	MBP2 MLP7 CB2 CB3 CB4 CB5	B2
3	MBP3 BLP5 BLP9 CB2 CB5 CB6 CB9	B3
4	MBP4 BLP2 CB9 CB10	B4
5	MTP1 BLP10 CB1 CB3 CB10	T1
6	MLP4 MLP6 BLP3 CB4 CB6	L1
7	MLP7 BLP1 BLP6 BLP10 CB1 CB6 CB7 CB10	L2
8	MLP8 MLP9 BLP2 CB8 CB9	L3
9	MLP2 BLP8 CB2 CB8	L4
10	MTP1 MLP8 MLP9 CB1 CB3 CB8 CB9	B4 L4

Table 2. Diagnosis outputs of the test samples.

Fault Diag. No	B_1	B_2	B_3	B_4	T_1	L_1	L_2	L_3	L_4
1	1.10264	0.03556	0.00235	0.13554	0.14536	0.06859	0.08012	0.06041	0.05911
2	0.01254	1.05971	−0.05984	0.12541	0.01485	0.08952	0.00594	0.0612	−0.01451
3	0.1477	0.02542	1.02481	0.21655	0.25412	0.15487	0.04587	0.09214	−0.01547
4	0.09521	0.24154	−0.02544	0.95484	0.14958	0.15986	0.05872	0.10254	−0.00241
5	0.12418	0.14554	−0.01272	0.09951	1.14523	0.14988	−0.01145	0.25012	0.02015
6	0.02950	0.21459	0.26489	−0.10015	0.02350	0.098452	0.14222	0.31431	−0.10205
7	−0.26010	0.01300	0.18421	−0.11981	0.02698	0.21547	1.05413	−0.11296	0.02899
8	0.03658	0.13547	0.24987	0.01956	0.268495	0.30999	0.25465	0.84124	0.25483
9	0.58432	0.25687	−0.34762	0.15476	−0.26874	0.26847	0.65247	−0.04793	1.03544
10	0.25479	−0.12946	0.48657	0.75412	0.32487	0.25684	−0.02547	0.11856	0.62145

configuration system, which includes ten breakers $CB_1 \sim CB_{10}$, the main line protection, the main bus protection MBP and the main transformer protection MTP[13]. $N = 32$ typical fault conditions are constructed as the training sample sets for the neural network. Under each fault condition, the states of all protections and breakers are inputs of the neural network. The state of a protection or a breaker is expressed with a binary number, 1 for a protection or breaker with action, and 0 for one without action. If an input of the neural network is close, the corresponding element is identified as faulty.

To test the generalization ability of the neural network designed, we select faults not in the training sample set as the test samples. Here, 10 faults listed in Table 1 are used as test samples. All the test samples show severe faulty conditions, the most severe having as many as two protections and breakers in malfunction or dual failure.

5.2 Simulation result and analysis

The test system and fault samples are tested. Table 2 presents the diagnosis result, in which each line corresponds to the input of one fault condition. If one component of the output vector is larger than 0.5, the corresponding element is identified as faulty. Of the test samples numbered $1 \sim 10$ in the table, the corresponding output test results larger than 0.5 are B_1, B_2, B_3, B_4, T_1, L_1, L_2, L_3, L_4, B_4 and L_4. The results fully agree with the pre-settings of Table 1. From the output vectors in Table 2, the improved RBF neural network is able to give correct diagnosis conclusions for all the test samples. It is also able to give correct diagnosis when σ varies. The simulation results indicate that the improved RBF neural network is well capable of diagnosing power grid faults.

6 CONCLUSIONS

An improved RBF neural network based on data derived from big data mining is applied to an intelligent power grid monitoring system. In this paper, big data mining processes are applied to intelligent power grid monitoring systems. The fault diagnosis using an improved RBF-ANN is described and analysis. The least square method is used to optimize this RBF neural network. Computer simulation indicates that this improved RBF-ANN is very effective in diagnosing power grid faults. An intelligent monitoring system based on this algorithm and TCP/IP communication is designed. According to our study, the improved RBF neural network is highly effective in diagnosing and restoring faults when applied to intelligent power grid monitoring systems.

REFERENCES

[1] Chen, W.H. & Jiang, Q.Y. (2005) Risk assessment of voltage collapse in power system. *Power System Technology*, 29 (19), 6–10.
[2] Li, S.G. (2006) Development of supervisory control and forewarning system for power grid. *Power System Technology*, 09, 77–82.
[3] Jiang, L.H., Wang, F.R. & Xu, T.S. (2002) Review on approaches of power system restoration. *Electric Power Automation Equipment*, 05, 70–73.
[4] Wang, Q. & Liu, Z.(2011) Research methods of power grid fault diagnosis. East China Institute for Electric Engineering (Electric). *Proceedings of the 19th Symposium on Power Transmission*. East China Institute for Electric Engineering (Electric), 6.
[5] Fu, Q. (2007) Turntable subsystem fault-diagnosing approach based on RBF neural networks. *Transducer and Microsystem Technologies*, 06, 26–28+32.
[6] Cui, H.Q. & Liu, X.Y. (2009) Parameter optimization algorithm of RBF neural network based on PSO algorithm. *Computer Technology and Development*. 12, 117–119+169.
[7] Lei, Z.W., Xu, Z.S. & Liu, F. (2010) The improvement of information neural network and its application. *Instrumentation Technology*, 05, 57–59.
[8] Tang, L., Sun, H.B., Zhang, B.M. & Gao, F. (2003) Online fault diagnosis for power system based on information theory. *Proceedings of the Chinese Society for Electrical Engineering*, 07, 5–11.
[9] Wen, F.S. & Han, Z.X. (1994) Fault section estimation in power systems using genetic algorithm and simulated annealing. *Proceedings of the Chinese Society for Electrical Engineering*, 03, 29–35+6.
[10] Li, E.G. & Yu, J.S. (2002) An integrated fault detection and diagnosis approach to sensor faults based on RBF neural network. *Journal of East China University of Science and Technology*, 06, 640–643.
[11] Wang, C.L., Wang, H.H., Xiang, C.M. & Xu, G. (2012) Generator leading phase ability model based on RBF neural network. *Transactions of China Electrical Society*, 01, 124–129.
[12] Guo, C.X., You, J.X., Peng, M.W., Tang, Y.Z., Liu, Y. & Chen, J. (2010) A fault intelligent diagnosis approach based on element-oriented artificial neural networks and fuzzy integral fusion. *Transactions of China Electrical Society*, 09, 183–190.
[13] Xiong, G.J., Shi, D.Y. & Zhu, L. (2013) Fault diagnosis of power grids based on multi-output decay radial basis function neural network. *Power System Protection and Control*, 21, 38–45.
[14] Zhang, Y.C., Yang, C.F., Wang, W.J. & Li, H.Q. (2007) The arithmetic study of fault restoration in distribution network based on artificial neural network and pattern recognition method. *Central China Electric Power*, 06, 1–4+7.

Engineering Technology and Applications – Shao, Shu & Tian (Eds)
© 2014 Taylor & Francis Group, London, ISBN 978-1-138-02705-3

Study on the effect of nonlinear load on grid energy quality and its optimization

Nan Yao, Daojun Luo & Xingshi Zhang
Nanyang Power Supply Company, State Grid Henan Electric Power Company, Nanyang, Henan, China

ABSTRACT: Grid energy quality criteria are summarized to derive important indicators for grid energy quality and provide a basis for establishing energy optimization goals. Next, the mechanism underlying the effect of nonlinear load on grid energy quality is investigated before the method for calculating one of the influencing factors, distorted power, and the algorithm for extracting distorted signals based on adaptive notch filter are presented. Finally, a solution to optimizing grid energy quality by compensator control of distorted signals of nonlinear load is provided, and the optimization effect is simulated and examined to arrive at the anticipated grid energy optimization effect. In the final discussion, some of the drawbacks of the optimization solution discussed herein are pointed out and the outlook on future work on grid energy optimization is provided.

Keywords: grid energy quality; nonlinear load; distorted power; adaptive notch filter; compensator control

1 INTRODUCTION

National economy is growing rapidly under the current favorable atmosphere. Economy and technology are inseparable from each other whilst in technology itself, electricity, electronics and automation equipment make up a considerable proportion. As equipment precision and complexity yell for better power, the provision of high-quality grid energy quality becomes a prerequisite. However, as the nonlinearity of equipment as load plays a role in affecting energy quality, how to sort out a rational solution amid the conflict between the provision of high-quality electric energy and the mass use of nonlinear load is becoming an area of interest. In this paper, the algorithm for acquiring distorted load signals and solution to minimizing the effect of nonlinear load on energy quality are discussed from the perspective of energy quality evaluation indicators and the distortion of nonlinear load, with a view to providing theoretic basis and direction for optimizing the energy quality in China.

Many researchers have made efforts in grid energy quality and its optimization. It is these efforts that provide scientific basis for supplying high-quality electric energy in China. Among the many examples, Zhang Y.Y. (2013) examined a range of interference sources in power grid and presented a calculation method suited for evaluating the percentage of harmonic current[1]. Liu S.M. et al (2012) characterized the power use of some ten typical energy quality interference sources present in medium or high-voltage distribution systems and provided effective measures for energy management of electric power authorities and energy quality control of power users[2]. Chen J.L. et al (2011) presented a novel solution to realizing energy calculation according to the power properties of nonlinear load and contributed to the measurement of nonlinear load[3]. Zheng W. et al (2012) described the typical energy quality of special loads, simulated the energy quality problems of a special load concentration region in Gansu power grid using PSCAD/EMTDC, analyzed the simulation results under steady extreme conditions and transient different variations and provided reference for energy quality control[4]. Brennan M. (2010) characterized the energy quality intrinsic to electric railroad and possible control measures, and reported some achievements[5]. Qin B.Y. et al (2010) studied the effect of nonlinear load on the measuring performance of energy meters by using both experimenting and network operation, examined rational ways of energy measurement under nonlinear load and provided optional suggestions and basis for energy meters under nonlinear load[6]. Oztir Salih Mutlu (2009) carried out a lot of studies on energy quality properties intrinsic to wind power and control measures, having yielded mature results[7].

In this paper, the distortion of nonlinear load and the extraction of distorted signals are analyzed using previous findings to provide basis for optimizing grid energy quality; a solution to optimizing grid energy based on compensator control is designed to provide theoretical basis and direction for controlling high-quality energy supply.

2 SUMMARY GRID ENERGY QUALITY CRITERIA AND ENERGY CALCULATION PRINCIPLE FOR NONLINEAR LOAD

Accompanying the extensive use of electric and electronic equipment in electric power systems are concerns for the current energy quality. By category, the seven expressions of affected energy quality are tabulated below.

The sources of these energy quality problems all point to the load nonlinearity of electric and electronic equipment presently used. To locate the exact causes of energy quality problems, it is necessary to characterize the load nonlinearity of electric and electronic equipment, or in other words, only with an accurate algorithm for measuring nonlinear load energy is it possible to find out a scientific approach to optimizing energy quality. This section summarizes energy quality criteria with a view to providing energy quality optimization goals, and then discusses conventional measuring ways for nonlinear load energy and the essentials of measuring energy using adaptive notch filter with a view to providing theoretic basis for more accurate energy measurement.

2.1 Summary grid energy quality criteria

Energy quality means supplying high-quality electric power. This subsection looks at energy quality in terms of supply quality, use quality, voltage quality and current quality, with a view to providing improvement direction for energy measurement and quality optimization.

1) Supply quality: Supply quality can be discussed in its technical sense and non-technical sense. The former refers to voltage quality and supply reliability while the latter means the harmony level between suppliers and individual users.
2) Use quality: Use quality can also be discussed from current quality in its technical sense and from payment credibility in its non-technical sense.
3) Voltage quality: This is generally reflected by the deviation of the given voltage from the ideal level. The larger deviation is an indication of poorer voltage quality and vice versa.
4) Current quality: Current quality is generally discussed from current harmonic, interharmonic, subharmonic and noise, and reflects the deviation of the current from the ideal sinusoidal waveform. In the same way, the deviation of the given current from the ideal sinusoidal waveform current is also a measure for current quality.

Table 1. Expressions of affected energy quality.

1	2	3	4	5	6	7
Voltage interruption	Voltage swell	Voltage sag	Voltage constant fault	Waveform distortion	Voltage fluctuation	Frequency variation

On these grounds, energy quality can be defined as "deviation of voltage, current or frequency that causes equipment failure or malfunction", in which "deviation" includes both the steady and transient behaviors of a power supply system. As tasks designated by energy quality criteria are measured with permissible deviations from certain energy quality indicators, we have to know how to judge energy quality criteria properly. Following are base guidelines for energy quality criteria:

1) Ability to maintain safety, continuous supply and economic operation of the electric power system;
2) Ability to provide normal power load for electric equipment of individual users; and
3) Conformity to the permissible deviation from energy quality indicators for the current technology level.

Main indicators for energy quality evaluation include voltage deviation, frequency deviation, harmonic content, voltage fluctuation and flicker, and 3-phase voltage imbalance. Harmonic refers to the component sine wave of the electric quantity of one cycle and causes of harmonic include current sources shown in expression (1), current at the ends of steady load under nonlinearity shown in expression (2) and the current properties of time-varying load shown in expression (3).

$$i_s(t) = \sum_{h=0}^{m} \sqrt{2} I_h \sin(h\omega_1 t + \varphi_h) \tag{1}$$

$$i(t) = \frac{3}{2}\sqrt{2}kU_1^3 \sin \omega_1 t - \frac{1}{2}\sqrt{2}kU_1^3 \sin 3\omega_1 t \tag{2}$$

$$i_s(t) = \frac{\sqrt{2}}{2} U_1 G_0 [\cos(\omega_1 - \omega_g)t - \cos(\omega_1 + \omega_g)t] \tag{3}$$

In expression (2): ω_1 stands for the frequency of the input voltage; in expression (3): when ω_g is an even multiple of ω_1, the frequency in $i(t)$ is an odd multiple of ω_1; when ω_g is an odd multiple of ω_1, the frequency in $i(t)$ is an even multiple of ω_1. From the causes of harmonic, the essential source of harmonic in electric power systems is the nonlinearity intrinsic to some loads or equipment. In other words, the relation between voltage and the current induced is nonlinear.

2.2 Energy calculation principle for nonlinear load based on adaptive notch filter

In response to the increasing nonlinear load in the nation's power grid, Zhang X.B. (2007) introduced a new method of measuring electric energy under nonlinear load more accurately, suggesting that the energy consumed should be the difference between total energy and distorted energy[8]. To investigate the AC grid signals under nonsteady distorted conditions, this subsection tries to simplify the grid model as illustrated in Fig. 1, in which $u(t)$ is the sine voltage source voltage of the grid; $i(t)$ is the current of the grid; Z_L is the impedance of the grid circuit, in which Z is the impedance of nonlinear load.

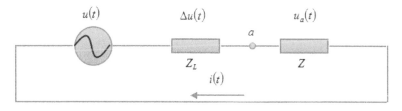

Figure 1. Simplified modeling chart of AC power grid.

In this chart, the voltage $u_a(t)$ at point a and the current $i(t)$ and point a can be expressed as (4):

$$\begin{cases} u_a(t) = u_B(t) + u_S(t) \\ i_a(t) = i_B(t) + i_S(t) \end{cases}$$ (4)

where: the subscript of fundamental wave is B; the subscript of distortion is S. From these, we get the expression of circuit impedance voltage drop $\Delta u(t)$ shown in (5):

$$\Delta u(t) = \Delta u_B(t) + \Delta u_S(t)$$ (5)

According to the principle of voltage times current for power, we get the expressions for the transient power $P_a(t)$ and average power $\overline{P}_a$ at point a shown in (6), in which variable are described below:

$\overline{P}_B$: Average power of fundamental wave absorbed by nonlinear load;
$\overline{P}_S$: Average power of distortion absorbed by nonlinear load;
$\overline{P}_{BS}$: Average power generated by fundamental voltage and distorted current absorbed by nonlinear load;
$\overline{P}_{SB}$: Average power generated by distorted voltage and fundamental current absorbed by nonlinear load.

$$\begin{cases} P_a(t) = u_B(t)i_B(t) + u_B(t)i_S(t) + u_S(t)i_B(t) + u_S(t)i_S(t) = P_B(t) + P_{BS}(t) + P_{SB}(t) + P_S(t) \\ \overline{P}_a = \frac{1}{T}\int_0^T P_a(t)dt = \frac{1}{T}\int_0^T (P_B(t) + P_{BS}(t) + P_{SB}(t) + P_S(t))dt = \overline{P}_B + \overline{P}_{BS} + \overline{P}_{SB} + \overline{P}_S \end{cases}$$ (6)

From the four average powers indicated in (6), the following conclusions can be drawn according to their direction of tide:

Conclusion 1: If $\overline{P}_a \leq 0$, the power generated by fundamental voltage and fundamental current is a non-positive;
Conclusion 2: If $\overline{P}_{BS} < 0$, the power generated by fundamental voltage and distorted current is negative;
Conclusion 3: If $\overline{P}_{SB} > 0$, the power generated by distorted voltage and fundamental current is positive power that is fed into the grid in the form of fundamental current and could affect the grid quality.
Conclusion 4: If $\overline{P}_S > 0$, the distorted power is non-negative power that is fed into the grid in the form of distorted current and does not affect the grid energy quality.

From these conclusions, we get the following formula for the average energy power $\overline{P}$ from the grid as shown in (7):

$$\overline{P} = \overline{P}_B + \overline{P}_{BS} + \overline{P}_{SB} = \overline{P}_a - \overline{P}_S$$ (7)

To study distorted signals, it is important to filter fundamental signals. In our study, adaptive notch filter is used for this purpose before distorted signals are yielded, and these distorted signals are examined to find out how nonlinear load affects the grid quality, with a view to establishing workable recommendations for optimizing the grid energy quality.

In our study, LMS algorithm is used for adaptive notch filter. Supposing $d(k)$ is the sample signal obtained by sampling the original input signals, $x_1(k)$ and $x_2(k)$ the reference input by discretizing a pure sine wave having the present fundamental frequency and the reference input after a 90° phase shift, and $e(k)$ is the final input, the weight vector equation for the final distorted signal $e(k)$ is

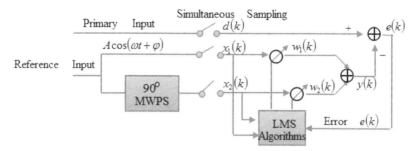

Figure 2.   Schematic adaptive notch filter.

expressed as (8) below [5]:

$$\begin{cases} w_i(k+1)= w_i(k)+2\mu e(k)x_i(k) \\ y(k)= \sum_{i=1}^{2} w_i(k)x_i(k) \\ e(k)= d(k)- y(k) \\ i=1,2 \end{cases}$$

(8)

where: $w_1(k)$ and $w_2(k)$ stand for two weights in order to equalize the sine wave amplitude and phase angle after functional combination to those of the interference component in the original input so that the inference with the frequency ω_0 in the final output $e(k)$ is eliminated to provide basis for obtaining more accurate distorted signals. In expression (8), μ is the parameter controlling the convergence rate and stability of the algorithm. If the sampling cycle is symbolized as T, we get the transmission function of notch filter in Z-transform.

$$H(Z)= \frac{E(Z)}{D(Z)}= \frac{1}{z^2 -2\cdot z\cdot \cos\omega_0 \cdot T\cdot(1-\mu c^2)+(1-2\mu c^2)}\cdot z^2 -2\cdot z\cdot \cos\omega_0 \cdot T+1$$

(9)

From the transmission function, we derive a schematic of adaptive notch filter shown in Fig. 2.

3   GRID ENERGY QUALITY OPTIMIZATION BASED ON COMPENSATOR

3.1   *Essentials of compensator control*

A compensator is used to eliminate the effect of local nonlinear load harmonic and imbalance in the grid, and to provide preset quantities of active power to nonlinear load on this basis so as to compensate for current and voltage stabilization. From the anticipated effect of compensation, the grid current and voltage after compensator will be balanced. The summation of 3-phase AC currents, in the case of household appliances, should be zero at nodes if balanced, and the average voltage would remain constant. The single-phase equivalent circuit of the converter is shown in Fig. 3.

In this diagram, $u \cdot V_{dc1}$ is the input voltage of the converter and u level, which is either $+1$ or -1, indicates the switching function. As the main purpose of a converter is to produce a switching function, a state vector shown in (10) is selected:

$$x^T = \begin{bmatrix} i_1 & i_{cf} & v_{p1} \end{bmatrix}$$

(10)

where: v_{p1} is the voltage at the point of common connection (PCC); and given $v_{p1} = v_{cf}$, the state vector space of the system can be expressed as (11). In expression (11) below, u_c is the form of

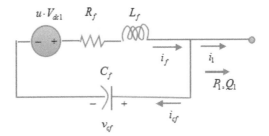

Figure 3.   Single-phase equivalent circuit of converter.

Table 2.   Simulation system parameter settings.

System quantities	Values	System quantities	Values
Systems frequency	50 Hz	Microgrid	$R_{01} = R_{02} = 0.2\,\Omega$
Source voltage	11 kV	DC votage	3.5 kV
Feeder impedance	$R_1 = 1.025\,\Omega$, $L_1 = 57.75$ mH	Tra rating	3/11 kV, 0.5 MVA, 2.5%
DG-1 and DG-2 local unbalanced load	$R_{1a} = 48.4\,\Omega$, $L_{1a} = 192.6$ mH $R_{1b} = 24.4\,\Omega$, $L_{1b} = 100.0$ mH $R_{1c} = 96.4\,\Omega$, $L_{1c} = 300.0$ mH	VSC losses Filter C m1	1.5 Ω 50 μF −0.1 rad/MVAr
DG-1/-2 local	3-phase R = 200 Ω, L = 100 mH	m2	−0.05 rad/MVAr
Common impedance load	$R_{1a} = 24.4\,\Omega$, $L_{1a} = 100.0$ mH $R_{1b} = 24.4\,\Omega$, $L_{1b} = 100.0$ mH $R_{1c} = 24.4\,\Omega$, $L_{1c} = 100.0$ mH	n1 n2 DGs and comeper	0.12 kV/MW 0.06 kV/MW /
Common load	Induction moror 40 hp, 11 kV	Droop coefficients	/

switching function under continuous time; $u_c(k)$ is the form of switching function under discrete time.

$$\left.\begin{array}{l}\dot{x} = Ax + Bu_c \\ x(k+1) = Fx(k) + Gu_c(k)\end{array}\right\} \Rightarrow u_c(k) = -K\left[x(k) - x_{ref}(k)\right] \tag{11}$$

By linear quadratic control shown in expression (12), we get the final payoff matrix:

$$\begin{array}{ll}\text{If} & u_c(k) > b \quad \text{then} \quad u = +1 \\ \text{Eles} \quad \text{If} & u_c(k) < -b \quad \text{then} \quad u = -1\end{array} \tag{12}$$

where: b is a very small setting that can be adaptively defined as required.

3.2   *PSCAD/EMTCD simulation environment and result*

PSCAD/ENTCD is used to monitor the voltage and power at two points of common connection where compensator is incorporated, with a view to identifying the optimization effect of compensated nonlinear load on the grid quality. Simulation parameters used are presented in Table 2.

With the system parameters listed in Table 2, we get the simulated 3-phase voltage at PCC1 and PCC2 shown in Fig. 4.

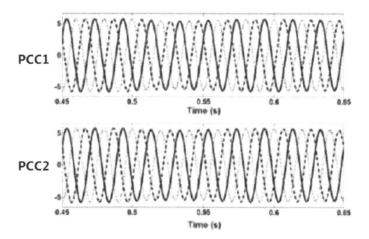

Figure 4.   Simulated 3-phase voltage at point of common connection.

Table 2.   System simulation parameters under PSCAD/EMTCD environment.

Voltage drop at PCC1			Voltage drop at PCC2		
Active power	Initial value, MW	Final value, MW	Reactive power	Initial value MVAr	Final value, MVAr
P_1	0.275	0.274	Q_1	0.282	0.281
P_{G1}	1.125	1.125	Q_{G1}	1.120	0.119
P_{L1}	1.401	1.392	Q_{L1}	1.401	1.400
P_2	0.710	0.695	Q_2	0.420	0.410
P_{G2}	0.693	0.691	Q_{G2}	0.182	0.173
P_{L2}	1.401	1.384	Q_{L2}	0.601	0.584

Voltage drop at PCC1 (%)			Voltage drop at PCC2 (%)		
3.15%			3.38%		
Active power	Initial value, MW	Final value, MW	Reactive power	Initial value MVAr	Final value, MVAr
P_1	0.142	0.711	Q_1	0.090	0.432
P_{G1}	0.568	0.000	Q_{G1}	0.363	0.000
P_{L1}	0.712	0.693	Q_{L1}	0.451	0.433

3.3   *Effect of grid energy optimization based on compensator*

The simulation environment for compensator-based energy quality optimization is PSCAD/EMTCD. The system simulation data at PCC1 and PCC2 when sharing load with the power grid of Tangshan are given in Table 2.

From the simulation data listed in Table 2, by controlling the grid energy quality parameters at PCC1 and PCC2 under variable common load with a compensator, we can optimize the energy quality within a specific area.

4   DISCUSSIONS

The essential mechanism underlying the effect of nonlinear load on grid energy quality is the voltage or current distortion during the use of electric power. To optimize grid energy quality, it is important to eliminate this distortion and to maintain stable active power of the electric power supplied. In this paper, an algorithm for extracting distorted signals and a solution of compensator control are presented, both of which are highly effective for optimizing electric energy but are fairly complicated in practical application. The compensator optimization solution discussed herein can serve as a guiding idea. Further studies will have to be carried out on equipment and solutions if we want to optimize grid energy on a larger scale.

REFERENCES

[1] Zhang, Y.Y. (2013) The study on the energy quality characteristics of typical interference sources and countermeasures. *China Electric Power & Electrical.*

[2] Liu, S.M., et al. (2012) Study of power quality characteristics of nonlinear electric user in medium-high voltage distribution networks. *Power System Protection and Control.* 40(15), 150–155.

[3] Chen, J.L., et al. (2011) Application of adaptive notch filter for power energy measurement of nonlinear load. *Electrical Measurement & Instrumentation.* 48(541), 10–12.

[4] Zheng, W., et al. (2012) Analysis of power quality problems in a regional grid with many types of disturbing loads. *Electrical Automation*, 34(1), 4–8.

[5] Brenna, M., et al. (2010) Electromagnetic model of high speed railway lines for power quality studies. *IEEE Transactions on Power System.* 25(3), 1301–1308.

[6] Qin, B.Y., et al. (2010) Discussion on the practical technology of electric energy measuring for nonlinear load. *Southern Power System Technology*, 4(1), 56–59.

[7] Oztir, S.M., et al. (2009) Power quality analysis of wind farm connected to Alacatl substation in Turkey. 34(5), 1312–1318.

[8] Zhang, X.B. (2007) Electrical network power flow analysis and research on new measuring methods of electric energy on the condition of distorted signals. Harbin, Harbin University of Science and Technology.

[9] Liu, H., et al. (2007) Accurate estimation of power system frequency using a modified adaptive notch filter. *Journal of Shanghai Maritime University*, (9), 24–27.

Engineering Technology and Applications – Shao, Shu & Tian (Eds)
© 2014 Taylor & Francis Group, London, ISBN 978-1-138-02705-3

Research on the detection of coal mine underground ventilation capacity based on cloud computing k-unth algorithm

Zhou Lv

Institute of Mechatronic Engineering, Taiyuan University of Technology, Taiyuan, Shanxi, China

ABSTRACT: Coal mine accidents are generally associated with the ventilation system used. Cloud computing k-unth algorithm has brought about major changes in the multi-variable prediction structure of coal mine underground ventilation capacity. This paper discusses the use of cloud computing k-unth algorithm for the detection of coal mine underground ventilation capacity and a spatial data fusion approach as well as a cloud computing information aggregation technique and four-dimensional nonlinear indexing for newly developed intelligent ventilation networks are discussed. With options for dynamic modifications of R+tree, the minimum k-unth algorithm prediction error is used to determine the optimal embedding dimension. Prediction customer error is used to predict wind waveform flicker and ventilation diagonal branch unbalance.

Keywords: detection of coal mine underground ventilation capacity detection; k-unth algorithm; cloud computing

1 INTRODUCTION

There are a wide variety of underground coal mine accidents in China, the causes of which, however, can be sourced according to certain regularities. Statistics has indicated that any major accident is generally associated with the ventilation system used, either because the ventilation system is not properly designed or a complete ventilation system is not established, leading coal methane explosion or coal dust explosion. This shows how important a properly designed ventilation system is for the production safety of an underground coal mine.

k-unth algorithm is capable of calculating the air distribution and resistance loss in all roadways of an underground coal mine. The resolution is performed for roadway conditions with and without a ventilator. By using the k-unth algorithm, we can decide whether a blade angle is suitable according to the characteristic curve of the ventilator and the load of the coal mine so that the right angle is selected. We can also find out about the air distribution and resistance loss in any roadway of the coal mine. Identification and stability analysis of diagonal airway are one of the core part of the stability theory of coal mine ventilation system[1]. When the ventilation system is unstable, the air in the airway at the diagonal branch will reduce and even flow backward. If at this time the airway at a certain diagonal branch is an air load point (e.g. a working face or the main air shaft for a working face), accumulation of coal methane or other hazardous gases will lead to unforeseeable disasters.

2 NEW TECHNOLOGY RESEARCH ON COAL MINE UNDERGROUND VENTILATION CAPACITY

Major changes have taken place in the multi-variable prediction structure cloud computing k-unth algorithm applied to coal mine underground ventilation capacity. A method of using cloud computing k-unth algorithm for the multi-variable quality detection of coal mine underground ventilation capacity, a spatial data fusion approach and a cloud computing information aggregation technique for newly developed intelligent ventilation networks are discussed. Mutual information is used to select the optimal delay time for the quality detection of each component, while the minimum k-unth algorithm prediction error is used to determine the optimal embedding dimension so as to predict the voltage fluctuation and flicker or three-phase unbalance caused by voltage waveform distortion, which may result in severe disturbance or "contamination" to the coal mine underground ventilation capacity.

3 NEW TECHNOLOGY FOR THE DETECTION OF COAL MINE UNDERGROUND VENTILATION CAPACITY

3.1 *Essentials of present detection of coal mine underground ventilation capacity*

Monitoring the underground ventilation capacity of a coal mine is a direct way of acquiring information on the coal mine underground ventilation capacity. Traditional methods like hash segment aggregation and directed boundary, are all based on, and improvements of, edge tracking. However, as edge tracking is quite time-consuming, its processing result hardly provides a topology and some pre-processing is needed under specific conditions. The need to mark the searching process, in particularly, has prevented it from being used for larger ventilation networks[2], making it impossible to become a practically applicable method.

3.2 *Application of new detection technology for coal mine underground ventilation capacity*

Application system platforms of hardware and software for coal mine ventilation capacity mainly include like digital information aggregation processing, cloud computing nonlinear aggregation, and four-dimensional nonlinear index. As new technologies like the dynamic modification of R+tree structure and new k-unth algorithms like wavelet transform have been developed, a new requirement has emerged for the detection of coal mine underground ventilation capacity in addition to system timeliness: it has to support complex algorithms. Changing raster images into vector images using the idea of nonlinear topology is not faster than traditional methods, but is also able to process very large images without the need of pre- or post-processing.

1) Capturing point objects

Exact way of capturing point objects: to increase the computation speed, the system uses logical operation as a substitute for powers and roots operations, which greatly increases the computation speed.

A capture point is extended to a capture box. The screen coordinates of the capture box are turned to the same logic coordinates as the target point (top left (x1, y1); top right (x2, y2)). In this way, the distance calculation is turned to logic judgment[3] so that the only thing you need to do is to determine whether the object point falls into the capture box. That is: if $x1 < xi < x2$ and $y1 < yi < y2$, then (xi, yi) is the object point. Otherwise the capture is unsuccessful.

2) Capturing line objects

To increase the computation speed, the system uses distance approximation to eliminate the powers and roots operation, which makes the computation much faster.

3.3 *Essentials of new detection technology for coal mine underground ventilation capacity*

In processing the mass data of coal mine underground ventilation, a topographical map usually involves a tremendous quantity of vector data. One roadway can involve hundreds of trillions of data. Under the traditional ventilation network display method, the speed is untolerably slow. Some of the typical examples, all of which are less than satisfactory, are listed below.

Quad tree: This is a very simple and easy way, but the division of the tree is not flexible enough and is more suitable for the spatial distribution of individual objects rather than for area search.

R+tree: This is suitable for spatial area inquiry[4]. But as it is a two-dimensional nonlinear index, with the dynamic modification of the R-tree structure, it behavior will deteriorate markedly.

Q-tree, linear Q-tree: These methods are able to reflect the dynamic changes of an index, but they have a complex architecture and huge storage. In our study, an indexing method is used so that the mass data of coal mine ventilation networks are well managed and detected. Grid indexing mechanism of points and curve segments.

This method divides a spatial regional area laterally and longitudinally into a number of rectangular cells, each having a unique ID, and then stores the IDs of the points and segments falling into this cell together with this ID in the form of variable-length record. To address the multi-scale nature of the system, details reflecting the scales have to be stored when establishing spatial index.

The grid indexes of points and curve segments are established in the following steps. First, for a large data file, we establish a longitudinal and lateral rectangular index grid, the length and width of which can vary. Then, we establish index files of the grid, possibly in the following format:

Index file header:

m_Xmin	Min of x-axis in the grid;
m_Ymin	Min of y-axis in the grid;
m_Xmax	Max of x-axis in the grid;
m_Ymax	Max of y-axis in the grid;
m_Row	Number of rows in the grid;
m_Col	Number of columns in the grid;

Index file body:

m_GridCode	Grid area ID
m_ScaleCode	Scale ID
m_ObjectNumber	Total objects
m_ObjectPos1	Coordinate pointer of object 1
: :	
m_ObjectPosn	//Coordinate pointer of object n

The ID of a grid cell is expressed as a four-digit integer of any number from 0–9999[5]. The first two stand for the row, and the last two the column. The calculation formula is: grid cell ID = line number × 100 + column number. Finally, calculate the number of grids involved in the display, extract and prioritize data according to the position provided by the corresponding index file, and cut out the display so that the data processing speed is twice as fast.

4 NEW TECHNOLOGY FOR CAPTURING CLOUD COMPUTING OBJECTS IN THE RESEARCH OF COAL MINE UNDERGROUND VENTILATION CAPACITY

Real-time object capture is indispensable for graphic editing and information inquiry. This includes point objects, line objects, areal objects and notes which, in concrete operation, can be grouped into point, line and areal objects. Notes can be attributed to point objects. These objects are captured in the following way.

When capturing a point object, we compare the distance of the capture point to all the point objects. The one having the shortest distance is the point object we need. The calculation formula is:
Here (x0, y0) are the coordinates of the capture point; (xi, yi) are the coordinates of the object point; Di is the distance of the capture point to point i.

$$D_i = \sqrt{(x_i - x_0)^2 + (y_i - y_0)^2} \quad (i = 0, \cdots, n) \tag{1}$$

When capturing a line object, we compare the distance of all line segments to the capture point. The one having the shortest distance is the line object we need. The calculation formula is:

$$D_i = \frac{|(x_0 - x_{i1})(y_{i2} - y_{i1}) - (y_0 - y_{i1})(x_{i2} - x_{i1})|}{\sqrt{(x_{i2} - x_{i1})^2 + (y_{i2} - y_{i1})^2}} \qquad (i = 0, \cdots, n) \qquad (2)$$

Here (x0, y0) are the coordinates of the capture point; (xi1, yi1) and (xi2, yi2) are the coordinates at the end points of line segment i; Di is the distance of the capture point to line segment i.

5   SIMULATION SYSTEM AND OUTPUTS

Three-dimensional simulation experiment of an underground coal mine is to establish a three-dimensional model with the mine ventilation data as shown in Fig. 1. This mainly includes the

Figure 1.   Roadway wind quantity and dynamic wind flow.

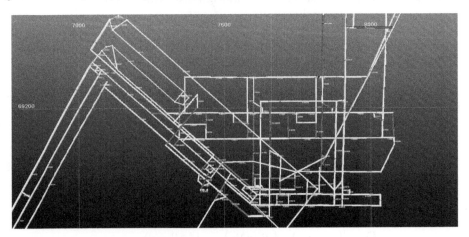

Figure 2.   Simulation of key routes on the ventilation network map.

138

mean wind speed	friction loss Pa	overall loss Pa	The friction loss	specific friction Ns2/m8	The total drag Ns2/m8	coefficient of shock resistance	friction resistance of unit length R/100m	coefficient of frictional resistance Ns2/m4	number
0.5	0	10.05	0.59	0.0061	0.00734	0.9014	0.01551	0.00665	37
1.48	0.01	1.14	0.12	0.00022	0.00227	0.90758	0.01168	0.00273	80
2.05	0.17	3.41	3.45	0.0035	0.00351	0.00414	0.17487	0.03953	81
0.66	0.02	0.37	0.34	0.00329	0.00329	0.00015	0.17486	0.03953	82
0.83	0.03	0.64	0.74	0.00455	0.00455	0.00015	0.17489	0.03953	83
0.3	0	35.51	0.01	0.00057	1.4834	0.67088	0.01221	0.0034	89
1.72	0.12	24.95	27.37	0.03947	0.03959	0.04857	0.17495	0.03953	90
6.66	0.14	34.78	7.89	0.0008	0.00366	1.16963	0.01447	0.00239	91
0.76	0	0.55	0.04	0.00029	0.00287	1.14191	0.01058	0.00247	92
4.84	0.07	2.33	1.34	0.00018	0.00031	0.0755	0.00996	0.00329	93
1.35	0.04	2.31	0.87	0.00193	0.00436	1.10644	0.09733	0.02392	98
0.72	0	0.81	0.03	0.00021	0.00225	0.9044	0.01089	0.00254	99
2.02	0.1	2.18	2.12	0.00207	0.00214	0.02781	0.09733	0.02392	100
0.03	0	750.26	0	0.00213	3156...	0.92355	0.05264	0.01228	101
0.03	0	748.08	0	0.0014	3144...	0.93393	0.03478	0.01348	102
0.03	0	0	0	0.00068	0.00212	0.94201	0.03215	0.01228	103
3.02	0.13	132.27	145.23	0.05113	0.05113	0	0.04546	0.01425	104
1.55	0.1	15.74	16.41	0.02896	0.02896	0	0.17494	0.03953	133
0.11	0	8.64	0	0.00072	1.50998	0.33517	0.01145	0.00665	134
2.87	0.1	44.29	35.8	0.01209	0.01346	0.90228	0.03479	0.01348	156
4.45	0.11	71.67	63.68	0.00666	0.00672	0.05359	0.01201	0.00665	161

Figure 3.   Network resolution result.

automatic establishment and management of the topology of the mine ventilation system. The system provides a variety of three-dimensional graphic tools. AutoCAD DXF graphs can also be directly imported to establish the topological diagram of the ventilation system. Our method is able to identify the maximum ventilation resistance route (or key route) of the mine using k-unth algorithm with relevant data and resolution results, and produce the maximum resistance slope map that provides power basis for reducing the mine ventilation resistance. The resulting graphs and data can be shared with Microsoft Word (as shown in Fig. 2).

According to the mine ventilation management method, the type of the main surface ventilator and the combined operation of multiple ventilators were examined. The mine power consumption and the maximum capacity of the ventilation system were analyzed as shown Fig. 3. On the basis of the analysis result, we were able to establish the rational adjusting position of the ventilation system. Fig. 4 shows our observation and evaluation of the roadway wind speed distribution.

6   CONCLUSIONS

Two aspects of applying cloud computing k-unth algorithm to the new detection, analysis and monitoring technologies of coal mine underground ventilation capacity are studied. One is full-intelligent control which automatically identifies and processes data of the underground ventilation capacity of a coal mine to achieve unattended operation. The other is remote communication, which incorporates the use of mutual information method to select the optimal delay time for the chaotic time series of each component, and the use of the minimum k-unth algorith prediction error to determine the optimal embedding dimension[6]. This decides that remote communication is flexible

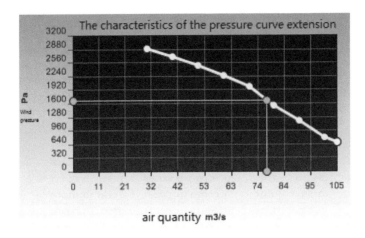

Figure 4.   Ventilation hydrostatic evaluation curve.

to the monitoring of different levels. This makes it possible to place the underground ventilation capacity monitoring points anywhere in the ventilation network and provides good online behavior at the same time.

REFERENCES

[1] Yang, G. & Sui, Y.L. (2009) Adaptive approach to monitor resource for cloud computing platform. *Computer Engineering and Application*, (29), 14–17.
[2] Neal, L. (2009) Is Cloud Computing Really Ready for Prime Time? *IEEE Computer*, (1), 15–20.
[3] Li, B.H., Chai, X.D., Hou B., Li T. et al. (2009) Networked modeling & simulation platform based on concept of cloud computing – cloud simulation platform. *Journal of System Simulation*, (17).
[4] Hu, X.P. (2010) The concept of cloud library. *Information Studies: Theory & Application*, (6), 29–32.
[5] P'erez, P., Hue, C., Vermaak, J., et al. (2009) Color-based Probabilistic Tracking [C]//Proc. of Eur. Conference on Computer Vision. Copenhagen, Denmark: [s. n.], 2002. Cloud Definition [J]. ACM SIGCOMM Computer Communication Review, 2009, 39(1): 50–55.
[6] Buyya R; Yeo CS; Venugopal S. Market-oriented cloud computing vision, hype, and reality for delivering IT services as computing utilities. In: Proceedings of the 10th IEEE International Conference on High Performance Computing and Communications, 2008.

Engineering Technology and Applications – Shao, Shu & Tian (Eds)
© 2014 Taylor & Francis Group, London, ISBN 978-1-138-02705-3

A privacy protection model for location-based social networking services

Rong Tan & Wen Si
College of Information and Computer Science, Shanghai Business School, Shanghai, China

Jian Wang
Cyber Physical System R&D Center, The Third Research Institute of Ministry of Public Security, Shanghai, China
School of Electronic Information and Electrical Engineering, Shanghai Jiao Tong University, Shanghai, China
Shanghai Chenrui Information Technology Company, Shanghai, China

ABSTRACT: Since the location-based social networking services emerged, the location privacy issue has gained widespread concerns from consumers and researchers. While the location history plays an important role in knowledge discovery, many people worry that their privacy will be disclosed unsafely. In this paper, we introduce a spatial-temporal cloaking model which follows the k-anonymity principle to protect the location history. The corresponding algorithm which can find a minimum cloaking 3D box covering the actual location and its temporal information is proposed as well. Moreover, the algorithm is able to prevent the modeled records against the continuous attacks which infer the actual location by computing the overlapping spatial region. The experimental results show that the proposed algorithm has a high efficiency.

Keywords: privacy model; location privacy protection; privacy protection; mobile computing; location-based social networking services; location-based services

1 INTRODUCTION

Recently, the location-based social networking services (LBSNS) have gained considerable prominence in social computing. Applications like Foursquare[1] and Gowalla[2] are closely associated with geographical information and aim to provide the platforms on which people expand their social ties mainly on the basis of their locations. Furthermore, with the huge amount of location history collected, it is able to further study the people's offline social behaviors[3,4,5,6] based on which personalized services such as recommendation and advertising are provided. Although LBSNS has made great commercial success, it has to face a lot of challenges among which the location privacy issue raises widespread concerns from consumers and researchers[7,8,9].

Locations can reveal more than longitude and latitude. With the help of advanced data mining technologies, LBSNS applications can compile detailed profiles of people's lives such as the places they live, the places they work, and the people they meet, etc. by analyzing their location histories. While it is easy to locate friends, it is far more difficult to provide strong and clear privacy protections for these LBSNS applications. As a result, many applications promise that they do not collect or maintain users' location history. For example, Loopt[10] maintains only a user's most recent location for necessary usages. Although such promises help building users' trust to some extent, the diversity of LBSNS is certainly decreased.

This paper introduces a k-anonymous spatial-temporal cloaking model (KSTCM) for the purpose of protecting the privacy of users' location histories. The model perturbs the records on both the spatial and temporal dimensions by reducing the granularity of related information. In addition, algorithm which is able to find a minimum cloaking box covering the actual location and its temporal information is proposed. And it can prevent actual location from continuous attack as well. The experimental results show that the algorithm performs well with respect to reasonable settings.

2 RELATED WORK

Most location privacy protection models essentially aim to make users' identities unmatchable with their accurate locations such that the adversary cannot infer the sensitive information by binding location and identity. Using a fake identity called pseudonym instead of user actual identity is generally adopted in the early stage, as well as replacing the genuine location with a fake one called dummy[11,12]. While these two methods are easy to implement, lack of modeling standards and poor quality of services make them far from satisfaction.

Spatial cloaking is utilized to perturb the locations on the spatial dimension by covering the actual location with a spatial region (e.g. circle or rectangle)[13]. Based on spatial cloaking, spatial-temporal cloaking which further perturbs the location information with respect to temporal dimension. Furthermore, k-anonymity which is originally used to preserve privacy in database system is modified to location cloaking[14]. Although these models upgrade the location privacy protection, problems like how to find the minimum cloaking box, how to define the value of k, and how to handle the attack models, etc. have to be solved[15].

3 K-ANONYMOUS SPATIAL-TEMPORAL CLOAKING MODEL

A typical location record in LBSNS can be organized as (u_{id}, t, [x,y], l_{id}) where u_{id} stands for the identification of user, t stands for timestamp when it is collected, [x,y] represents the location coordinate pair and l_{id} is the identification of the place user located. For a location record, the k-anonymous spatial-temporal cloaking model (KSTCM) is defined as follows.

Definition 1. *Let S be the set of location records, δ and θ represents the spatial and temporal tolerance respectively. A record in S is said to be modeled by KSTCM if and only if it satisfies:*

- *It is indistinguishable from other k-1 records with respect to spatial and temporal dimensions.*
- *For any two records of above related k records, $r_1 = (u_1, t_1, [x_1, y_1], l_1)$ and $r_2 = (u_2, t_2, [x_2, y_2], l_2)$, they should meet $\|[x_1, y_1] - [x_2, y_2]\| <= \delta$, $\|t_1 - t_2\| <= \theta$ and $u_1 \neq u_2$.*

As we can see, KSTCM follows the k-anonymity principle proposed in [14]. A record modeled by KSTCM is called a KSTCM object whose data structure is defined as:

$$K_i^{uid} = (u_{id}, [ts_{begin}, ts_{end}], [x_{min}, y_{min}], [x_{max}, y_{max}], A) \tag{1}$$

The u_{id} is the user identifier. [ts_{begin}, ts_{end}] and ([x_{min}, y_{min}], [x_{max}, y_{max}]) represent the temporal interval and spatial cloaking box which cover the timestamps and locations of the k original records on the temporal and spatial dimensions respectively. And [x_{min}, y_{min}] and [x_{max}, y_{max}] are the lower-left and upper-right coordinate pair of the spatial cloaking box. A is the semantic annotations that describe the types of KSTCM object.

4 MODELING ALGORITHM

To find the minimum cloaking 3D box with re spect to spatial and temporal dimensions is the key issue of our model. It is possible that the k records spin in large area and extended time period, in which case that the cloaking box is very large resulting in a degradation of quality of services. Even we define the spatial and temporal tolerances, if there are more than k records, how to select those k-1 records is still a challenge. On the other hand, the potential threats arise from associated attack models (e.g. continuous queries attack) [18] should be taken into consideration as well.

4.1 Similarity between records

In order to find the minimum cloaking box, we define the similarity measurements between two records with related to the spatial and temporal dimensions.

- *Spatial similarity.* The similarity between the locations of two records, measured by their Euclidean distance. (δ is the spatial tolerance)

$$sim_{l1,l2} = \frac{\|[x_1, y_1] - [x_1, y_1]\|}{\delta} \tag{2}$$

- *Temporal similarity.* The similarity between the timestamps of two records, measured by their differences in minute. (θ is the temporal tolerance)

$$sim_{t1,t2} = \frac{\|t_1 - t_2\|}{\theta} \tag{3}$$

Therefore, the similarity between two location records is defined as follows

$$sim_{r1,r2} = 1 - w_t * sim_{t1,t2} - w_s * sim_{l1,l2} \tag{4}$$

where w_t and w_s is the weight of temporal and spatial dimension respectively, and $w_t + w_s = 1$.

4.2 Attack model

Although with the KSTCM model, a location record is indistinguishable from other k-1 ones on spatial and temporal dimensions, the actual location still can be extracted by continuous attacks. For example, in Fig. 1 assuming location A has been visited twice by user 1, as a result, there are two cloaking boxes with respect to location A of user 1. If the overlapping spatial region of these two cloaking boxes is very small, then the location A may be extracted. And it tends to that the more cloaking boxes of a same location with respect to a particular user, the more likelihood it can be found.

Therefore, it is specified when modeling location records, if two records of a same user have the same coordinate pair, then their spatial cloaking box should be the same.

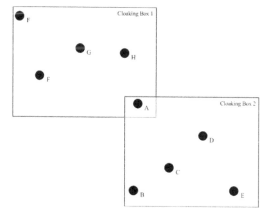

Figure 1.   Continuous attack model.

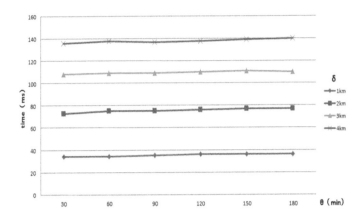

Figure 2. Impacts of spatial and temporal tolerances on efficiency of KSTCM modeling algorithm.

4.3 *Modeling algorithem*

For a set of location records with the data structure of $(u_{id}, t, [x,y], l_{id})$, the pseudo code of KSTCM modeling algorithm is given in Algorithm 1.

Algorithm 1. KSTCM modeling

$M = Map<u_{id}, l_{id}List>$ is a map structure for recording the l_{id} of distinct users.

(1) **for each** record r in record set S **do**
(2) **while** not exist KSTCM(r)
(3) find other records according to tolerances δ and θ;
(4) **if** more than k-1 records are found **then**
(5) compute their similarity;
(6) select the k-1 most similar records to build KSTCM(r);
(7) **if** the l_{id} of r is in M **then**
(8) get the KSTCM(r_i) with respect to u_{id} and l_{id};
(9) compare spatial parts of KSTCM(r) and KSTCM(r_i) and pick the smaller one as new spatial part;
(10) **else** add u_{id} and l_{id} to M;
(11) **else** build KSTCM(r) with default tolerances;

In particular, to build KSTCM(r) with default tolerances means that if we cannot find more than k-1 records, the algorithm will randomly generate a spatial cloaking box and a time interval covering the accurate location and timestamp with respect to the tolerance δ and θ.

5 EXPERIMENT RESULTS

The experimental dataset is the check-in data collected from Gowalla between Feb. 2009 and Oct. 2010 in California State. There are 667,821 original records and 15,039 users.

Our first experiment evaluates the impacts of the spatial and temporal tolerances on the efficiency of the KSTCM modeling algorithm. We choose 5 as the default value of k, four spatial tolerance 1,2,3 and 4 kilometers and six temporal tolerance ranging from 30 minutes to 180 minutes. Moreover, we specify that spatial and temporal tolerances have the same weight when computing the similarity. Fig. 2 shows the average time to build a KSTCM object under different conditions. While temporal tolerance has little influence on the efficiency, spatial tolerance impacts a lot. On the other hand, the time consumption is satisfied in most cases.

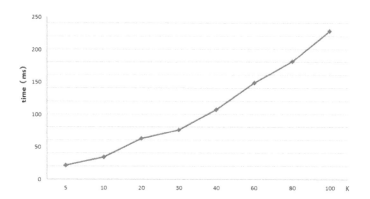

Figure 3.   Impact of k on efficiency of KSTCM modeling algorithm.

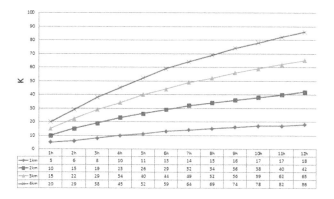

Figure 4.   Relationship between k and tolerances.

The second experiment examines the impact of the value of k on KSTCM modeling algorithm. We set default temporal tolerance to 360 minutes, and do not limit the spatial tolerance. As we can see in Fig. 3, with the value of k increases, the time consumption increases accordingly.

Next experiment investigates the relationship between k and tolerances. We select four spatial tolerances: 1, 2, 3 and 4 kilometers, and twelve temporal tolerances ranging from 1 hour to 12 hour. Fig. 4 shows the average k found under different conditions of tolerances. It is obvious that spatial tolerance influence the value of k more than the temporal one.

By summarization of the experimental results, it can be found that spatial tolerance impacts more than temporal one with respect to our KSTCM. Therefore, it is desirable to define a small spatial tolerance and a relative big temporal one in order to obtain a high efficiency. Assuming it is satisfied that the value of k ranges from 5 to 10, the spatial tolerance is suggested to be specified as 1 km and the temporal tolerance as 60 minutes.

6   CONCLUSION

In this paper, a k-anonymous spatial-temporal cloaking model to protect location history in location-based social networking services is introduced, as well as its corresponding algorithm. Our algorithm proposed can find a minimum cloaking 3D box covering the actual location and its temporal information by measuring the similarity between two location records with respect to the spatial and temporal dimensions. Furthermore, the algorithm can ensure that the accurate location won't be inferred by continuous attacks. The experimental results show that our proposed algorithm has a high efficiency with a reasonable value of k.

ACKNOWLEDGMENT

This paper was sponsored by following projects:

– National High-tech R&D Program of China ("863 Program") (No. 2013AA01A603);
– National Science and Technology Support Projects of China (No. 2012BAH07B01);
– Program of Science and Technology Commission of Shanghai Municipality (No. 12510701900);
– 2012 IoT Program of Ministry of Industry and Information Technology of China.
– Young University Teachers Training Plan of Shanghai Municipality under Grant (No. SXY12003).
– Management department of Shanghai Business School (NH1-2-1-1316).

REFERENCES

[1] Foursquare. [Online]. Available: http://www.foursquare.com/
[2] Gowalla. [Online]. Available: http://www.gowalla.com/
[3] H. Cao, N. Mamoulis, D.W. Cheung, "Discovery of Periodic Patterns in Spatiotemporal Sequences", IEEE. Transactions on Knowledge and Data Engineering, Vol. 19, No. 4, pp. 453–467, 2007.
[4] V.W. Zheng, Y. Zheng, X. Xie, and Q. Yang, "Collaborative Location and Activity Recommendations With GPS History Data", In Proceeings of the 19th International Conference on World Wide Web, pp. 1029–1038, 2010.
[5] E. Cho, S.A. Myers, J. Leskovec, "Friendship and mobility: user movement in location-based social networks", In Proceedings of the 17th International Conference on Knowledge Discovery and Data Mining, pp. 1082–1090, 2011.
[6] X. Xiao, Y. Zheng, Q. Luo, and X. Xie. "Inferring Social Ties between Users with Human Location History". Journal of Ambient Intelligence and Humanized Computing, 2012.
[7] R. Tan, J.Z. Gu, J. Yang, "Designs of Privacy Protection in Location-Aware Mobile Social Networking Applications", In Proceedings of 5th International Conference on Pervasive Computing and Application, pp. 62–68, 2010.
[8] X. Lin, S.P. Li, C.H. Yang. "Attacking algorithms against continuous queries in LBS and anonymity measurement". Journal of Software, vol. 20, no. 4, pp. 1058–1068, 2009.
[9] M. Prabaker, J. Rao, I. Fette, P. Kelley, L. Cranor, J. Hong, and N. Sadeh, "Understanding and capturingpeople's privacy policies in a mobile social networking application". In Proceeds of the Workshop on Ubicomp Privacy. pp. 1–14, 2007.
[10] Loopt. [Online]. Available: http://www.loopt.com/
[11] L. Liu. "From data privacy to location privacy: models and algorithms", In Proceeding of The 33rd International Conference on Very Large Data Bases, pp. 1429–1430, 2007.
[12] H. Kido, Y. Yanagisawa, T. Satoh. "An anonymous communication technique using dummies for location-based services", In Proceedings of IEEE International Conference on Pervasive Services, pp. 88–97, 2005.
[13] P. Kalnis, G. Ghinita, K. Mouratidis, et al. "Preserving anonymity in location based services", Technical Report TRB6/06, Department of Computer Science, National University of Singapore, 2006.
[14] B. Gedik, L. Liu. "Protecting location privacy with personalized k-anonymity: Architecture and algorithms". IEEE Trans. on Mobile Computing, vol. 7, no. 1, pp. 1–18, 2008.
[15] R. Cheng, Y. Zhang, E. Bertino, et al. "Preserving user location privacy in mobile data management infrastructures", In Proceedings of Privacy Enhancing Technology Workshop, 2006.

Engineering Technology and Applications – Shao, Shu & Tian (Eds)
© 2014 Taylor & Francis Group, London, ISBN 978-1-138-02705-3

Marine controlled source electromagnetic acquisition station with low power consumption technology

Di Zeng, Hongmei Duan, Ming Deng, Kai Chen & Shaocong Bu
School of Geophysics and Information Technology, China University of Geosciences, Beijing, China

ABSTRACT: MCSEM requires several data acquisition stations located at the bottom of the sea to gather weak electromagnetic signals for long-term continuous synchronous observation. During the observation, the station should be under controlled within an extremely low self-noise level, the time synchronization error between the acquisition stations should be very small, and their power consumption should be very low to sustainable work for dozens of days. In order to achieve low-power goal, guarantee prerequisites for low noise and low drift, we developed the acquisition station hardware which is consisting of on-duty circuit, high efficiency and low noise power management circuit and low power data acquisition circuit, and we also lowered the power consumption of the preamplifiers and the magnetic sensors. After indoor test, land test and marine tests, it is proved that the low-power design specifications can meet the actual requirements, according to the desired purpose.

Keywords: marine CSEM, CSEM acquisition station, low power, low noise

1 INTRODUCTION

Recent years, MCSEM (Marine controlled-source electromagnetic methods) plays an increasingly important role in the deep water oil and gas hydrate exploration area, especially in the area of offshore drilling success rates [1–2]. This method has gained the attention of numerous international oil exploration companies. A number of professional marine magnetic exploration companies were set up and have conducted hundreds of commercial operations for exploration so far in major offshore oil and gas fields of the world. And they become the focus of many research institutions and universities. The marine electromagnetic methods have also been listed into the national 863 Program. China National offshore oil company has been introducing foreign troops in the South China Sea for trials. Sinopec invested in equipment and technology research or introduction, indicating the awareness and attention of relevant departments of the State to electromagnetic exploration techniques for marine.

According to the skin effect, the thick conductive seawater layer leads to the natural field source characteristics of MT signal with weak amplitude and low frequency, posing great challenges to the practical submarine signal observation. MCSEM method compensates for the lack of high-frequency signals in deep water condition by means of exciting artificial source signal 0.1 Hz–10 Hz in deep water, making the medium shallow subsea electrical structure probing possible, therefore the method in offshore oil and gas exploration and other aspects has been widely used. The observation of mixed source signal of natural and artificial sources by the multiple acquisition stations generally requires the stations in the submarine to work for dozens of days [3–6]. Limited by the volume of the pressurized chamber and the buoyancy of acquisition stations, the stations are not allowed to carry more batteries pack into the water. So we may only reduce the power consumption of each unit in the circuit, in order to make the circuit can continuously work longer.

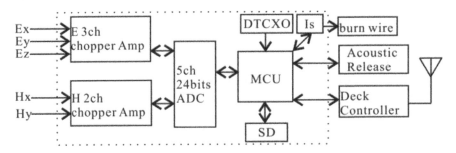

Figure 1. The overall block diagram of the circuit of marine controlled source acquisition station.

2 HARDWARE DESIGN

MCSEM acquisition station observes undersea electromagnetic five components of the signals (Ex, Ey, Ez, Hx, and Hy), synchronizes with GPS for precise time, and records their attitude information. The process of offshore construction includes clock program, release, positioning, recycling, and clock calibration. Acquisition station circuit design includes the electromagnetic field sensor part, data acquisition part and release control part. The sensors use low-noise Ag/AgCl electrode and the low-power consumption and small size CAS-10M inductive magnetic sensor. The data acquisition circuit consists of AD conversion unit, control unit and deck unit. Under the condition of low noise, low power consumption design goals, we choose the proper chips and design an effective power management system. The release part is related to the overall structure of the framework, and contains floating ball, counterweight, releaser and hook device to adapt to different submarine topography and ensure the safety of recycling the station.

2.1 Overall design

According to the special underwater work environment and the characteristics of the electromagnetic signal, we designed the hardware circuit as shown in figure 1. The acquisition circuits are completely sealed in pressurized chamber and connected to the control box through watertight connectors, and mainly include the front three-channel EAmp (electric preamplifier), two-channel HAmp (magnetic preamplifier), five-channel ADC (analog-digital conversion), MCU (master control unit) and lithium batteries. The electric preamplifier has the characters of 0.01 to 40 Hz −3 dB bandwidth, 60 dB to 80 dB optional gain, less than 1 nV/rt (Hz) @ (1–100 Hz) typical noise. The magnetic preamplifier has the characters of 0.01 to 40 Hz −3 dB bandwidth, 1, 8, 64 optional gains, 6 nV/rt (Hz) @ (1–100 Hz) typical noise. The ADC unit integrate 5-channel, 24-bit large dynamic range ADCs, and complete high-precision conversion of 5-channel analog signals, the typical dynamic range of which is 120 dB @ (DC–100 Hz). The control unit integrates MCU, DTCXO (digital temperature compensated crystal oscillator), CPLD (complex programmable logic device), attitude sensors, RTC (real time clock), constant current source circuit and power conversion circuits. The deck control unit integrate GPS module, signal generating circuit and so on, in order to achieve the communication between the acquisition circuit in the pressurized chamber and PC, Power Switch delay measurement, and self-test signal generation.

2.2 Low-power control circuit

Low-power control circuit is designed primarily in four aspects: the processor, software platform, clock, memory.

The MCU control unit use AT91SAM9G45 embedded microprocessor. It's a 32-bit processor operating at up to 400 MHz, and has an efficient power management controller including clock gating and backup battery part in activities which reduces the maximum power consumption in standby mode. The power consumption is usually less than 300 uW/MHz At 400 MHz rate and 1.0 V core voltage. According to the requirements, the processor integrates NOR Flash, NAND Flash, SDRAM, SD memory card, CPLD and other devices and achieves the user's control and data transfer via Ethernet and serial ports.

And the processor transplant embedded Linux operating system, which is streamlined, strong real-time, open source, portable, can be cut and has other advantages compared to WinCE, uCOS systems, while supporting FTP, FAT32 file system, USB Host, Telnet, advanced applications such as multi-threading. To the applications in the marine CSEM acquisition station, it has good practicality.

Traditional clock program use GPS + OCXO, but in view of the status quo that underwater GPS signals cannot be received, we program clock before entering the water, eliminating unnecessary power consumption of the GPS module. Besides, we use DTCXO instead of OCXO, and the frequency stability declines from 10 ppb to 50 ppb in the range of allowable error, but the power consumption reduces from 1250 mW to 10 mW.

In terms of data storage, large-capacity storage can choose IDE hard drives, U disk, CF cards, NANDFLASH and SD card, taking low power, pluggable, small-footprint, read and write speed indicators into account, we chose the 16 GB SD as the memory of acquisition station, read and write speed of which is up to 10 MB/s.

2.3 *Low-power chopper amplifier*

The conventional low-noise amplifier is unable to meet the requirements because its 1/f noise is high, and the corner frequency is mostly between 1 Hz and 100 Hz. While the chopper amplifier chopper the slowly varying signal and modulate it to a high frequency signal, thus avoiding the 1/f noise at low frequency, obtaining the best noise performance. Available chopper amplifier CS3301, whose noise level can reach 8.5 nV/rt (Hz) @ (DC–2 kHz), but still does not meet the requirements, therefore we need to design and develop the chopper amplifier on our own. If the chopper amplifier is achieved through integrated operational amplifier, the overall power consumption will be great, so the design of low-power chopper amplifier we design uses CMOS tube structures. The block diagram of the chopper amplifier is shown in figure 2, the demodulators are placed at low impedance nodes, the cascade circuit which utilizes low impedance nodes can modulate input signal to high frequency band without extra power [7–9]. The final design indicators of the three-channel amplifier field in shown in Table 1.

2.4 *Low-power magnetic sensor customization*

Table 2 lists the parameters of current main MT induction type magnetic sensor and the customized CAS-10M, to meet the special needs of marine controlled source electromagnetic methods

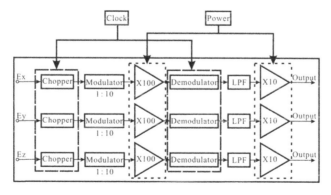

Figure 2.   Block diagram of the amplifier circuit.

Table 1. Three-channel electric amplifier circuit indicators.

NOISE	GAIN	CONSUMTION	DR	−3 dB BW	CROSSTALK
1 nV/rt(Hz) @ (1/100 Hz)	60/80 dB	56 mW/ch	>110 dB	0.01–40 Hz	>110 dB

Table 2. Magnetic sensor data table.

Model	BF-4	MFS-06	MTC-80	MTC-50H	CAS-10M
Manufacturer	EMI	Metronix	Phoenix	Phoenix	CAS
Sensitivity	300 mV/nT	800 mV/nT	50 mV/nT	500 mV/nT	300 mV/nT
Noise	0.1 pT/rt(Hz) @ 1 Hz	0.1 pT/rt(Hz) @ 1 Hz	0.2 pT/rt(Hz) @ 1 Hz	0.1 pT/rt(Hz) @ 1 Hz	0.1 pT/rt(Hz) @ 1 Hz
Corner Frequency	0.2 Hz	3 Hz	0.5 Hz	0.2 Hz	0.06 Hz
Size	1420 mm*60 mm	1250 mm*75 mm	970 mm*60 mm	1440 mm*60 mm	880 mm*46 mm
Weight	7 kg	8.5 kg	5 kg	8.2 kg	4.5 kg
Power Dissipation	288 mW	600 mW	150 mW	unknown	100 mW

and ensure the noise performance, we redesign the sensor in size, power consumption, weight and other parameters. Test results showed that, compared to other models, CSA-10M has a high sensitivity and lowest power consumption, and has a minimum volume, the minimum weight.

2.5 On-duty circuit

On-duty circuit is designed for the purpose of achieving power switch without open pressurized chamber, which requires a very low static current. The circuit uses MSP430G2553 as the microcomputer, which has a remarkable feature of low current consumption of 0.1 uA in a dormant state and less than 200 uA at full speed state. As it integrates digital calibration crystals, integrated flash, serial ports, and other functions, the circuit eliminates some of the peripheral devices, reduces the power consumption of peripheral devices. On-duty circuit accepts PC's control through the deck unit, works in low power mode with other modules turned off, and consumes only 40 mW.

2.6 Power management

Power conversion circuit designs following the balance between efficiency and noise, the analog power requires low ripple but must sacrifice some efficiency but the digital circuits are the opposite. The power system is shown in Figure 3. The battery power supply is converted to different power voltages for various parts through power conversion chips as follows. The power conversion chips both have wide power supply range and high efficiency, especially the advantage of the isolation of TEN 8-2413WI for the corrosion power, high driving ability of TPS54040 and the low noise level of LT1962 and LT1964 for the analog power.

In order to reduce power consumption, achieving dynamic power management, power supply system according to the practical conditions of the acquisition station can realize to shut off part of the module. As shown in Figure 3, R0 region represents the whole power supply system, R1 region represents the analog part power, R2 region represents the power for electric corrosion, and both regions are controlled by the MCU unit, R3 region is under control of on-duty circuit (Watch Power) unit. The base power of marine controlled-source electromagnetic data acquisition station uses customized, 27.5 AH battery, which consists of 6 series of 18 standard 18,650 2.5 AH, 3.7 V Li batteries, and is configured overcurrent overcharge protection circuit.

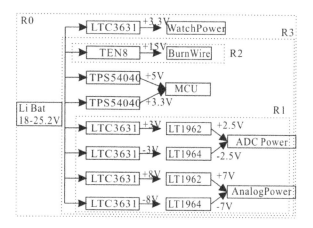

Figure 3. Power conversion topology.

Table 3. Power consumption statistics form.

Module	M. Sensor (2)	E. AMP (3ch)	ADC (5ch)	MCU
Power Supply [V]	±7	±7	±2.5	+5
Current [mA]	8	±12	40	160
Power Dissipation [mW]	224	168	200	800

3 TEST RESULT

To test the power consumption indicator, we conducted a series of indoor, land and marine field tests.

3.1 *Indoor test*

Indoor we use the Agilent E3646A as the power supply to test the power consumption of the magnetic sensor, electric preamplifier circuit, ADC and MCU module separately, the results is shown in table 3.

Under the condition of 10 v power supply, 150 Hz sampling, 5 channels working together testing the whole system, the supply current is 180 mA, power consumption is 1.8 W, the total power consumption of each part calculated from table 3 is 1.392 W, then the actual power efficiency is 77.3%. Under the same conditions, the system powered by customized 27.5 AH, 25.2 V Li battery, can work continuously for 12 days.

3.2 *Land test results*

During July to August in 2013, our research group did two weeks of land magnetotelluric field test near the Mantou Village, Zhangbei Country, Hebei province, finally processed the apparent resistivity curve as shown in the figure below.

Land test results show: the instrument can measure frequency range from 0.5 mHz to 316 Hz, and apparent resistivity curves are continuous. It proves that the instrument can provide reliable MT data, better performance.

3.3 *Marine field tests*

During October to November in 2013, our research group went on the "Ocean IV Research and Survey" ship and tested the self-developed MCSEM acquisition station near the northern waters of the South China Sea and selected four points to put in.

151

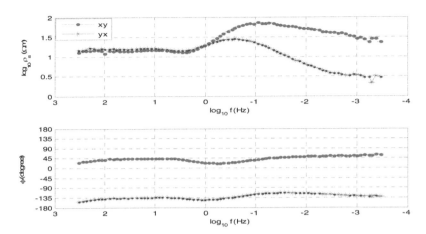

Figure 4.   Apparent resistivity curve in Zhangbei.

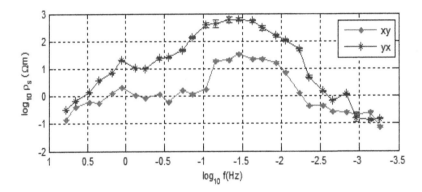

Figure 5.   Apparent resistivity curve of marine test.

Figure 5 shows the ocean test results: the system can measure frequency range from 0.56 mHz to 5.6 Hz of marine electromagnetic signals, the apparent resistivity curve is relatively flat, the acquisition data is stable and reliable, system performance can reach the indicator, and it also can meet the demand for marine controlled source electromagnetic hydrate survey work.

4   CONCLUSION

Long-term operation of marine controlled source acquisition station makes high demands of power consumption level of hardware circuit, our self-developed marine controlled source acquisition station do not affect overall system performance prerequisite condition, put the focus on the performance of low-power system including from the front sensor to the back end control circuit. Through indoor, land and marine experiments, it proves that the acquisition station can operate a long-term stable work and can provide reliable marine controlled source measurement needs.

ACKNOWLEDGEMENT

Thanks for all the staff of the "Ocean IV Research and Survey" ship during sea trials to hard work.

REFERENCES

[1] He, Z.X. & Yu, G. (2008) Marine electromagnetic survey techniques and new progress. *Progress in Exploration Geophysics*, 31(1).

[2] Summer field P J. Marine CSEM Acquisition challenges [J]. Expanded Abstracts of 75th Annual International SEG Meeting, 2005, 538–541.

[3] Chen Kai1, Deng Ming1, Wu Zhong-liang, Jing Jian-en1, Luo Xian-Hu, Wang Meng. Low Time Drift Technology for Marine CSEM Recorder [J]. Geoscience, 2008, 26(6).

[4] Chen Zu-Bin, Teng Ji-Wen, Lin Jun, Zhang Lin-Hang. Design of BSR-2broad band seismic recorder. Chinese Journal of Geophysics, 2006, 49(5): 1475–1481.

[5] Lin Pin-rong, Guo Peng, Shi Fu-sheng, Zheng Cai-jun, Li Yong, Li Jian-hua, Xu Bao-li. Electromagnetic method comprehensive detection system research [J]. Acta Geologica Sinica, 2006, 40(10): 1539–1538.

[6] Lin Pin-rong, Guo Peng, Shi Fu-sheng, Zheng Cai-jun, Li Yong, Li Jian-hua, Xu Bao-li. A Study of the Techniques for Large-depth and Multi-functional Electromagnetic Survey [J]. Institute of Geophysical and Geochemical Exploration, Chinese Academy of Geological Sciences, Langfang, Hebei 065000.

[7] Bilotti A, Monreal G. Chopper-stabilized amplifiers with a track-and-hold signal demodulator. IEEE Journal of Solid 2 state Circuits, 1999, 46(4): 490–495.

[8] Enz CC, Temes G C. Circuit techniques for reducing the effects of op2 amp imperfections: Auto zeroing, correlated double sampling, and chopper stabilization. Proc. IEEE, 1996, 84: 1584–1613.

[9] Yoshihiro Masui, Takeshi Yoshida, Atsushi Iwata. Low power and low voltage chopper amplifier without LPF. IEEE Electronics Express, Vol. 5, No. 22, 967–972.

Engineering Technology and Applications – Shao, Shu & Tian (Eds)
© 2014 Taylor & Francis Group, London, ISBN 978-1-138-02705-3

Research on automatic monitoring system of medium and small size reservoirs based on internet of things

Weiqun Cui, Shuili Zhang, Xianwei Qi & Jiyu Yu
Department of Information Engineering, Shandong Water Polytechnic, Rizhao, Shandong, China

ABSTRACT: Based on the technology of internet of things, the system can detect the parameters of medium and small size reservoirs such as water regime, rainfall regime and dam structure behavior remotely, automatically and on real-time. Moreover, it can also control the sluice gate automatically in remote or local. It realizes a reliable, opening structure, low cost automatic monitoring system for medium and small size reservoirs. It has played an important role in promoting modernization on management level, improving the safety and capacity of disaster prevention and reduction for medium and small size reservoirs.

Keywords: internet of things; medium and small size reservoirs; automatic monitoring

1 INTRODUCTION

Medium and small size reservoirs account for more than 99 percent of all Chinese reservoirs and they are an important part of flood control system and water infrastructure. But currently, the means of safety monitoring for most medium and small size reservoirs in China are far behind, they lack necessary measuring and monitoring instruments such as water regime, rainfall regime and dam structure behavior and there are a lot of hidden troubles, so the automatic monitoring of them must be solved as quickly as possible.

The conventional means of safety monitoring for most medium and small size reservoirs are setting multi rainfall observation points to monitor the water regime that flows into the reservoir and observing the seepage water level of each observation point on saturation line artificially, then make the safety analysis of the dam saturation line manually. This method is far behind. So, to realize the low coast automatic monitoring system is an important means to improve the safe operation level for medium and small size reservoirs. At present, the distributed monitoring systems for medium and small size reservoirs developed by China have reached the international advanced level in the system structure of modularity etc., but they need to be improved further in system overall performance, data transmission mode, power supply diversity, reliability and long-term stability etc.

According to the management situation and service demands of medium and small size reservoirs, aiming at the problems of the informatization and digitalization for medium and small size reservoirs, based on the idea of one unified network platform, one unified database platform and one unified application platform, we establish a reliable performance, practical function, opening structure and low cost system that integrate the information of water regime, rainfall regime, dam structure behavior, sluice gate automatic control, information query and publish. So it's the base of informatization and modernization for medium and small size reservoirs management, and it can play an important role in promoting modern management level, improving the reservoirs' safety and the capacity of disaster prevention and reduction.

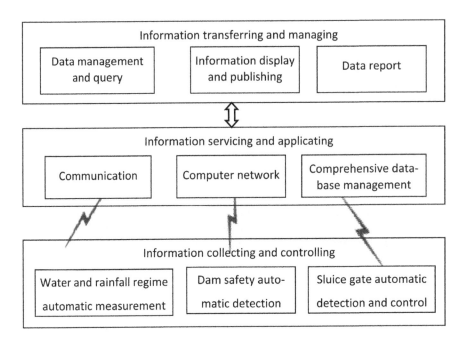

Figure 1. System logic structure.

2 THE SYSTEM STRUCTURE

2.1 *The system logic structure*

The system logic structure is as figure 1. It mainly includes information collecting and controlling, information transferring and managing, information servicing and applicating. The information collecting and controlling includes subsystems of water and rainfall regime automatic measurement, dam safety automatic detection, sluice gate automatic detection and control etc. Information transferring and managing includes subsystems of communication, computer network and comprehensive database management. Information servicing and applicating includes subsystems of data management and query, information display and publishing, data report etc.

2.2 *The system physical structure*

The system physical structure is as figure 2. It mainly includes the field measurement and control system of each reservoir, data communication system and monitoring and management center.

2.2.1 *The reservoir field measuring and controlling system*

According to different reservoirs in size, each reservoir includes several rainfall regime, water regime and dam structure behavior automatic detecting points and one set of sluice gate automatic monitoring system. Each detecting point uses mature industry sensors to collect information and complete analog to digital conversion by the DTU, then the DTU transfer the converted data to data convergence point through ZigBee, WIFI or ethernet network. The data convergence point communicates with the computers of monitoring and management center bidirectionally through short message, GPRS or wireless bridge etc.

Water regime detection is mainly completed by ultrasonic stage gauges or put-into liquid level gauges, and each reservoir must set two sets of water level detecting gauge at least, one in head water and the other one in tail water. Rainfall regime detection is mainly completed by tipping bucket

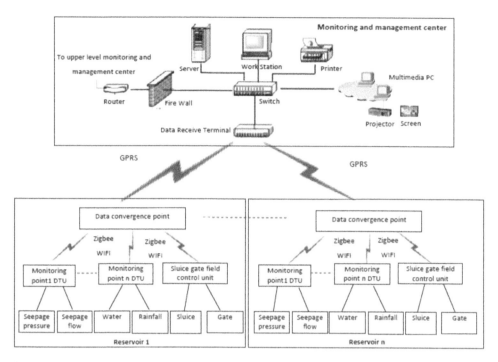

Figure 2.   System physical structure.

rain gauges and they transform rainfall information into pulse signal, then the DTU transforms the pulse signal into data.

The dam structure behavior detection mainly includes seepage pressure and seepage flow. Seepage pressure detection is completed by the float-type stage gauge or put-into liquid level gauge that has been buried in the seepage pressure pipe. Seepage flow detection is completed by measuring the water level of the measuring weir and then calculating the seepage flow discharge through empirical formula. The water level of the measuring weir is measured by float-type stage gauge. The power is provided by battery.

The sluice gate automatic monitoring system is placed in the gate hoist room. It's responsible for controlling and opening height measurement of the gate. The control mode can be automatic or manual and automatic mode includes field control as well as remote control. Field control equipments mainly include control cabinet, touch screen, PLC, gate opening height detector, contactor and industry configuration software etc. These equipments have extremely good reliability and anti-interference ability.

2.2.2 The data communication system

Each reservoir set one data convergence point. The data of water regime, rainfall regime and dam structure behavior for every detection point communicates with data convergence point through internet of things such as ZigBee or WIFI. The data of sluice gate monitoring system communicates with data convergence point also through ZigBee or WIFI. The data convergence point of every reservoir communicates with the computers of monitoring and management center bidirectionally through short message, GPRS or wireless bridge etc.

2.2.3 The monitoring center

The monitoring center mainly includes data receiving terminal, switch, server, workstation and large screen projection etc. It receives, processes, analyses, displays, publishes data of every reservoir,

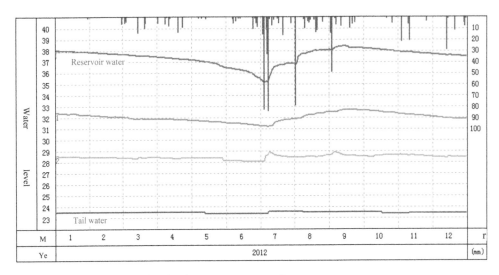

Figure 3. Water level processing line for seepage pressure pipe.

it also creates and prints the data reports and communicates with upper level monitoring center. Computers of monitoring center and local field are all Industrial Computer.

3   THE SYSTEM SOFTWARE DESIGN

The key point of software design is the software of monitoring and management center. It mainly includes data receiving and storing, data analyzing and processing, data displaying and publishing, data report and remote sluice gate controlling.

Data receiving uses the data receiving terminal based on GPRS, short message or wireless bridge. We design the background software specially to receive and send data between the monitoring center computers and each information detection point of every reservoir, and store the received data. The data stores in Microsoft SQL server database which has high reliability. The software of data analyzing, processing, displaying, publishing and data report is designed on Microsoft VS.Net and C#. Data displaying and publishing can running on both structures of C/S and B/S, this increases the system flexibility. The data report is designed on the base of Microsoft crystal report. It can display or print the information such as water level, seepage pressure and seepage flow etc. by both modes of graph and data report based on different time unit such as year or month. Figure 3 is an example of water level processing line for seepage pressure pipe by one year.

4   CONCLUDING REMARKS

Based on the technology of internet of things, WEB, database, solar power and battery, the system realizes automatic monitoring of medium and small size reservoirs. It makes the medium and small size reservoirs' safety monitoring system come to wireless, automation, networking and actualize the data sharing through Internet. It overcomes many deficiencies such as traditional manual monitoring and operation, cable power supply, cable network transmission etc. It can make managers know well or control the water regime and rainfall regime of medium and small size reservoirs or the operation situation of the dam rapidly and accurately. It can also provide the timely and accurate data for basin flood forecast and evaluation of pre-disaster. Meanwhile, it can provide important basis for flood control and drought resistance.

The system has been applied to medium and small size reservoirs such as Rizhao Maling reservoir, Hubuling reservoir etc. Practical applications show that the system has some advantages such as working steadily and reliably, fast response, low cost etc. Furthermore, it realizes the automatic monitoring and untended operation of medium and small size reservoirs. It ensures safety of medium and small size reservoirs strongly and has good application and promotion value.

REFERENCES

[1] Stankovic J A, Real-Time communication and coordination in embedded sensor networks [J], Proceedings of the IEEE, vol. 7, 2003, pp. 1002–1022.
[2] Amardeo C, and Sarma J G, Identities in the Future Internet of Things [J], Wireless Personal Communications, vol. 8, 2009, pp. 353–363.
[3] Akyildiz L E, Wireless sensor networks:A survey [J], Computer Networks, vol. 3, 2002, pp. 393–422.
[4] Junshui Jin, The safety monitoring system based on the Mixed Architecture of B/S and C/S [J], China Flood & Drought Management, vol. 21, 2011, pp. 120–124.
[5] Tao Ma, Zaixu Liu, and Liang Song, Study on automation system for the management of Medium and Small size size reservoirs [J], Technical Supervision in Water Resources, vol. 16, 2008, pp. 61–65.
[6] Xiaowen Du, Jinkui Chu, Xinying Miao, and Qing Guo, WSNs node design for dam safety monitoring based on Zig Bee technology [J], Transducer and Microsystem Technologies, vol. 28, 2009, pp. 101–105.
[7] Wenjiang Li, and Juan Wei, Design of Low-Power Consumption ZigBee Wireless Sensor Networks Nodes [J], Journal of Chengdu University (Natural Science Edition), vol. 27, 2008, pp. 5–9.
[8] Jianyu Wu, Xi Liang, and Yuntao Zhang, Design of dam security information system based on GPRS [J], Journal of Ningbo Polytechnic, vol. 36, 2012, pp. 26–30.

Engineering Technology and Applications – Shao, Shu & Tian (Eds)
© *2014 Taylor & Francis Group, London, ISBN 978-1-138-02705-3*

An innovative framework for database web services

Ke Yan
School of Information Science and Technology, Hankou University, Wuhan, Hubei, China

Tianfa Jiang
College of Computer Science, South-Central University for Nationalities, Wuhan, Hubei, China

Qingfang Tan & He Liu
School of Information Science and Technology, Hankou University, Wuhan, Hubei, China

ABSTRACT: In this paper, we proposed and implemented an innovative framework for database web services. This framework consists of two components: services provider and services consumer. With services provider, SQL statements or stored procedures of database are mapped to web services, so database can be accessed by client in the form of web services. With services consumer, SQL statements in database server are extended to support external web services invoking, so external web services can be invoked directly by database. With this framework, the function of web services offering and web services consuming can be provided by database directly and conveniently.

Keywords: database, web services, SOAP

1 INTRODUCTION

Nowadays, most web-based applications are developed by different tools and deployed on different platforms. It is very important to integrate these applications and provide specific function. Web services are loose-coupled, language-neutral and platform-independent, so they offer a good approach [1] to integrate different applications amongst different organizations in pervasive space.

Database is one of the most important components of information system and stores all information for web applications. In Web services environment, database systems are not limited to data management. They need to interact with Web services. On the one hand, databases are accessed by Web services; on the other hand, databases invoke external Web services to obtain or manipulate the data outside. In the future, database will be integrated and interact with Web services inevitably.

There are two aspects of database Web services that support database to interact with Web services:

– Providing Web services: database as service provider. Client access database via Web services mechanisms, i.e. SQL statements or stored procedures of database are invoked by a client application through Web services.
– Consuming Web services: database as service consumer. External Web services are invoked by database, i.e. a SQL query in a database session consumes an external Web service.

Both DB2 and Oracle support database Web services. In Oracle [7–11], the Java files, which invoke SQL statements or stored procedures of databases, should be created before database operations are exposed as Web services. Thus, SQL statements or stored procedures aren't published as Web services directly. Similarly, Java files, which invoke external Web services, should be registered before Web services are consumed by Oracle. Before external Web services are

161

consumed by DB2 [12–14], a dynamic link library should be registered and the body element of SOAP message should be supplied to DB2 database. Soap body can't be constructed automatically. Therefore, neither DB2 nor Oracle is convenient to support database Web services.

In this paper, we proposed an innovative framework for database Web services and is implemented on a database management system. A service provider called DBWS (Database Web services) is developed for database to provide the function of Web services offering. Additionally, we extend the SQL statements for a database to consume external Web services.

The main contributions of this paper are:

– An innovative framework was proposed to support database Web services and implemented on a database. The framework consists of two component: services provider and service consumer;
– For services provider, relationships between SQL statements or stored procedures and Web services are built in databases as tables. These tables are called mapping tables. And then, a model of processing chain is presented to receive SOAP requests, invokes SQL statements or stored procedures according to the mapping tables, and returns results in the form of SOAP message;
– For services consumer, SQL statements are extended for a database to consume external Web services directly. Thus, SOAP requests can be constructed automatically, and external Web services can be invoked directly and conveniently.

The rest of this paper is organized as follows. We compare our work with related work in section 2. Section 3 presents the overall architecture of DBWS that is used to offer Web services for database. Section 4 explains how to consume Web services by extending SQL statements. We conclude the paper in section 5.

2 RELATED WORK

Based on the model of SaaS (Software as a Service) [2,3], Hakan Hacigumus, Bala Iyer and Sharad Mehrotra proposed the idea about providing database as a service [4]. They implemented the NetDB2 for enterprise to build, maintain and manage databases. NetDB2 provides the function of DB2 as a service.

E. Dogdu, Y. Wang and S. Desetty used the Web services approach for data access. They implemented the JDBC-WS (JDBC Web Service) [5] to allow universal access to data resources without complicated installation and maintenance issues. Applications that access to databases needn't to deal with drivers and leave the driver-oriented communication to JDBC-WS.

K. Michael etc. deployed database Web services on PostgreSQL database management system through GDS (Generic Database Service) [6]. Web services are used to process SQL query and the corresponding result sets are sent back as strongly typed XML documents. However, only database as service provider is supported by GDS.

The function of offering and consuming Web services is provided in DB2 [7–11]. For DB2, SQL statements and stored procedures are invoked as Web services using a framework called WORF. However, before external Web services are consumed, five functions in the form of DDL should be built and registered. Additionally, SOAP body should be provided. For a database, there is no need to register any DDL. Only the Web services that are invoked need to be register in the database. SOAP requests can be created automatically with the information of parameters and registered Web services.

For Oracle, Java files, which are used to invoke PL/SQL stored procedures, should be built and deployed as Web services before stored procedures are invoked as Web services. In our study, the relationship between SQL statements or stored procedures and Web services are built in database through DBWS. Both SQL statements and stored procedures can be deployed as Web services.

In Oracle, a Java file that invokes web service should be registered as stored procedures before Web services consuming. In our study, we extend SQL statements to support Web services consuming. Web services can be invoked by extended SQL statements directly and conveniently.

162

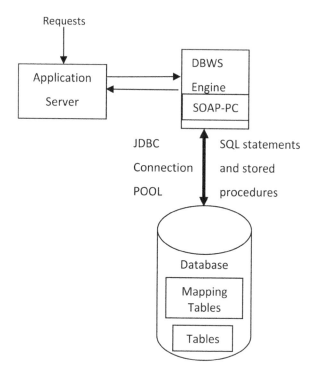

Figure 1. Architecture of database as web services provider.

3 DATABASE AS SERVICES PROVIDER

Web services offering mean that SQL statements or stored procedures of database are exposed as Web services and then can be invoked directly by client.

When a database as Web services provider, there is a key issues should be resolved: while client requests to invoke Web services of database, Web services should be mapped to real SQL statements or stored procedures that can be processed by database.

Additionally, SOAP messages are sent to database from client to request Web services invoking and sent back to client from database to return results. So, parsing the SOAP requests and encapsulating results from database as a SOAP response are also important problem.

3.1 *The architecture of database as web services provider*

The architecture of database as Web services provider is illustrated in Figure 1. It consists of three parts: Application Server, DBWS Engine and database.

Application Server monitors and examines the requests from client. If a request needs to invoke Web services of database, it will be sent to DBWS Engine. Also, Application Server returns the result from DBWS Engine to client.

In DBWS Engine, there is a SOAP processing chain (SOAP-PC) that processes the SOAP requests from Application Server. SOAP requests are parsed by SOAP-PC to specify which Web services need to be invoked. Then, the Web services that will be invoked are mapped to SQL statements or stored procedures according to the mapping tables and are processed by database. When results are returned from database, it is encapsulated as a SOAP message by DBWS Engine and returned to Application Server.

In this paper, a database is connected to DBWS Engine by JDBC. Mapping tables are stored in database.

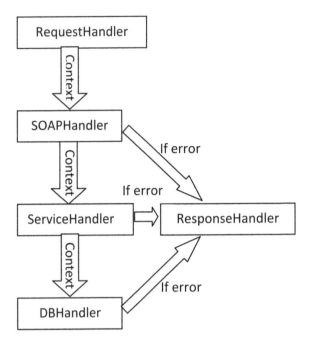

Figure 2.   SOAP processing chain.

3.2   *DBWS engine*

DBWS Engine is one of the most important components in the architecture of database as Web services provider.

With the consideration of flexibility, a model called SOAP processing chain (SOAP-PC) is proposed, which consists of five processors: RequestHandler, SOAPHandler, ServiceHandler, DBHandler and ResponseHandler. SOAP-PC is showed in Figure 2.

The first processor in SOAP-PC is RequestHandler. It receives HTTP requests from Application Server, extracts SOAP requests and other information, stores them to DBWSContext and sends the DBWSContext to SOAPHandler.

DBWSContext is used to store the result information of processors and transports the information between processors. Each processor receives DBWSContext from the previous processor and inserts result information to DBWSContext, and then, sends the new DBWSContext to the successor processor.

SOAP requests from RequestHandler are parsed by SOAPHandler to collect the information about names of Web services and methods that will be invoked, and parameters list. The result information is stored into DBWSContext and sent to ServiceHandler.

The SQL statements or stored procedures corresponding to the Web service being invoked will be found by ServiceHandler according to the DBWSContext from SOAPHandler and the mapping tables. SQL statements and information about parameters and database connection are stored into DBWSContext and sent to DBHandler.

DBHandler connects to databases with JDBC and the SQL statements or stored procedures stored in DBWSContext can be executed by databases. The results from database will be received by DBHandler, and then, sent to ResponseHandler.

In ResponseHandler, results are encapsulated as a SOAP response and returned to Application Server. Also, any processors make an exception, an error message will be returned to Application Server from ResponseHandler.

3.3 Mapping tables

In order to map Web services to corresponding SQL statements or stored procedures of databases, we create mapping tables in databases. There are four tables in databases to store the information of relationships between Web services and SQL statements or stored procedures.

Table Service is used to store the attributes of Web services. Table Method is used to store the information about methods of Web services. Table Parameter is used to preserve the information about parameters of methods. Table Properties is used to preserve the information about the database connections.

4 DATABASE AS SERVICES CONSUMER

Web services consuming means that external Web services can be invoked through a database session. So, database can obtain or manipulate the data outside directly.

When database as Web services consumer, there are three issues should be resolved. Firstly, external Web services should be registered in databases before they are invoked by databases. Secondly, external Web services should be invoked as simply and conveniently as possible. So, SQL statements should be extended to support Web services invoking. At last, the SOAP responses, in which the results of Web services are encapsulated, are parsed in databases to get the external data.

4.1 Registry of web services

Before external Web services are invoked, the definition information of them should be registered in databases.

In order to store the definition information of Web services, we extend the data dictionary of database. There are three data dictionary tables added to the database. The attributes of Web services, such as name and URL of Web services, are defined in table *syswebservices*. Methods of Web services are described in table *sysmethods*. The information about the parameters of methods is described in table *sysparameters*.

When a database starts up, corresponding data structure will be built in memory according to three extended data dictionary tables. The data structure about Web services is presented in figure 3, and the data structure about methods is presented in figure 4.

Web services can be registered in a database by extended SQL statements. We introduce the extended SQL statements in the next subsection.

4.2 Extension of SQL statements

By extending SQL statements, Web services and methods can be registered in database and deleted from database. Additionally, extended SQL expressions are used to invoke external Web services directly.The syntax for registry of Web services is:

CREATE WEBSERVICE <name> <URL of web service>;

After the execution of this statement, a Web service is registered in database and the information about the Web service, such as name and URL of the Web services, are stored in the database. The syntax used to delete a Web service is:

DROP WEBSERVICE <name >;

After the execution of this statement, information about a Web service is deleted from database. The syntax for methods of Web services to be registered is:

CREATE METHOD <name of method> (<parameter> {,< parameter>,......}) ON <name of web service>;
< parameter>::=<name> <type> <pattern>
<pattern>::=IN/RETURN

165

```
struct dict3_webservice_struct
{
    dict3_hdr_t hdr;        //the head structure of dictionary
                             table
    ulint id;              //ID of Web service
    sysname_t name;        //name of Web service
    char* url;             //URL of Web service
    dict3_db_t* db;        //point to the memory object of database
    dict3_schema_t* sch;   //memory object of the schema the Web ser-
                             vice belong to
    dbbool is_public;      // public symbol
    hash_node_t name_hash; //hash node of web service name
    hash_node_t id_hash;   // hash node of Web service ID
    DB_LIST_NODE_T(dict3_webservice_t)   link;
                      //link to memory object of web service
};
```

Figure 3.   Data structure in memory about web services.

```
struct dict3_method_struct
{
    dict3_hdr_t hdr;       //the head structure of dictionary
                            table
    ulint id;             //ID of method
    sysname_t name;       //name of method
    ulint service_id;     //ID of Web service the method
                            belongs to
    dict3_db_t* db;       //point to the memory object of
                            database
    dict3_schema_t* sch;  //memory object of the schema
                            the Web service belong to
    hash_node_t name_hash; //hash node of method name
    hash_node_t id_hash;   // hash node of method ID
    DB_LIST_BASE_NODE_T(dict3_parameter_t)
    lst_parameters;       //parameter list of the method
    DB_LIST_NODE_T(dict3_method_t) link;
//link to memory object of method
};
```

Figure 4.   Data structure in memory about methods.

There are two kinds of pattern in parameters: *IN* denotes the parameter is an input parameter and *RETURN* denotes the parameter is a result parameter of Web services.

The syntax used to delete the method of Web service is:

DROP METHOD <name >;

After registry of Web services and methods, details about how to invoke Web services are explicit. When input parameters are specified, the SQL expression in database needs to be extended to invoke external Web service. The basic structure of an SQL expression is:

select [top n1 [percent | ,n2]] < list>
from <table reference> [{,<table reference >}]
[<where>]
[<group by >]
[<having >]
{,union[all]<expression >}
[<order by>]
<table reference>::=[[<name of database>.]<name of schema>.]<name of base table>|<name of view>

There is no need to extend all clauses of SQL expression except *FROM* clause. We extend the *table reference* of *FROM* clause. The *table reference* is extended as follow:

<table reference>::=[[<name of database>.]<name of schema>.]<name of base table>|<name of view>
|<web service invoking>[[AS]<alias>]
<web service invoking>::=<name of method> (<input value>[,<input value>]) on <name of web service>

For example, we use the extended SQL expression presented below to invoke the method Get*TradePriceweb* of a Web service called *StockQuote*. The value of input parameter is *"DB"*.

SELECT price FROM GetTradePrice ('DB') on StockQuote;

4.3 *Extension of database engine*

The processing procedures of the extended statements are different from general statements. So, database engine need to be extended to support the extended SQL statement. In this paper, we extend the kernel of the database. The basic modules of the database include: *Lexical Analysis Module*, *Syntactic Analysis Module*, *Query Plan Module* and *Actuator*. We extend these basic modules to support extended SQL statements and add a new module called *SOAP Processor* to invoke external Web services. The framework of the extended database engine is showed in Figure 5.

For Web services registry, the structure of lexical analysis is:

KW_CREATE KW_WEBSERVICE webservice_name lt_string

Parsing module is in charge of lexical analysis is extended to create the syntactic trees for the extended SQL statements.

According to the syntactic tree, Query plan module is extended to create query plan for the extended SQL statements. *Actuator* is extended to execute the query plan for the extended SQL statements. External Web services are invoked through *SOAP Processor* with the information from *Actuator*. After SOAP responses are returned from external Web services, they are parsed in *SOAP Processor* and results are returned to *Actuator*.

4.4 *SOAP Processor*

Soap Processor is the key module for database as services consumer. When Web services are invoked through SQL statements or stored procedures, SOAP requests are sent to service provider through *Soap Processor*. When SOAP responses are received, they are parsed in *SOAP processor* and results are returned to *Actuator*. The processing procedures of *SOAP Processor* are illustrated in Figure 4. *SOAP Processor* consists of three submodules: *SOAP module*, *HTTP module* and *Network module*. Web Services Connection Pool is used to store the connections to Web services.

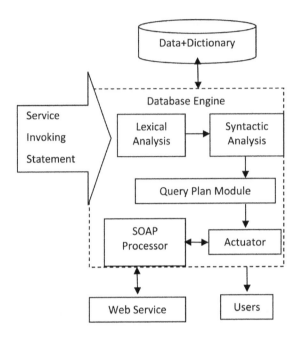

Figure 5. Framework of Extended Kernel.

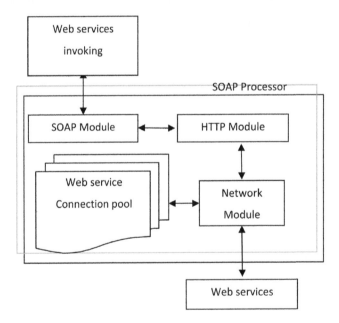

Figure 6. SOAP Processor.

The function of *SOAP module* is to create SOAP requests and parse SOAP responses. This module receives the registry information of Web services and the information about input parameters. SOAP requests are constructed in *SOAP module* according to the format of SOAP-RPC protocol and sent to *HTTP module*. SOAP responses from external Web services are received by *SOAP module* and parsed to obtain results.

The *HTTP module* is used to encapsulate SOAP requests into HTTP messages and get SOAP responses from HTTP messages. SOAP messages can be transported by many protocols, such as TCP, SMTP and MSMQ. HTTP is the most popular transportation protocol. In this study, we use HTTP to transport SOAP messages.

When the *HTTP module* receives the SOAP requests from the *SOAP module*, it encapsulates the SOAP requests into *HTTP Body* and creates *HTTP header* with information about the Web services. The *HTTP module* sends the HTTP requests to the *Network module*. When the *HTTP module* receives the HTTP responses from the *Network module*, they are parsed according to HTTP protocol and delivered to the *SOAP module*.

The *Network module* is used to communicate with Web services. It creates connections with providers of Web service and receives or dispatches messages. Connections are preserved in *Web Service Connection Pool* instead of releasing immediately.

5 CONCLUSION

In this study, we implement the functions of Web services offering and Web services consuming on database.

For the database as services provider, we built the relationships between SQL statements or stored procedures and Web services, and stored the relationships in the database in the form of tables. Both SQL statements and stored procedures can be published as Web services that can be invoked directly. The function of Web service offering for database is provided by DBWS, and DBWS can be deployed on other database.

We extended SQL statements of database to invoke external Web services directly and extended a database server to support the extended SQL statements. External Web services can be consumed directly in the database.

ACKNOWLEDGEMENTS

I would like to express my gratitude to all who helped me during the writing of this paper. My deepest gratitude goes first and foremost to our team members. Because of their effort and cooperation, we can complete this article successfully.

Our work was supported by science and technology research project of the Education Department of Hubei province (No. B2013006).

REFERENCES

[1] N. Milanovic, M. Malek. Current solutions for Web service composition. IEEE Internet Computing, 2004, 8(6): 51–59.
[2] Turner, M. Budgen, D. Brereton. Turning software into a service. Computer, 2003, 23(2): 38–44.
[3] M.P. Papazoglou. Service-oriented computing: concepts, characteristics and directions. IEEE Computer Society, 2003, 3–12.
[4] Hakan Hacigumus, Bala Iyer, Sharad Mehrotra. Providing Database as a Service. 18th International Conference on Data Engineering, 2002, 139–150.
[5] E. Dogdu, Y. Wang, S. Desetty. A Generic Database Web Service. SWWS, 2006, 117–121.
[6] K. Michael, H. Markus, M. Paul. A Generic Database Web Service for the Venice Service Grid. International Conference on Advanced Information Networking and Applications, 2009, 129–136.
[7] F.F. Donald, K. Rainer. WebSphere as an e-business server. IBM Systems Journal, 2001, 40(1): 25–45.
[8] Grant Hutchison. Developing XML Web services with WebSphere Studio Application Developer. IBM Systems Journal, 2002, 41(2): 178–197.
[9] S. Malaika, C.J. Nelin, R. Qu. DB2 and Web service. IBM Systems Journal, 2002, 41(4): 112–130.
[10] T.J. Brown, T. HS. Web Service for DB2 Practitioners. IBM Corp., 2004, 5–20.

[11] K. Beyer, R. Cochrane, M. Hvizdos. DB2 goes hybrid: integrating native XML and XQuery with relational data and SQL. IBM Systems Journal, 2006, 45(2): 271–298.

[12] Kuassi Mensah, Ekkehard Rohwedder. Database Web services. ORACLE White Paper, 2002, 5–16.

[13] Thomas Van Raalte. Application Server Web services. ORACLE Corp., 2003, 65–79.

[14] Thomas Van Raalte. Oracle Application Server Web services Developer's Guide. ORACLE 10g Release, 2004, 35–47.

Engineering Technology and Applications – Shao, Shu & Tian (Eds)
© *2014 Taylor & Francis Group, London, ISBN 978-1-138-02705-3*

Comparison of A* and Lambda* algorithm for path planning

Ping Li
*Zhongshan Institute, University of Electronic Science and Technology of China, Zhongshan,
Guangdong, China*

Junyan Zhu
*Technical Center, Zhongshan Entry-Exit Inspection and Quarantine Bureau, Zhongshan,
Guangdong, China*

Fang Peng
*Zhongshan Institute, University of Electronic Science and Technology of China, Zhongshan,
Guangdong, China*

ABSTRACT: Inspired by Visibility Graph and A*, a new path planning algorithm Lambda* is presented in this paper, which needs two list as A*. But differently, Lambda* algorithm's computation complexity is reduced by cutting down the number of vertices contained in list, this makes Lambda* algorithm find short and realistic path fast. Using Lambda* for path planning, as compared with A* algorithm, the run time decreased by 48.76% in 2D environment, 30.11% in 3D environment. Lambda* algorithm can find short and realistic looking path fast even in the complex 3D environment, and meets the demand of quickness for modern industrial robot path planning better.

Keywords: path planning; visibility graph; A*; Lambda*

1 INTRODUCTION

With increasing requirements for industrial robots, traditional technologies such as template matching[0], artificial potential field[0], map building planning[0] and artificial intelligence programming[0] [0], which are usually used in two dimensional path planning, cannot satisfy the needs of industrial robots anymore many times. There is an urgent need for a mature fast path planning algorithm can be used to three dimensional spaces[0]. Planning method based on free space structure is a kind of important path planning method[00], continuous environment with obstacles is described into a graph by geometric method firstly, and then a collision free shortest path from the starting point to the goal is searched from the graph by certain graph search method. Among all graph search methods, A* algorithm is a widely used best-first search method, and in 2D environments, it can be adopted if an appropriate heuristic function is selected[0]. That is, if the shortest path exists, A* can find it in any case. But it is time-consuming for extending many nodes. In order to raise the rapidity of path planning, A* algorithm was improved and a new planning algorithm named Lambda* is proposed in this paper, and the two algorithms were compared.

2 DESCRIPTION OF THE ENVIRONMENT

Developers in robotics area usually solve the generate-graph problem by discretizing the continuous environment into 2D grid composed of squares or hexagons, regular 3D girds composed of cubes (grid graphs)[0], visibility graphs[0], way point graphs, voronoi graphs, navigation meshes, probabilistic road maps (PRMs)[0] or rapidly exploring random trees[00] and so on. Visibility graphs contains the start point and the goal point and the corners of all obstacles, a point is connected to another point via a straight line if and only if they have line-of-sight, that is, the straight line between them does not pass through obstacles[0]. All paths along these lines from the starting point to goal point are collision free, and searching optimal path means to choose a shortest path from all these collision free paths. Without grid constraints, it is allowed to search paths in any direction while optimal path in grid graph can only be consisted of lines between adjacent nodes, so the shortest path in 2D and 3D grid graphs are respectively 1.08 and 1.12 times in length to the shortest path in the corresponding visibility graph[0].

But the time complexity for visibility graph is $O(n^2)^{0}$, n is the number of obstacles vertexes. Space obstacles with irregular geometry have many vertexes which are difficult to determine, and the calculation is complex, this affect the rapidity of path planning directly[0]. To simplify the path planning, we assume all obstacles can be enveloped by rectangles in 2D continuous environment; this expands the blocked region to some extent, but regularizes obstacles and makes the path more secure.

P represents the set of all path points including all obstacles vertexes. $p_{start} \in P$ is the start point and $p_{goal} \in P$ is the goal point, $succ(p) \subseteq P$ is the set of points which have line-of-sight to $p \in P$, $c(p,p')$ is the Euclidean distance between p and p', $lineofsight(s,s')$ is true if and only if they have line-of-sight.

3 LAMBDA* ALGORITHM

Lambda* (shown in Figure 1) maintains two global data structures open list and closed list as A*[0]. In A*, in order to avoid losing best node, the open list retains all vertices to be considered to for expansion, the closed list contains vertices that have already been expanded and ensures that each vertex is expanded only once. But in Lambda*, the closed list hold all path points that have been expanded and all these points will be composed into path point-by-point. The key different between Lambda* and A* is that the open list in Lambda* only contains unexpanded points which have line-of-sight to the last point in closed list. What is more, values for points in open and closed list are also different in Lambda*.

Unlike A*, Lambda* maintains g-value for point p only when it is inserted into the closed list, g-value is the length of the shortest path from the start point p_{start} to p found so far. But f-values for all points inserted into the open list need to be calculated when the last point p in the closed list is expanded. We uses $c(p_{goal},p')$ (Euclidean distance between p_{goal} and p') as the h-value for point p' in the experiments. The g-value for p' in open list is calculated by considering the length of path from the start point to $p[=g(p)]$, and the Eucliden distance between p and $p'[=c(p,p')]$, resulting in a length of $g(p') = g(p) + c(p,p')$. In Fig 1, $closed.insert(p_{start})$ inserts point p_{start} with into $closed$ firstly. When p_{goal} is not in $closed$, the last vertex p_{closed} is extended. $open$ list is cleared, if a successor p of p_{closed} has already been in $closed$, its corresponding node in $closed$ is marked as p'_{closed}, $p = p'_{closed} \cdot f = p_{closed}.gval + c(p,p_{closed}) + hval(p)$ while $p'_{closed}.fval = p'_{pre}.gval + c(p'_{closed},p'_{pre}) + hval(p') = p'_{pre}.gval + c(p,p'_{pre}) + hval(p)$, and the father node p'_{pre} of p'_{closed} is extended earlier than p_{closed}, it is a senior of p_{closed}. Without loss of generality, assuming p'_{pre} is the parent of p_{closed}, then $p_{closed}.gval + c(p,p_{closed}) = p'_{pre}.gval + c(p_{closed},p'_{pre}) + c(p,p_{closed})$. For $c(p_{closed},p'_{pre}) + c(p,p_{closed}) > c(p,p'_{pre})$, so $f \geq p'_{open}.fval$, it does not need to update p. Else if successor p of p_{closed} is not in $closed$, update p by $UpdateVertex(p,p_{closed})$, and insert p into $open$ by $open.insert(p)$. If there is no successor for p_{closed}, $open$ is empty means that no path is found, else a vertex px in $open$ with minimum cost $fval$ is insert into $closed$ by $closed.insert(px)$. This extension progress is repeated until p_{goal} is inserted into $closed$ and an optimal path is found. Open list in Lambda* algorithm retains only successors of the extending vertex and discards its parent and brother vertices, this can largely decrease the number of vertices in open, so that to reduce calculation and time consumption, but this will lead to the suboptimal of path planning scheme. Those discarded vertices may have line-of-sight with the expanding vertex p_{closed}, and can be reinserted into $open$, in the path $p_{start}, p_1, p_2, \ldots, px$, discarded p as a successor is reinserted

```
1.     main
2.     | for all points
3.     |  | fval := ∞;   gval := 0;   pre := [];
4.     | closed := [];     closed.insert( p_start );
5.     | while p_goal ∉ closed
6.     |     open := [];
7.     |     for each p ∉ succ( p_closed ) do
8.     |       | if  p ∈ closed
9.     |       |  | UpdateVertex( p, p_closed ); open.insert( p );
10.    |     if  open = ∅
11.    |       | return "no path found!";
12.    |     else
13.    |       | px = Pop( open ); closed.insert( px );
14.    |     Smooth( closed )
15.    | return "path found!";
16.    end
```

Figure 1.   Pseudo code for Lambda*.

into *closed*, but *p* and the parent of *px* (without loss of generality, assume it is p_1) have line-of-sight, vertices between *p* and p_1 can be deleted and the path can be smoothed, so that the path can be further optimized, so we add smooth progress after searching to optimize the path planning scheme, but cannot ensure to get the optimal scheme still. On the other hand, the adding of smooth progress increases time consumption correspondingly. Fortunately, *Smooth*( ) processes only once after the path is founded, and deals with vertices on the path which are less than vertices in open list. So Lambda* can find short and realistic path in less time when compared with A* algorithm.

4 EXPERIMENTAL RESULTS

Aforementioned Lambda* algorithm is a path planning method for environments represented by 2D visibility graphs, but can be extended to 2D or 3D environments represented by other graphs, as long as vertices are depicted as required data structures.

In order to compare the performances of A* and Lambda* algorithms, path planning environments are represented by 2D or 3D environment with obstacles. The size of 2D field is set to be 100×100, starting point is (0,0), and goal is (100,100); the size of 3D field is $100 \times 100 \times 100$, starting point is (0,0,0) and goal is (100,100,100). Obstacles represented by rectangles (cuboids when in 3D environments) are generated randomly according to certain proportions (assumed to be 0%, 5%, 10%, 20%, 30%, 40%). Two path planning algorithms are implemented in Matlab R2012a, and executed on a 2.10 GHz core 2 Duo with 2 Gbyte of RAM, our implementations are not optimized for performance and can possibly be improved.

Paths are planned 100 times in each case of certain obstacle proportion, taking the average value of the path length, time consumption, number of total vertices, number of extended vertices and number of total vertices in open list as indicators, performances of A* and Lambda* algorithm are shown in table 1 and table 2.

Experimental results demonstrate that two path planning algorithms took more time to find longer path as obstacles added, meanwhile, the number of extended vertices and total vertices in open list of Lambda* are obviously less than their respective value of A*, so that the time consumption is reduced.

In 2D environment with obstacles, when the obstacle proportion is 0%~5%, total vertices are relatively less and path planning is simple, the shortest paths found by A* and Lambda* are almost

173

Table 1. Experimental results in 2D.

Obstacle Proportion	Number of total vertices	Path length	Time consumption	Number of extended vertices	Number of total vertices in open	Algorithm
0%	2.00	141.4214	0.1150	2.00	2.67	A*
		141.4214	0.0232	2.00	3.00	Lambda*
5%	12.57	144.7118	0.3471	3.43	18.93	A*
		144.8073	0.1635	3.43	13.86	Lambda*
10%	18.40	146.7484	0.4879	3.85	28.80	A*
		147.1628	0.3098	3.70	19.35	Lambda*
20%	33.33	153.8671	0.9820	4.42	49.96	A*
		154.2514	0.7434	4.08	29.42	Lambda*
30%	58.00	161.1964	7.4232	8.63	174.96	A*
		161.9125	2.5560	5.50	35.71	Lambda*
40%	72.09	161.2122	4.8584	7.57	138.04	A*
		162.4313	3.2288	6.00	43.09	Lambda*

Table 2. Experimental results in 3D.

Obstacle Proportion	Number of total vertices	Path length	Time consumption	Number of extended vertices	Number of total vertices in open	Algorithm
0%	2.00	173.2051	1.9853	2.00	2.00	A*
		173.2051	0.0648	2.00	1.00	Lambda*
5%	82.00	177.6354	4.3022	3.13	107.19	A*
		177.3686	3.0953	3.25	82.56	Lambda*
10%	144.55	183.3138	13.9156	3.86	235.32	A*
		183.3817	11.3499	3.68	155.86	Lambda*
20%	317.33	192.4383	33.3411	4.17	393.29	A*
		194.3397	31.7022	4.67	251.33	Lambda*
30%	533.13	198.2998	76.5913	4.17	473.04	A*
		198.2030	67.7328	4.13	267.91	Lambda*
40%	711.33	192.3206	265.6775	11.00	719.67	A*
		193.6356	209.9539	8.67	562.33	Lambda*

equal, with the increasing proportion of obstacles, the path planning scheme of Lambda* is slightly worse than that of A*, the average cost of shortest paths found by Lambda* is 1.003 times to that of A*, but meanwhile, the time consumption decreased by 48.76%.

In 3D environment with obstacles, A* cannot ensure to get the optimal path neither. When the obstacle proportion is 0%, the shortest paths found by A* and Lambda* are almost equal; when the obstacle proportion is 5%~30%, the cost of shortest path found by Lambda* even less than that of A*. The average cost of shortest paths found by Lambda* is 1.0025 times to that of A*, and the time consumption decreased by 30.11%.

5 CONCLUSIONS

A new path planning algorithm Lambda* is presented in this paper, which needs two list as A*, but the computation complexity is reduced via cutting down the number of vertices retained in open list, so that short and realistic paths can be found fast. Experimental results show that Lambda* can obtain path planning schemes which are almost equal to that obtained by A*. What is more, it can get short and realistic paths fast even in complex environment, this can be better meet the rapidity requirement of practical application.

ACKNOWLEDGEMENTS

This work was financially supported by the special funds for university discipline and specialty construction in Guangdong province (2013LYM0103), research fund for the doctoral program of Zhongshan institute, university of electronic science and technology of China (410YKQ01), scientific research team construction project for study and application of intelligent system in Zhongshan institute, university of electronic science and technology of China (412YT01).

REFERENCES

[1] Vasudevan, C. & Ganesan, K. (1994) Case-based path planning forautonomous underwater vehicles. *IEEE Int Symposiumon Intelligent Control. Columbus*, pp. 160–165.

[2] Fujimura, K. & Samet, H. (1989) A hierarchical strategy for path planning among moving obstacles. *IEEE Trans on Robotic Automation*, 5(1), 61–69.

[3] Oh, J.S., Choi, Y.H., Park, J.B., et al. (2004) Complete coverage navigation of cleaning robots using triangular-cell-based map. *IEEE Trans on Industrial Electronics*, 51(3), 718–726.

[4] Rajankumar, B.M., Tang, C.P. & Venkat, K.N. (2009) Formation optimization for a fleet of wheeled mobile robots: A geometric approach. *Robotics and Autonomous Systems*, 57(1), 102–120.

[5] Aybars, U. (2008) Path planning on a cuboid using genetic algorithms. *Information Sciences*, 178, 3275–3287.

[6] Chen, Y., Zhao, X.G. & Han, J.D. (2010) Review of 3D path planning methods for mobile robot. *Robot*, 32(4), 568–576.

[7] Cheng, W.M., Tang, Z.M., Zhao Chun-xia et al. A Survey of Mobile Robots Path Planning Using Geometric Methods [J]. Journal of Engineering Graphics, 2008(4), 12–20.

[8] Zhu Daqi, Yan Mingzhong. Survey on technology of mobile robot path planning [J]. Control and Decision. 2010, 25(7), 961–967.

[9] Zhou Xiaojing. Research of routing in the gamemap based on improved A* algorithm [D]. Chongqing: Southwest University, 2011, 22–23.

[10] Yu Chong, Qiu Qi-Wen. Hierarchical robot path planning algorithm based on grid map [J]. Journal of University of Chinese Academy of Sciences, 2013.7, 30(4), 528–538, 546.

[11] Oommen B, Iyengar S, Rao N, et al. Robot navigationin unknown terrains using learned visibility graphs, part I: the disjoint convex obstacle case [J]. IEEE Journal of Robotics and Automation, 1987, 3(6), 672–681.

[12] Hsu D, Latombe J, Kurniawati H. On the probabilisticfoundations of probabilistic roadmap planning [C]. 12th Int. Symp. on Robotics Research, 2005, 627–643.

[13] Choset, Howie M., ed. Principles of robot motion: theory, algorithms, and implementations [M]. MIT press, 2005.

[14] Björnsson Y, Enzenberger M, Holte R, et al. Comparison of different grid abstractions for pathfinding on maps [C]. IJCAI. 2003, 1511–1512.

[15] Hua Jianning, Zhao Yiwen, Wang Yuechao. A new global path planning algorithm for mobile robot [J]. Robot. 2006.11, 28(6), 593–597.

[16] Nash A, Koenig S, Tovey C. Lazy Theta*: Any-angle path planning and path length analysis in 3D [C]. Third Annual Symposium on Combinatorial Search. 2010.

[17] Zhang Ying, Wu Chengdong, Yuan Baolong. Progress on path planning research for robot [J]. Control Engineering of China. 2008 (z1), 152–155.

[18] Mu Zhonglin, Lu Yi, Ren bo, et al. Study on path planning for UVA Based on Improved A* Algorithm [J]. Journal of Projectiles, Rockets, Missiles and Guidance, 2007, 27(1), 297–300.

[19] Jia Zhenhua, Siqingbala, Wang Huijuan. Path planning based on heuristic algorithm [J]. Computer simulation. 2012.1, 29(1), 135–137.

Engineering Technology and Applications – Shao, Shu & Tian (Eds)
© *2014 Taylor & Francis Group, London, ISBN 978-1-138-02705-3*

Research on the comparison testing technique of induction coil sensor

Xianhu Luo & Zhongliang Wu
Guangzhou Marine Geological Survey, Guangzhou, Guangdong, China

Ming Deng, Meng Wang & Kai Chen
China University of Geosciences (Beijing), Beijing, China

ABSTRACT: Induction coil sensor plays an extremely important role in the magnetotelluric field. Compared with the high coil sensor technology outside China, our research in this regard is still some distance behind. Newly developed coil sensors have to receive comparison testing against their overseas state-of-the-art counterparts before their performance can be determined to allow further improvement. In this paper, the newly developed CAS-10M induction coil sensors are compared with an MTC-80 series coil sensors produced by Canada's Phoenix. A set of comparison including frequency response, noise and actual measurements was carried out. The results show that the comparison testing method proposed herein is able to identify the performance difference of the newly developed coil sensors from its mainstream counterparts and can be used as important basis for evaluating the performance of new sensors.

Keywords: induction coil sensor; noise; correlation detection; comparison testing

1 INTRODUCTION

Induction coil sensors are the most widely used magnetic field receivers for magnetotelluric prospecting. They are widely applied to the geophysical exploration operations like underground energy and resource exploration, engineering and environmental exploration, marine and space electromagnetic survey for their lightweight, wide coverage, high efficiency, low cost and high resolution.

Some leading European and American geophysical instrument manufacturers have made intensive researches on the development of coil sensors with the purpose to improve the performance of magnetotelluric instrumentation, resulting in higher technical level and rapid product updating. Representative products include GMS-06, MFS-07 of Metronic, and MTC-80, MTC-50 of Phoenix. China is fairly a late-comer in the magnetotelluric research and is some distance behind other countries with respect to coil sensor studies. Domestic academies conducting researches on induction coil sensors include Jilin University, CAC Institute of Electronics and Institute of Geophysics and Geochemistry. Some achievements have also been reported.

For newly developed sensors, they have to receive comparison test to compare them against the performance indices of their overseas state-of-the-art counterparts. The comparison result can serve as important basis for assessing the performance of the new sensors as well as important reference for improving the design of the sensors. In this paper, the newly developed CAS-10M induction coil sensors are compared against the indices of the MTC-80 coil sensors produced by Canada's Phoenix. A set of comparison testing including frequency response, noise level and measurements is carried out. The performance of the newly developed coil sensors is accurately evaluated.

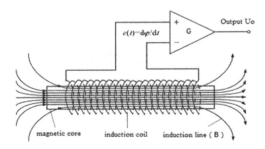

Figure 1. Working schematic of the coil sensor.

2 PRINCIPLE OF THE COIL SENSOR

The working principle of the induction coil sensor is the Faraday law of electromagnetic induction. The induction coil sensor is also called an induction coil rod that consists of a high-conductivity magnetic core wrapped with certain turns of coil. Any change in the magnetic flux in the point of the coil will result in induced voltage at the coil ends. Assuming that the external field changes according to $H(t) = H_0 \sin \omega t$ and is parallel to the axes of the core, the magnetic flux through the coil $\phi = NS\mu_e H$ will change with the time. Here N *is the turn of the coil; S* is the cross section of the core; μ_e is the effective conductivity of the flux material in response to the uniform external field. The induced electromotive force produced at the coil ends is:

$$e = -\frac{d\phi}{dt} \tag{1}$$

As the phase and direction of the external field is normally ignored in the design of the induction coil rod, the electromotive force of the induction coil rod is expressed as follows:

$$|e| = 10^{-9} \cdot 2\pi f N S \mu_e H_0 \sin(\omega t) \tag{2}$$

Fig. 1 shows the working schematic of the induction coil sensor.

3 COMPARISON TESTING

Important technical indices for measuring the performance of an induction coil sensor include frequency response, noise level, power consumption, size and weight. Now we are going to compare the newly development CAS-10M induction coil sensor with the MTC-80 produced of Phoenix against some of these indices.

3.1 *Frequency response comparison*

The mainframe of the testing system is the MTU5A of Phoenix. Data of three channels, i.e. Hx, Hy and Hz were used. Channel Hx was connected to a CAS-10M 1010 sensor, to CAS-10M 1013, and Hz to MTC-80 8080. The three coil sensors were placed parallel to each other about 2 m apart and about 10 m from the mainframe as shown in Fig. 2.

Comparison of the frequency response is shown in Fig. 3.

The comparison result of frequency response indicates that the induction coil sensors are normally operational. Error statistics was performed on the frequency response result: CAS-10M 10 vs CAS-10M 13 is 1.98%; CAS-10M10 vs MTC-80 is 13.82%; CAS-10M13 vs MTC-80 is 12.15%.

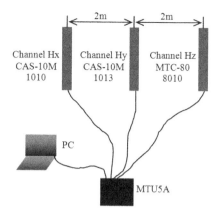

Figure 2. Equipment arrangement.

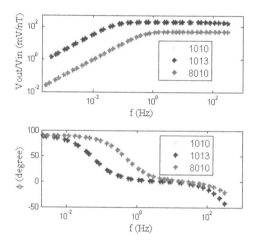

Figure 3. Comparison frequency response of CAS-10M vs MTC-80.

3.2 *Parallel noise comparison, error statistics and mutual coherence coefficient comparison*

Order that the observation mainframe works under MT mode and acquires data continuously for 2 h. The observation data was parallel-noise processed. The processing result is shown in Fig. 4, in which the upper part is the PSD curve and the lower part is the calculated correlation coefficients.

The comparison result indicates very good agreement in the PSD curve though the CAS-10M13 coil sensor shows a marginal large signal at the 0.01 Hz point. Full-range error statistics was made on the PSD calculation result: CAS-10M10 & MTC-80 is 7.40%; CAS-10M13 & MTC-80 is 8.12%. The latter indicates fairly large error, which should be associated by the difference in the PSD in the range nearby 1 Hz. Correlation coefficient of CAS-10M10 vs CAS-10M13 is largely fluctuated within the 0.1 hz–3 hz range, close to 1 in other ranges and abnormal at the 0.01 Hz point. Correlation coefficient curve of CAS-10 M vs MTC-80 is consistent with that of CAS-10 M vs CAS-10 M, though it is marginally smaller.

3.3 *Correlated noise comparison*

The noise level of a sensor is an important gauge for measuring its performance. The acquisition of magnetotelluric data also involves a variety of noise interferences. To measure the noise level of the sensor itself accurately and efficiently, correlation detection technique was used to compare

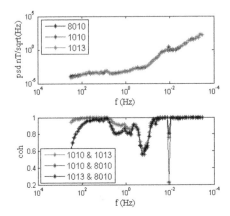

Figure 4. Comparison parallel testing result of CAS-10M and MTC-80.

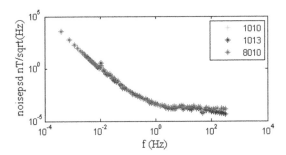

Figure 5. Noise evaluation result of the coil sensors.

the noise level of the coil sensors.

$$s_i = x + n_0 + n_i = x_0 + n_i \qquad (i = 1, 2, 3) \tag{3}$$

Data acquired from all the three channels include input signal x and noise N, the latter of which contains the ambient background noise n_0 and the noise of itself n_i ($i = 1, 2, 3$). The signals acquired from the channels are recorded as:

$$s_i = x + n_0 + n_i = x_0 + n_i \qquad (i = 1, 2, 3) \tag{3}$$

As the data were acquired from the channels synchronously, we assume that the input signal x and the ambient background noise n_0 are correlative, while the noise of the sensor itself x_0 is not correlative with x_0 or the noise o any other sensor. Hence, according to the nature of correlation function, we get:

$$\begin{cases} R_{n_1 n_1}(\tau) = R_{s_1 s_1}(\tau) - R_{s_1 s_2}(\tau) \\ R_{n_2 n_2}(\tau) = R_{s_2 s_2}(\tau) - R_{s_2 s_1}(\tau) \\ R_{n_3 n_3}(\tau) = R_{s_3 s_3}(\tau) - R_{s_3 s_1}(\tau) \end{cases} \tag{4}$$

The correlated noise of CAS-10M10, CAS-10M13 and MTC-80 coil sensors was analyzed using this method. Fig. 5 shows the noise evaluation result of coil sensors.

From the resulting curve, the noise level of the newly developed coil sensors is roughly as high as that of the MTC-80 in the lower-than-10 Hz range and marginally lower than that of the MTC-80

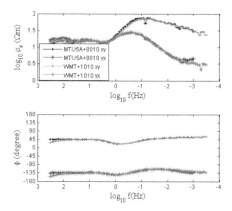

Figure 6.   Comparison actual application result of WMT+CAS-10 M vs MTU5A+MTC-80.

Table 1.   Comparison main indices of CAS-10 M vs MTC-80

Model	MTC-80	CAS-10 M
Producer	Phoenix	CAS
Sensitivity	50 mV/nT	300 mV/nT
Noise	0.2pT/rt(Hz)@1 Hz	0.1pT/rt(Hz)@1 Hz
Corner frequency	0.5 Hz	0.06 Hz
Volume	970 mm*60 mm	880 mm*46 mm
Weight	5 kg	4.5 kg
Power	150 mW	100 mW

in the mid-to-high range. Our correlation detection method is able to obtain the noise evaluation result of coil sensors successfully.

3.4 *Actual testing result*

After the comparison testing, the actual application result of the newly developed sensors was tested and compared with that of the MTC-80 sensor. The MTC-80 coil sensor was connected to the MTU5A of Phoenix while the CAS-10 M sensor was connected to proprietary mainframe WMT. The two systems acquired data synchronously for about 14 h. Fig. 6 compares the actual application result of the two coil sensors.

The comparison result shows that the apparent resistivity curve measured by the newly developed coil sensor virtually agrees with that measured by the MTC-80 coil sensor and is fairly smooth.

4   COMPARISON TESTING RESULT

Table 1 lists the comparison testing result of the newly developed CAS-10 M induction coil sensor with the MTC-80 coil sensor against some important indices. Using the comparison result, we were able to evaluate the performance of the new coil sensor and conclude that its performance is satisfactory and comparable with the technical level of the MTC-80 coil sensor.

5   CONCLUSIONS

To evaluate the performance of the newly developed induction coil sensor CAS-10 M, it is compared with the MTC-80 coil sensor of Phoenix. By performing 1) frequency curve comparison;

2) parallel noise comparison; 3) mutual coherence coefficient comparison; 4) correlated noise comparison; and 5) actual application result testing, we prove that our comparison testing method is able to provide correct performance evaluation of newly developed coil sensors.

REFERENCES

[1] Chen, X.P., Song, G., Zhou, S., Xi, Z.Z. & Wang, H. (2012) Development of magnetic sensor in audio-frequency magnetotelluric sounding. *Transactions of Nonferrous Metals Society of China*, 22(3), 922–927.
[2] Deng, M., Du, G. & Zhang, Q.S. (2004) The characteristic and prospecting technology of the marine magnetotelluric field. *Chinese Journal of Scientific Instrument*, 6(25), 742–746.
[3] Zhang, F. (2012) *Research on Amplification Technology of Wideband Inductive Magnetism Transducer*. Jilin, Jilin University.
[4] Wei, W.B. (2002) New advance and prospect of magnetotelluric sounding. *Progress in Geophysics*, 17(2), 245–254.
[5] Jiang, A.L., Zhou, S.H., Zhang, X.B. & Chen, Z.Y. (2011) Design of induction coil sensors. *Avionics Technology*, 17(31), 35–37.
[6] Chen, K. & Deng, M. (2010) Study and implementation of key technology for magnetotelluric signal measurement. *Modern Scientific Instruments*, 2, 7–9.
[7] John, G.P. & Dimitris, G.M. (1996) Digital Signal Processing Principles, Algorithms, and Applications, Prentice Hall PTR, London, pp. 118–131.

Engineering Technology and Applications – Shao, Shu & Tian (Eds)
© 2014 Taylor & Francis Group, London, ISBN 978-1-138-02705-3

Applying data mining to improve faculty evaluation for educational organizations

Wen Si & Rong Tan
College of Information and Computer Science, Shanghai Business School, Shanghai, China

Jian Wang
Cyber Physical System R&D Center, The Third Research Institute of Ministry of Public Security, Shanghai, China
School of Electronic Information and Electrical Engineering, Shanghai Jiao Tong University, Shanghai, China
Shanghai Chenrui Information Technology Company, Shanghai, China

Hongmei Liu
Donghua University, Shanghai, China

Jianye Li
The Third Research Institute of Ministry of Public Security, Shanghai, China

ABSTRACT: One of the challenges of educational organization improvement is to manage talented teachers effectively, i.e. to assign the right teacher for given courses. For providing scientific performance evaluation, some factors related to a teacher, such as age, experience, professional, etc. usually play important roles. Therefore, it is crucial to find the underlying associations between teachers' personal information and their working behaviors. In this paper, data mining technique is employed to improve the faculty evaluation. An algorithm called Quick Apriori, which is based on the classic Apriori algorithm, is proposed. The results not only provide the decision rules related to personal characteristics with working performance, but also show that the obtained rules are consonant with existing experience.

Keywords: data mining; knowledge management; Quick Apriori algorithm; human resource management; talented management

1 INTRODUCTION

The evaluation of teaching quality is one of most important part of faculty management in educational organizations. Dependent on the evaluation results, educational organizations are capable of assessing, selecting, training and rewarding the right faculties. For example, teachers' evaluation results can be generally classified into four ranks: excellent, competent, qualified and unqualified. The "excellent" teachers deserve all the praises, whereas the "unqualified" teachers have to be trained and enhance their abilities. The evaluation plays a vital role in teachers' individual growth and career development[1,2].

It is obvious that the evaluation results are related to the factors such as age, experience, professional, occupation, performance, etc. However, the underlying associations among above factors are still unclear which may lead to inefficient faculty management. How to manage talented teachers is one of the challenge of education institutions, especially to ensure the right teacher for the right students and courses[3,4]. Like most companies, many educational organizations nowadays have established human resource information systems and accumulated large amounts of personnel data which could be valuable for further exploration. Actually in the past few years, there has been an increasing tendency for researchers to investigate various data mining methods for educational organizations to improve their management.

The Apriori algorithm, which is proposed by Rakesh Agrawal in 1993, is the most classical algorithm for mining association rules among data mining technologies. Association rules of commodities can be found by Apriori algorithm in the process of commodity sales in retail industry. As a result, the relevant demands of customers can be found and cross-selling can be processed. With large database, the process of mining association rules is time consuming. Many variations of Apriori algorithm which focus on the efficiency improvement have been proposed[6,7,8,9].

This paper aims to apply data mining to explore the association rules between personnel characteristics and work behaviors in educational organizations. In order to improve the efficiency, an algorithm named Quick Apriori (QA) which is based on the classic Apriori algorithm is proposed. The QA algorithm is divided into three phases: candidate frequent item sets generation; candidate frequent item sets filtering, and association rules production. And different from the Apriori algorithm, QA put confidence in the first place, so that the association rules with very high confidence will be mined firstly. Our study shows that the consequences of QA are consistent with existing experience. The remainder of this paper is organized as follows: Section 2 describes the QA algorithm. Section 3 illustrates the faculty evaluation system. Section 4 shows the evaluation and concludes the paper.

2 ASSOCIATION RULES AND QA ALGORITHM

Due to the large data collections in the database, the efficiency of the data mining algorithm is crucial. Based on the classic algorithm Apriori, the improved algorithm named Quick Apriori (QA) will be introduced in this section. It will mainly filter more than one database to speed up the candidate frequent item sets searching, and thus improve the efficiency.

In the classic Apriori algorithm, supportiveness may determine quite as much. With respect to QA algorithm, the confidence is put in the first place so that the association rules with very high confidence will be mined first. The whole algorithm is divided into three phases: create candidate frequent item sets, filter candidate frequent item sets, and produce association rules. The second step is the key. After the scan of database, the original candidate frequent item sets are generated, supportiveness is calculated, and the low supportive data item is found out. Then the confidence of the data item sets in the corresponding original database is calculated with the methods based on the probability. If the confidence is too high, the corresponding data item sets will be filtered. Therefore, when scanning the database secondly, the scope will be reduced and the search time will be shortened accordingly. At last, the biggest selected set is generated and the association rules will be provided[10,11].

Although the missing rates of QA algorithm may be increased, the space and time efficiency is gained. Through the experiments, the efficiency of QA is better than that of Apriori[12,13]. And the algorithm also can produce the same useful association rules.

3 REALIZE THE FACULTY EVALUATION SYSTEM

The faculty evaluation system is focusing on assessing teachers' performance in the educational organizations. The whole system contains four main functionalities: information initialization, data acquisition, data pretreatment, and data analysis and mining.

3.1 *Information initialization*

Information initialization is essential to system operations. Its main goal is to collect and manipulate the huge data collections required by the evaluation correctly and timely. Even some data are unnecessary in the collection of the evaluation, but very important in the data analysis and mining.

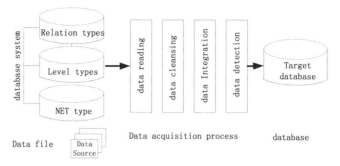

Figure 1. Data collection process.

Table 1. Teachers' evaluation data.

NO.	Age	Schooling	Major	Seniority	Dept.	Grade	Leader Y/N	results
21110009	30	PhD	Communication	1	Computer	lecturer	N	Good
21110010	45	M.E	Control theory	20	electronic	A-prof	N	Excellent
21110011	34	M.E	Computer	7	Computer	lecturer	Y	Good
21110012	55	B.E	Machine	30	Computer	experimenter	N	Good
21110013	43	M.E	Computer	15	E-business	A-prof	N	Good
21110014	56	B.E	Computer	32	E-business	prof	Y	Excellent

For example, when the system is analyzing and mining data, it is necessary to analyze whether the evaluation order is related with age, education experience, professional title, and working experience or not. This requires the system to store complete archival data of the teachers. However, this above information is not required in the evaluation information collection. As a result, the data initialization module needs to take into account the requirements of the subsequent modules, and initialize data processing. Information initialization mainly needs to complete following tasks:

1) Integrate data from different data sources correctly. The contestant unit data required by the evaluation (generally the latest information in the units), the teacher's performance data in original unit (generally evaluated by colleagues each other), as well as teacher detailed files required by the data mining are in the different application systems. This process need integrate these data sources from the different systems into the system database.

2) After the data integration, all kinds of information tables are generated and edited, including teacher's basic information table, unit's information table, etc. The main difficulties are to ensure the accuracy and reliability of the data tables in this process. This system requires absolutely accurate data. Inaccurate data will directly affect the reliability of the system. For example some professional teacher should be A, but due to incorrect data entry for B. So a evaluation information has become B evaluation information. Once such situation exists, the credibility of the system will be cut down.

3.2 *Data acquisition*

The target of the data acquisition is to realize complete and accurate collection of the large-scale evaluation data. Data acquisition steps are shown in figure 1. Teacher evaluation system uses the computer data transfer function to open the information database directly and acquire data. The contents of the data acquisition include teacher evaluation data and teacher archival data. The data of teacher evaluation collection is shown in Table 1.

3.3 Data pretreatment

Data pretreatment is an important part of the system operation. Its main purpose is to prepare data for data analysis and mining module. It is necessary to process data in advance according to the algorithm of the system data mining. The main tasks to be completed in data pretreatment are:

1) Data warehouse and data extraction. Data extraction is the act or process of transferring all kinds of data from transaction-oriented real-time operating database into data mining oriented warehouse, including teachers' information, the evaluation standards and so on. These data are from different departments.
2) Data cleaning. Data cleaning usually includes inconsistent data processing to fill vacancy and data transform. The inconsistent data and noise data in the system has been processed in the information initialization module. It focuses resolving fill vacancy and data transform.
3) Generate affairs table. Because QA algorithm is suitable for the data mining of transaction database, the system need convert the data of the teachers' information tables to transaction tables to provide the whole data mining the data mining object. The above data pretreatment operation enables the system complete, clean, and suitable for mining data output, and ensure that the following data mining can generate some useful rules. This paper regards a piece of record of the teacher evaluation table as a transaction. The values of the fields convert corresponding items with the code table (Table 1). These are shown in figure 3 below.

3.4 Data analysis and mining

Data analysis and mining is the core of the system. It analyses and mines the obtained evaluation data with the technology of the data mining, and finds the useful knowledge from the data sources. For example it finds out and extracts the association between the situation of the teacher evaluation and teacher's age, working age, education experience, professional, working experience and others for managers and organs to provide scientific decision information. The decision information promotes the universities comprehensive development and improves the quality.

The data mining of the system mainly depends on the analysis data, which user selected. Then it extracts the corresponding decision rules from these data and presents to the user. The user can make the decision according to the rules. On the collected evaluation data, it can be dug out the following information: the characteristics of the assessment of different teachers (different education background, position, age, etc.); what features are related with the evaluation results (age, sex, professional, education experience, etc.). The specific implementation procedure is as follows.

3.4.1 Generate frequent item sets

Using the mentioned mining technology, the system digs out the teachers who have excellent or good evaluation results. Firstly the system searches the original database and gets 36 pieces of records with excellent evaluation results, 74 pieces of records with good evaluation results. Given the minimum supportiveness, the system looks for the frequent item sets with QA algorithm.

3.4.2 Generate association rules

After generating the frequent item sets, according to the algorithm of the generated association rules, all possible proper subset will be found out from any frequent item sets, and be regarded as the conditions of the association rules. Then the confidence of the corresponding rules will be calculated. When the confidence of one of the rules is greater than the given minimum confidence, the rule will be output. The confidence of the data items in the original database will be calculated on the probability method from those data items with low supportiveness. If the confidence is very high, the rule will be output. Then the further data mining will be done till the end. The association rules with excellent achievement (min_sup $= 10\%$, min_conf $= 40\%$) are shown in figure 2; the association rules with good achievement (min_sup $= 10\%$, min_conf $= 40\%$) are shown in figure 3.

Table 2. Teacher evaluation table.

Project	Value	Code
Age	20–25	A1
	26–30	A2
	31–36	A3
	37–45	A4
	46–60	A5
Political Status	Party Member	P1
	Democratic Party	P2
	Public people	P3
Schooling	B.E	S1
	M.E	S2
	PhD	S3
Major	Engineering	M1
	Arts	M2
Seniority	1–10	Y1
	11–20	Y2
	21–40	Y3
Dept.	Electronic	D1
	E-business	D2
	Computer	D3
Grade	Lecturer, A-prof. to prof.	R1 to R4
Leader Y/N	Y	L1
	N	L2
Results	poor	W1
	Middle	W2
	good	W3
	Excellent	W4

Table 3. Transaction table.

NO.	Age	Schooling	Major	Seniority	Grade	Leader Y/N	results
21110009	A1	M1	Y1	D1	R1	L1	W4
21110010	A3	M2	Y1	D3	R4	L2	W2
21110011	A3	M1	Y1	D1	R5	L1	W3
21110012	A2	M2	Y1	D2	R3	L2	W3
21110013	A1	M1	Y1	D2	R4	L1	W3
21110014	A2	M1	Y1	D3	R1	L1	W3

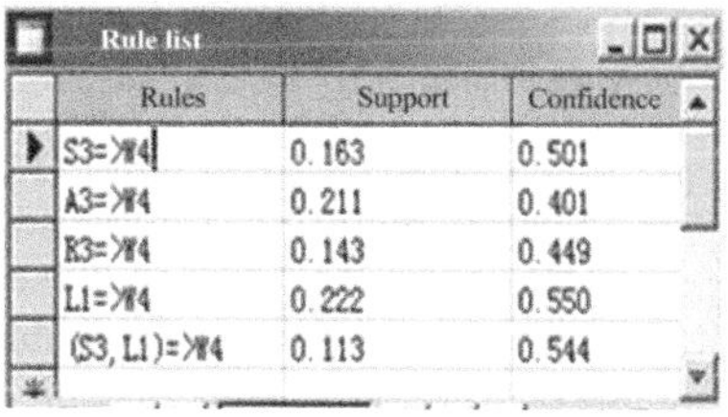
Rule list

Rules	Support	Confidence
S3=>W4	0.163	0.501
A3=>W4	0.211	0.401
R3=>W4	0.143	0.449
L1=>W4	0.222	0.550
(S3, L1)=>W4	0.113	0.544

Figure 2. The Association Rule with Excellent Achievement (min_sup = 10%, min_conf = 40%).

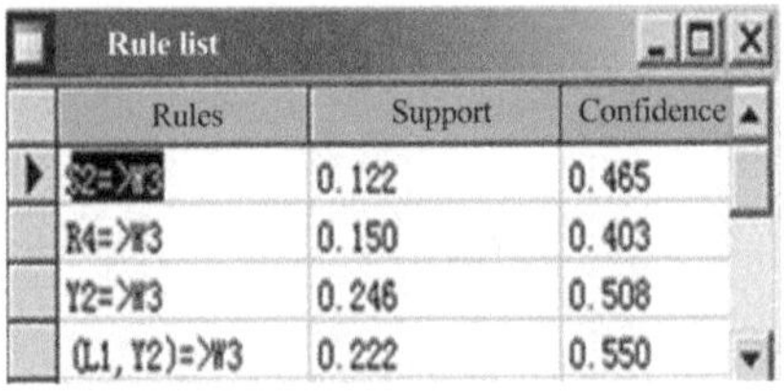

Rule list		
Rules	Support	Confidence
A2=>W3	0.122	0.465
R4=>W3	0.150	0.403
Y2=>W3	0.246	0.508
(L1, Y2)=>W3	0.222	0.550

Figure 3.   The Association Rule with Good Achievement (min_sup = 10%, min_conf = 40%).

3.4.3 *Examples of related meaning*

The rule: A3=>W4 converts to age $\in [31, 36]$ => the evaluation result is excellent, and the supportiveness is 21.1%. It shows there are 10% records "age $\in [36, 49]$ with excellent achievement" in the teachers evaluation tables; and the 45% confidence shows that the 50.1% teachers, whose ages $\in [36, 49]$, get the excellent achievement.

The rule: (L1, Y2)=>W3 converts to leaders $\cap$ working age in 11–20 years => the evaluation result is good, and supportiveness is 22.2%. It shows there are 22.2% records "as leaders, working age in 11–20 years with good achievement"; and the 55.0% confidence shows that the 55.0% leaders, whose working age in 11–20 years, get the good achievement.

4 EVALUATION AND CONCLUSION

1) Through the evaluation of the data mining, we conclude some suggestions on the teachers' development:
2) The school supports the on-the-job teachers enhance their educational backgrounds in the spare time; strengthens the trainings of the teachers, who have high educational backgrounds; improves the past employing rules; believes these teachers; and gives them enough opportunities to work.
3) The school attaches importance to the cultivation of the grassroots leaders in the teacher teams; creates the opportunity for teachers in the office to train in the other places at the same time.

The school gives priority to encourage teachers in the peacetime work; is patient and tolerance to the new teachers.

In this paper, we apply the data mining to improve the faculty evaluation for educational organizations. The Quick Apriori (QA) algorithm which is based on the classic Apriori algorithm is proposed. The usage of QA algorithm in evaluation system is illustrated step by step as well. The results show the association rules relating teachers' personnel information with their work performance. It is promising that our work can provide a valuable reference for talented teachers management.

ACKNOWLEDGEMENT

This paper was sponsored by following projects:

- National High-tech R&D Program of China ("863 Program") (No. 2013AA01A603);
- National Science and Technology Support Projects of China (No. 2012BAH07B01);
- Program of Science and Technology Commission of Shanghai Municipality (No. 12510701900);
- 2012 IoT Program of Ministry of Industry and Information Technology of China.
- Young University Teachers Training Plan of Shanghai Municipality under Grant (No. SXY12003)
- Management department of Shanghai Business School (NH1-2-1-1316).

REFERENCES

[1] Ji Qingbin. Development of Personnel Management System in Colleges. Mechanical Management And Development. Vol 24, pp. 127–128, April 2006 (In Chinese).

[2] Zhu Jie. Discussed shallowly the university educates the personnel system reform. China Water Transport. Vol. 5. pp. 229–230, December 2007 (In Chinese).

[3] Wang Xuan-wen. An Application of Association Rules Mining In Personnel System. Journal of Xi'an Institute of Posts and Telecommunications. vol. 1, pp. 21–23, 2011 (In Chinese).

[4] Chen Li, Chen Gencai. Research of Data Mining Based on Personnel Databases. Computer Engineering. vol. 26. pp. 117–119, 140, 2000 (In Chinese).

[5] Agrawal R. Mining association rules between sets of items in largedatabases. Proceeding of the 1993 ACM SIGMOD Conference, Washington, pp. 207–216, November 1993.

[6] Li Qingfeng; Yan Luming; Zhang Xiaofeng; Long Yanjun. Aneffective apriori algorithm for association rules in data mining. Computer applications and software, Vol. 21, No. 12, pp. 84–86, Jan. 2004.

[7] Liu Jing, Qiu Chu. An Improved Apriori Algorithm for Early Warning of Equipment Failue Computer Science and Information Technology. pp. 8–11 Aug. 2009.

[8] Yuntao Liu, Qing Liu, Danwen Zheng. Application and Improvement Discussion about Apriori Algorithm of Association Rules Mining in Cases Mining of Influenza Treated byContemporary Famous Old Chinese Medicine. 2012 IEEE International Conference on Bioinformatics and Biomedicine Workshops (BIBMW) pp. 316–322.

[9] U.M. Fayyd. From Data Mining to knowledge: discovery Advances in Knowledge Discovery and Data Mining. AAAI/MIT Press. pp. 101–106, 1996.

[10] D.W. Cheung, S.D. Lee, B. Kao. A General Incremental Technique for Maintaining Discovered Association Rules. In Proceedings of the Fifth Int. Conf. on Database Systems for Advanced Applications, Melbourne, Australia, pp. 185–194, April 1997.

[11] Han Jiawei. Mining frequent patterns without candidate generation. Proceedings of the 2000 ACM SIG-MOD, Dallas, pp. 1–12, August 2000.

[12] Zhao Chun-ling; NING Hong-yun. Improvement in Apriori algorithm and using in logistics information mining. Journal of Tianjin University of Technology, Vol. 23, No. 01, pp. 30–33, Jan. 2007.

[13] Liang Zhi-rui; CHEN Peng; SU Haifeng. Association rule miningin fault monitoring of power plant equipment. Electric Power Automation Equipment, Vol. 26, No. 06, pp. 17–19, June 2006.

Engineering Technology and Applications – Shao, Shu & Tian (Eds)
© 2014 Taylor & Francis Group, London, ISBN 978-1-138-02705-3

Research on the underground power quality optimization and control based on cloud computing TaskScheduler algorithm

Zhou Lv

Institute of Mechatronic Engineering, Taiyuan University of Technology, Taiyuan, Shanxi, China

ABSTRACT: Changes have taken place in the multi-variable prediction structure of cloud computing TaskScheduler algorithm applied to coal mine underground ventilation capacity. This paper discusses the use of cloud computing TaskScheduler algorithm to detect the underground power supply capacity quality, the use of FPGA to sample 16-way AC signals and perform 64 harmonic analysis fusion, as well as a cloud computing information nonlinear aggregation technique and four-dimensional nonlinear indexing for newly developed intelligent ventilation networks. With options for dynamic modifications of R+tree, the minimum TaskScheduler algorithm prediction error is used to determine the optimal embedding dimension, and prediction customer error is used to predict wind waveform flicker and power supply diagonal imbalance.

Keywords: energy quality; cloud computing; TaskScheduler; distribution optimization; ARM

1 INTRODUCTION

Over the past years, the extensive application of electric and electronic appliances in coal mines has added many nonlinear loads to the power load systems. The voltage and current waveform distortion as a result of these nonlinear loads, in turn, has led to severe deterioration of the current quality. The underground power systems in coal mines are facing austere challenge[1]. In the face of severe backup power shortage and power resource deficiency, it is important that a complete power quality monitoring system be established to keep the power management department of a coal mine always updated about the exact quality conditions of the power supply system, and enable them to take technical steps to guarantee good quality. However, to obtain the correct transient quality indicators, power data processing capability and high sampling rate are the most essential for a processor. The data acquisition and processing system of a normal CPU is not suitable for analyzing power quality data due process, structure, timer and bus limitations, the limited instruction feature, small addressing space and low computation ability that could affect its ability to process digital signals. In this paper, a dual-processing approach integrating cloud computing TaskScheduler and the embedded processor ARM9 is used to enable the best use of properties of the former and the feature of the latter, and achieve high cost performance. Our design includes hardware and software modules. The hardware design of the system comprises a data acquisition module and a data processing and display module[2]. The latter uses high-performance signal processor cloud computing TaskScheduler and an embedded ARM9 microcontroller as the core. In the software design of the system, cloud computing TaskScheduler is responsible for acquiring digital power quality signals, filtering these signals and processing data using wavelet algorithm. The embedded processor ARM9 is responsible for man-machine interaction on its software platform.

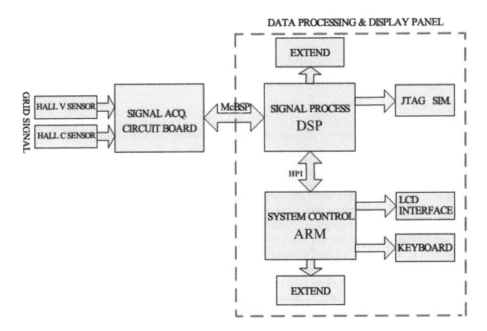

Figure 1.   General design diagram of the system hardware.

2   SYSTEM STRUCTURAL DESIGN

2.1   *System hardware design*

The hardware of the system comprises two modules: a signal acquisition circuit, and a data processing and display circuit. As an existing signal acquisition circuit board is used, our hardware design is focused on the data processing and display circuit board.

As a power quality monitoring system has to acquire the power supply system signals immediately, quickly and accurately and obtain processable digital signals after A/D conversion, the data acquisition circuit board has to be able to accurately convert the grid signals into digital signals that can be subsequently processed by cloud computing TaskScheduler[3].

At enable quick, real-time mass computation, a high-speed digital signal processor TMS320VC5402 is used as the data processing core. A large-capacity memory is used to record historic data. Besides, a JTAG interface for internal chip testing of cloud computing TaskScheduler is also incorporated. The embedded processor AT91RM9200 serves as the control kernel of the system that coordinates the work of the system. A keyboard module and an LCD display module are combined to provide man-machine interaction so that the user can control the system with the keyboard and view the needed parameters on the LCD. The Flash memory and synchronous dynamic memory (SDRAM) are extended[4]. Besides, the power circuit, reset circuit and keyboard circuit are also designed. Fig. 1 shows the general design diagram of the system hardware.

In the monitoring system, cloud computing TaskScheduler communicates data with A/D via a multi-channel buffer serial port (McBSP), and communicates with ARM via a parallel host interface (PHI).

The TMS320F2812 serial port SCI is generally regarded as UART. The SCI interface has mutually independent Enable bits and Interrupt bits that allow it to work both under half duplex and full duplex modes. As the two chips have different working voltages, they are optically isolated by an optical isolator 6N137. Fig. 2 shows the schematic diagram of the serial communication circuit.

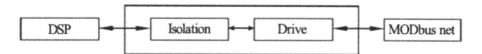

Figure 2.   Schematic diagram of serial communication.

2.2 *System software design*

In our power quality monitoring system, the software is a modular design composed of a data acquisition module, a data processing module, a data communication module and a man-machine interface module. The design of these modules is completed in cloud computing TaskScheduler and ARM.

2.2.1 *Software design in cloud computing TaskScheduler*

Data acquisition module. As the A/D data acquisition is controlled by initializing the serial port and the control register to open the receive interrupt of the serial port, the control register and serial port of A/D and cloud computing TaskScheduler is initialized first. Then, the system starts sampling. After that, A/D is closed. Data are sent to cloud computing TaskScheduler where they are processes. The result and some necessary data are then sent to the HPI interface area. After that, A/D is opened again to perform a new round of sampling.

Cloud computing TaskScheduler communicates with the host via the Modbus protocol RS-485 level. When the host program comes into the send interrupt service process, cloud computing TaskScheduler sends communication request to the host. If yes, it asks if the communication state is free. If yes, cloud computing TaskScheduler will generate a communication frame and starts send, and set the communication bit to BUSY. After receiving the data, the host will clear the receive bit and perform CRC check[5]. If yes, it will analyze the data frame and then set the communication bit to FREE and gets ready for the next receive. The data flow is as shown below.

2.2.2 *Software design in embedded system ARM*

MiniGUI is a graphic interface supporting system designed for embedded or real-time systems and typically works on Linux platform. It has good portability and an inventory file containing all functions of around 300 K. TaskScheduler is packed with cloud computing and provides a clear program framework for the developer to write his own customer user interface program. It has its own graphic engine that allows direct operation of the graphic drive frame buffer on the bottom[6]. Main function writing in TaskScheduler.

The application program of TaskScheduler is typically composed of header files, sources files and main functions. The program generally contains header files of all Qt components used, sometimes with additional customer header files. In some simple programs, the content of the header file can be written in the source file. The program writing of TaskScheduler also contains these three parts.

The TaskScheduler program normally contains the main function int app (int argc, char **argv), which is the cut-in point of the application program that performs some initialization work. Now, let's look some of the details of the main function writing framework.

For the main function, we first create the object of Displacing apps. The command is:

Displacing apps a (argc,argv);

a is the object of this program that is created and processes some come line parameters; agrc is the serial number of a command line parameter; agrv is a command line parameter group. After Displacing apps is created, the object of the components is created. For a pushbutton component QPushButton named "hello" is created, for example, the word is:

QPushButton hello ("hello world", 0);

193

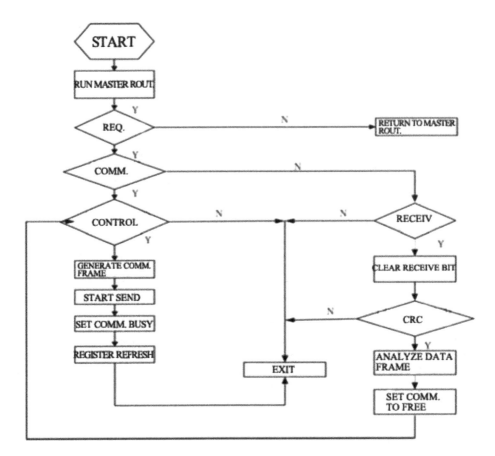

Figure 3.　Data flowchart.

The second parameter is the parent component of the component. We may also select a parentless parameter, when the parameter is 0. After creating the component, we set the main component of the application program to h. The word is:

a.setMainWidget (&h);

As the most principal mechanism of TaskScheduler is its signal/slot mechanism, the signal-slot connecting function connect () is also the core function of TaskScheduler. It connects the two TaskScheduler objects together, each of which is able to send or receive messages.

QObject::connect(& h,SIGNAL(clicked()),& a,SLOT(quit()));

As the initial state of the create window components are invisible, we have to use the show () function to display them.

hello.show();

Bedore displaying a component, we can set its attribute like the size, color, font and caption of the text displayed on the component. For example: h.setCaption ("hello world") means the caption of the component is set to be "hello world".

Finally, app() hands over the control to TaskScheduler. When the application program exists, the function exec() returns.

return a.exec();

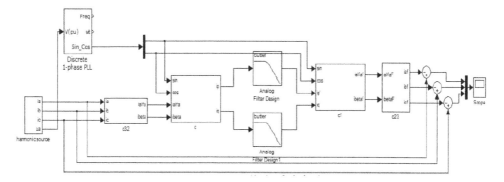

Figure 4.　Schematic diagram of simulation system.

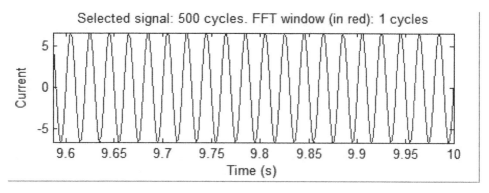

Figure 5.　Phase fundamental current.

3 SIMULATION EXPERIMENT AND RESULT

As nonlinear loads are the principal harmonic sources in a coal mine, and the AC-side rectifier circuit with resistance-inductance load is the most representative nonlinear load, a three-phase bridge rectifier circuit with resistance inductance load is used to simulate the harmonic sources. We assume the resistance is $50\,\Omega$ and the inductance is $0.002\,H$. Fig. 4 shows the simulation circuit diagram.

Using the traditional method, as part of the LPF, the phase fundamental current derived is shown in Fig. 5. The pre-process FFT result of the signal source harmonic is shown in Fig. 6. The post-process FFT result of the signal harmonic is shown in Fig. 7.

4 CONCLUSIONS

Two aspects of applying cloud computing TaskScheduler to practical problems in connection with the underground power supply capacity of coal mines are studied. One is the design of system data processing and display circuit boards using TaskScheduler+ARM as the core, which allows communication between ARM and TaskScheduler, extends the ARM and TaskScheduler memories and incorporates a control circuit, independent displaying and storing hardware circuits and some other hardware circuits. The other is the investigation of the signal/slot mechanism and the program writing of TaskScheduler, including the use of QtDesigner, the culture in Qt and the main function writing in QtDesigner TaskScheduler. Finally, we are able to display the base parameters of power quality and the waveform characteristics on the interface to allow direct control

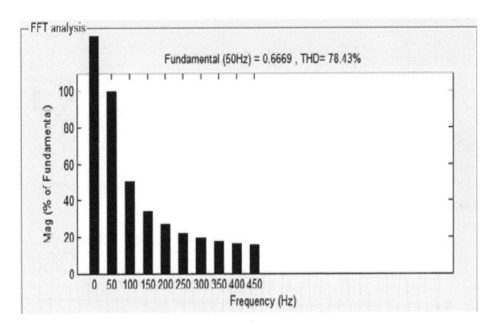

Figure 6. Pre-process signal source harmonic FFT.

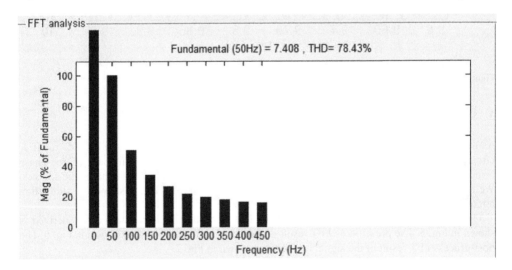

Figure 7. Post-process signal harmonic FFT.

of the monitoring system. This not only makes it possible to distribute the monitoring points of underground power supply capacity anywhere in the power supply network with satisfactory online behavior.

REFERENCES

[1] Xiao, X.N. (2004) *Analysis and Control of Power Quality*. Beijing, China Electric Power Press.
[2] Chen, X.P. (2006) *The Research and Application of Embedded Power Quality Monitoring Systems*. Hunan, Hunan University.

[3] The design of HPI interface and PCI bus interface of TMS320VC5402 processor [J] http://www.ic37.com/htm_tech/2008-1/12086_445320.htm.

[4] Mohammad Salem, Ramizi Mohamed. Development of a DSP-Based Power Quality Monitoring Instrument for Real-Time Detection of Power Disturbances [J] IEEE Power Electronics and Drives Systems. 2005: 78(03): 92–96.

[5] Wang M; Ochenkowski P; Mamishev A. Classification of power quality disturbances using time-frequency ambiguity plane and neural networks [J] IEEE Power Engineering Society Summer Meeting, 2001, 12(2): 1246–1251.

[6] Panteleymonov A. Interoperable thin client separation from GUI applications [J]. Software Maintenance and Reengineering, 2002. Proceedings. Sixth European Conference on, 2002.

[7] Yang S.Q. ARM embedded Linux system development techniques explanation [M]. Beijing: Publishing House of Electronics Industry, 2008.

Engineering Technology and Applications – Shao, Shu & Tian (Eds)
© 2014 Taylor & Francis Group, London, ISBN 978-1-138-02705-3

Evaluation modeling in English teaching based on frequent itemset mining algorithm

Yuexia Cui

Office of Academic Affairs, Weifang University of Science and Technology, Weifang, Shandong, China

ABSTRACT: As of the end of 2013, China's higher education enrollment was already higher than 30%, signaling a shift from elite education to mass education in terms of student enrollment. A competitive intelligence capabilities in English evaluation model (AKA) was established using frequent itemset mining algorithm with data collected on undergraduates' competitive intelligence capabilities in English. The model uses modified association rules mining algorithm of Boolean database (Ad-Apri) and succeeds in achieving higher efficiencies by trimming the database effectively and reducing pressures on both the processor and the memory by calculating Boolean vector during the computation, resulting in significant reduction of the time conventionally used for calculating high-order frequent itemsets. The performance of the algorithm was tested in the experiment with self-constructed data and the algorithm was applied to the English scores of undergraduates of a university during 2009–2012 in support of education and teaching processes.

Keywords: competitive intelligence capabilities evaluation; association rules mining; frequent itemset mining; modified apriori algorithm

1 INTRODUCTION

As of the end of 2013, China's higher education enrollment was already higher than 30%, signaling a shift from elite education to mass education in terms of student enrollment[1]. The former focuses on thick-foundation, wide-caliber, broad-spectrum adaptation cultivation while the latter places more emphasis on the professional skills of students and the correct definition among school, profession and student cultivation.

As a result, most talent cultivation systems in higher education institutions tend to define their target systems according to the classification of their school and profession. In other words, they try to adapt their targets and strategies to their specific type of school and disciplines, and plan and provide the education to students to achieve profession-specific education[2].

It is also noted that, in the context of current education, this pattern of talent education based on generic professional classification is exhibiting two obvious shortfalls: One is that is it fails to address individual needs for different occupations or individual variations, and the other is that it does not provide for real-time response of the professional cultivation targets of higher education institutions to local economic or social demands. It is therefore imperative that talent education processes start with the career planning of students, and implement classified or hierarchical teaching for different student communities, addressing their professional orientation, scores and personal growth throughout the teaching process from cultivation planning, course construction to probation and practical training. In a word, a new teaching model of classified cultivation targeted at the occupational skills of students should be implemented.

In this paper, a data mining model for evaluating students' competitive intelligence capabilities in English was built, in which reduced attributes from rough set modeling were inputted and examined for different capability groups using neural networks to examine how capabilities relate to grades and interpersonal relationship, and how this relationship, together with paid competitive intelligence service, contribute to the students' capability differences.

2 RELATIVE RESEARCHES

Competitive Intelligence is an information and research on the competitive environment, competitors, competitive situation and competitive strategy that emerged in the 1980s. Providing competitive intelligence education in universities and improving their competitive intelligence awareness help relieve their employment pressure and enable them to adapt their periodical targets to internal and external changes before finally achieving their lifetime goals.

John E. Prescott, University of Pittsburgh has a fairly complete and defined outline of the idea "Competitive Intelligence", saying that Competitive Intelligence System is an ongoing evolution in the formal and informal operating procedures combined enterprise management subsystem, its main function is to assess key trends, corporate organizations and members of the continuous change tracking emerging grasp of the business structure evolution, as well as the ability to analyze existing and potential competitors and trends to help maintain and develop competitive advantage[3].

Using data mining algorithms for competitive intelligence processing allows for automatic analysis of inputs, mining potential knowledge from them and automatic and intelligent knowledge discovery. While competitive intelligence system uses a range of algorithms to establish a competitive intelligence capabilities evaluation model, our study is focused on data mining and discovery in connection with relational databases. Among progresses already made in the research on dada mining and knowledge discovery of relational databases and transaction databases so far, the most influential algorithms include the Concept Hierarchy algorithm of Prof. J. Han, Simon Fraser University; the Association Rules algorithm of R. Agrawal, IBM; the Classification algorithm of Prof. J. R. Quinlan, Australia and the Genetic algorithm of Erick Goodman, Michigan State University. IBM, GTE, SAS, Microsoft, Silicon Graphics, Integral Solutions, Thinking Machines, DataMind, Urban Science, AbTech and Unica Technologies have also developed some useful KDD commercial systems and prototype systems including BehaviorScan, Explorer and MDT (Management Discovery Tool) for market analysis, Stock Selector and AI (Automated Investor) for the financial investment sector, and Falcon, FAIS and C10nedetector for fraud warning.

Apriori algorithm was introduced by Rakesh Agrawal et al in 1994. One of its main functions is to find a set of itemsets from a given transaction database that appears at a specific frequency[5]. As a classic algorithm generating frequent itemsets, Apriori marks a milestone for data mining. As research deepens, however, its shortfalls appear too. Two of the fatal performance bottlenecks in Apriori are the time consumed for processing tremendous candidate itemsets and the need for repeated scanning of the transaction database.

Improvements made to the performance of Apriori include reducing the I/O cost by reducing the scans of the transaction database, reducing the time for calculating the support degree of the candidate itemsets, and increasing the speed of calculation by using parallel algorithm. Park et al introduced a hash-based algorithm[6]. Savasere et al designed an algorithm based on data partition[7]. Toivonen suggested and improved a sample-based algorithm[8].

It is noted that, considering the temporal and spatial complexity intrinsic in classic Association Apriori, Jiawei Han et al proposed a new frequent pattern growth algorithm based on FP-tree, also known as FP-growth algorithm, in 2000 based on an important nature of Apriori: all non-empty subsets of a frequent itemset should also be frequent. This enables the generation of all frequent itemsets without producing candidate itemsets[9].

However, there are some disadvantages in FP-growth too, since this algorithm uses a recursion of pattern growth, the conditional FP-tree will become very large if a large number of itemsets are involved in the mining process and the FP-tree has many long branches, which is not only time consuming, but also occupies a lot of space. To achieve better mining effect, data partition could be incorporated so that FP-growth operates over a smaller database.

3 COMPETITIVE INTELLIGENCE CAPABILITIES EVALUATION MODELING BASED ON FREQUENT ITEMSET MINING ALGORITHM

Based on the Boolean idea of vector operation, we are now introducing an association rules mining algorithm based on Boolean database (Ad-Apri), which first converts the transaction database from English teaching to a relational database, and then discretizes and converts this relational database to Boolean database D following a specific rule. This allows for direct bitwise AND operation of rows to generate frequent itemsets and significant reduction of the time used for calculating high-order frequent itemsets by trimming the database effectively.

3.1 Definition

Definition 3.1 Boolean database D: The only value in the two-dimensional database is either "TRUE" or "FALSE".

Definition 3.2 Transaction density: The ratio of the total number of "TRUE"s to the total number of data present in Boolean database D, i.e. the total number of TRUEs/(lines*rows).

Definition 3.3 Minimum number of supports: As for a given minimum support degree min_sup, the number of transactions contained in Boolean database D (i.e. the recorded number of the database) is m, the minimum number of support min_sup_num = min_sup*m.

Definition 3.4 Number of k-dimension supports: In the bitwise (elements in the same line) AND operation of any k-row in Boolean database D, the number of "TRUE"s in the operation result is called the k-dimension support degree of vectors in this k-row.

From the definition of the number of k-dimension supports, we have inference 3.1.

Inference 3.1 If the number of "TRUE"s in a Ri line of Boolean database D is smaller than k, Ri line may be ignored in the operation when calculating the k-dimension support degree.

Proof: From the nature of Boolean AND operation, the result of the AND operation is not a "TRUE" unless all the elements involved in the AND are "TRUE". So, when the number of "TRUE" is in R_o is smaller than k, at least of the elements in R. is "FALSE".

Inference 3.2 If the number of "TRUE"s in one Lj row of Boolean database D is smaller than the minimum number of supports, Lj row may be ignored in the operation when calculating the k-dimension support degree.

Proof: According to the transcendental theory of Association Algorithm, all subsets of any frequent itemset are frequent itemsets, and all supersets of any non-frequent itemset are non-frequent itemsets [4]. As the number of "TRUE"s in line Lj is smaller than the minimum number of supports, Lj is a non-frequent itemset and any of its supersets is a non-frequent itemset. Thus Lj row may be ignored in the operation when calculating the number of k-dimension supports.

Inference 3.3 Lk, for the set of frequent k-itemset, Lk, if $|L_k| < k + 1$, the frequency of the maximum frequent itemset in the transaction database being mined is k, thus ending the operation. Where: $|L_k|$ is the number of frequent k-itemsets in Lk.

Proof: For a frequent (k+1)-itemset $X = \{I1, I2, \ldots, Ik + 1\}$, there is always k+1 k-frequent subsets. If the number of elements in the set of k-itemsets, Lk is smaller than k+1, there is no frequent (k+1)-itemset in the transaction database being mined.

3.2 Algorithm realization: steps and description

The algorithm is detailed below.

4 EXPERIMENT AND OBSERVATIONS

4.1 Performance testing of algorithm

To facilitate our experiment, three artificial databases were constructed and sized 100000*20. Data in the table are for Boolean database. The difference among the three databases is the transaction

Input: transaction database D, minimum support degree min_sup
Output: all frequent itemsets L

(01) min_sup_num = min_sup*m;

(02) $A \leftarrow D$;//convert transaction database D to Boolean database A

(03) $L_1 = \varnothing$;

(04) for each Field do begin

(05) if count(Ai) $\geq$ min_sup_num then

   //count(Ai) calculate the number of TRUEs in row i of database

(06) $L_1 = L_1 \cup \{I_i\}$

(07) else delete Ai from A;

(08) end

(09) for each recorder Aj of A do begin

(10) if count(Aj)<2 then

   //count(Aj)calcualte the number of "TRUE"s in line j of database A

(11) delete Aj from A;

(12) end

(13) for(k=2; $\left| L_k - 1 \right| > k - 1$;k++) do begin

(14) $L_k = \varnothing$;

(15) Produce k-vectors field for all columns of A;

   //combine the k-dimensions in all rows of database D

(16) for each k-vectors field{Ai1,Ai2,..,Aik} do begin

(17) $B = \bigcap_{j=1}^{k} A_{ij}$ //derive the number of supports of k-dimensions

(18) if sum(B) $\geq$ min_sup_num then

(19) $L_k = L_k \cup \{I_{i1}, I_{i2}, ..., I_{ik}\}$;

   // $\{I_{i1}, I_{i2}, ..., I_{ik}\}$ is the itemset corresponding to $\{A_{i1}, A_{i2}, ..., A_{ik}\}$.

(20) end

(21) for each item Ii in Lk do begin

(22) if $\left| L_k(I_i) \right| < k$ then

(23) delete the field Ai from A;

   //Ai is the corresponding row of Item Ii in database D

(24) end

(25) for each recorder Aj of A do begin

(26) if sum(Aj)<k+1 then

(27) delete Aj from A;

(28) end

(29) k=k+l;

(30) end

(31) return $L = \bigcup_{k=1}^{i} L_k$;

Figure 1. Description of Ad-Apri.

Table 1. Accumulated times of computing ABBD and Apriori algorithms under different supporting degrees.

		0.1	0.2	0.25	0.3	0.4
D1	Ad-Apri	11310	835	334	140	32
	Apriori	137950	60451	21456	6148	1340
D2	Ad-Apri	36185	5210	1941	453	53
	Apriori	264021	137983	60521	21672	1340
D3	Ad-Apri	47703	5424	2174	530	67
	Apriori	264087	136997	60521	21699	6201

density, which is 0.26, 0.29 and 0.31 respectively. To enable comparison, the minimum support degrees were also defined to be 0.1, 0.2, 0.25, 0.3 and 04. The experiment compared the efficiency of our association algorithm based on Boolean database for databases with different transaction densities under different minimum support degrees, and also compares its efficiency with Apriori.

From Table 1, the times of calculation for databases D1, D2 and D3 are 11310, 36185 and 47703 when the minimum support degree is 0.1; and are 32, 53 and 67 when the minimum support degree was 0.4.

From Fig. 1, the times of computing by Apriori are all larger than those by Ad-Apri for all the three databases under all minimum support degrees. As the minimum support degree diminishes, the increase in the times of computing by Apriori is far larger than that by Ad-Apri, especially for databases with a larger transaction density.

As can be concluded, Ad-Apri requires fewer times of computing than Apriori regardless of the density of the database or the minimum support degree; Ad-Apri is even more superior over Ad-Apri as the transaction density increases and the minimum support degree diminishes.

4.2 *Applying association rules to credibility test model*

The evaluation system for English in our school consists of five components: listening, speaking, reading, writing, and overall rating. The full score for the first four components is 100 and that for the overall rating is 40. A test was conducted using 126756 records obtained from the evaluation test papers of the latest three years. The test papers were evaluated using credibility method. Credibility is defined by the stability and reliability level of a test paper. It reveals the degree of deviation between measured and true values, and reflects how close samples are to the population. Theoretically, higher credibility will mean higher credibility of the measurement results. The most widely used and operable method for calculating credibility so far is the Cronbach's alpha[10], expressed as:

$$A = \frac{k}{k-1}\left|1 - \frac{\sum_{j=1}^{n} S_j^2}{s^2}\right| \tag{1}$$

where: A is the credibility coefficient of the test; K is the number of questions; s^2 is the variance of the test paper; S_j^2 is the variance of the questions. A is between 0 and 1. The closer it is to 1, the smaller is the deviation of the student's score from his true level. The closer it is to 0, the more incredible is the student's score and the more contingency factors there are.

According to this analysis, the credibility of the writing component of the 126756 records from the English evaluation system over the recent three years was tested. The correlation coefficient between the score of writing and other score is Table N below. Furthermore, we may also try to calculate the correlation coefficient between the score writing and that of listening, reading or dialog to identify the credibility of writing. Table N+1 show the calculated correlation coefficients.

Table 2. Correlation coefficient of writing and other Score.

Item 1	Item 2	Correlation
Score of writing	Total score	0.89
Score of writing	Other score than writing	0.71

Table 3. Correlation coefficient and calibration of writing and other score.

Item 1	Item 2	Credibility
Writing	Listening	0.30
Writing	Reading	0.44
Writing	Dialog	0.33

5 CONCLUSIONS

Apriori is the most classic mining algorithm that generates frequent itemsets from candidate itemsets. It produces enormous candidate itemsets and requires repeated visits to the database in each stage of calculation as well as huge memory space and high I/O cost, which make up bottlenecks for its extensibility and efficiency. To solve these shortfalls, hash-based, data partition-based or sample-based algorithms could be used to achieve better operation efficiencies.

By combining the nature of Association Rules and the Boolean vector operation rules, we propose an Association Rules Mining algorithm based on Boolean database (Ad-Apri). This algorithm is an improvement in efficiency by trimming the database effectively based on the transcendental theory of Association Rules that all subsets of any frequent itemset are frequent itemsets, and all supersets of any non-frequent itemset are non-frequent itemsets, and reduces pressures on both the processor and the memory by using Boolean vector calculation during the operation. The algorithm was validated with actual data and proves reliable and highly efficient.

We also discovered that students have to improve their competitive intelligence capabilities and provide some recommendations for this consideration. Given the austere employment challenges facing students nowadays, it is important that students themselves as well as schools and the society be fully aware of their responsibility to improve students' capability of obtaining and utilizing competitive intelligence so that they will be more capable of self-learning, self-adjustment and self-improvement. This will help students to grow on the one hand, and alleviate pressures for the society on the other hand.

REFERENCES

[1] http://www.chinadaily.com.cn/hqgj/jryw/2013-08-19/content_9893995.html
[2] Jiang Zongli. Targeted positioning, scientific teaching: On computer science in relation to the teaching implementation of core technical courses [A]. Proceedings of the University Computer Courses Report Forum [C]. Beijing: Higher Education Press. 2008: 3–6.
[3] Qiu Junping, Duan Yufeng. On Knowledge management and competitive intelligence [J]. Library and Information Service, 2000(4): 11–14.
[4] LUO Laipeng, LIU Ergen. Lottery prediction based on Apriori algorithm [J]. Statistics and Decision, 2007(3): 48–49.
[5] H. Xiong, P.N. Tan, V. Kumar. Mining Strong Affinity Association Patterns in Data Sets with Skewed Support Distribution. In Proc. of the Third IEEE International Conference on Data Mining. Melbourne, Florida, USA, November 2003.
[6] Li Xiongfei, Li Jun. Data Mining & Knowledge Discovery, 2003, 1st edition.

[7] (Canada) Jiawei Han, Micheline Kamber. Translated by Fan Ming and Meng Xiaofeng. Data Mining: Concepts and Techniques. Mechanical Industry Press, 2001–08.

[8] Liu Hongyan, Chen Jian, Chen Guoqing. Review of classification algorithms for data mining. Journal of Tsinghua University (Natural Sciences), 2002: 42.

[9] (U.S.) Mehmed Kantardzic. Translated by Shan Siqing et al. Data mining: Concepts, models, methods and algorithms. Beijing: Journal of Tsinghua University, 2003.

[10] He Yuantao, Sun Xifeng, Fang Jiqian et al. The Study of Statistical Methods Used for Item Selection. Chinese Journal of Health Statistics, 2004, 4(21): 209–211.

Engineering Technology and Applications – Shao, Shu & Tian (Eds)
© 2014 Taylor & Francis Group, London, ISBN 978-1-138-02705-3

Study on highway fire detection based on multi-information fusion technology

Songping Liu
Hunan Communication Polytechnic, Changsha, Hunan, China

ABSTRACT: Based on basic principle of data fusion technology and analysis of characteristics of highway tunnel fires, a highway tunnel fire detection algorithm based on data fusion was proposed. The structure of highway tunnel fire automatic detection fusion system was built and the intelligent algorithm of highway tunnel fire detection system was proposed. The algorithm proposed can accurately determine the tunnel fire, and be able to identify interrupt signal that may easily cause false alarm, which can improve the performance of the fire detection system. The fire can be detected early by the system, so it can relief the damage of the tunnel fire and ensure the sage operation of the highway tunnel.

Keywords: highway tunnel, fire detection, data fusion, neural network, fuzzy logic

1 INTRODUCTION

As throat section of the highway, tunnel has the characteristics of narrow pipeline space. Once fires occur in tunnel, it would greatly threat life safety of passengers because the temperature, smoke and concentrations of toxic gas would dramatically increase in tunnel. What's more, it can easily cause chain reaction that vehicles combust or explode continuously, which may fatally damage ventilation, lighting and other facilities of the tunnel, or even cause heavy casualties and significant economic losses. So, how to detect fires in tunnel fast and accurately gets a great deal of attention nowadays [1].

The term of data fusion appears in the 1970s, and developed into a specialized technology in the 1980s. It is the result that human mimic information processing capability of themselves. Data fusion was first used in military. In 1973, research institutions in American began to carry out research on the sonar signal interpretation system. Currently, multi-sensor data fusion technology is widely used in control, communication, artificial intelligence and other systems. It has become a hot study field to military, industrial and other development of high technology [2–4].

This paper does a lot of researches on means and methods of fire detection in tunnels. The multi-sensor data fusion method was applied into tunnel fire detection in the paper. An automatic tunnel fire detection system based on data fusion technology was built and implemented by information analysis, feature fusion and fuzzy decision technology. It can effectively improve the speed and accuracy of tunnel fire detection and reduce false alarm rate.

2 STRUCTURE OF MULTI-INFORMATION DATA FUSION SYSTEM

The basic principle of multi-information data fusion is to take full advantage of multiple sensors resources of different time and space. By reasonably distribute and use the sensors and their observation information, redundant or complementary information of multiple sensors in time or space is analyzed, synthesized, distributed and used, to get consistency interpretation or description. The highway tunnel automatic fire detection fusion system based on data fusion technology was

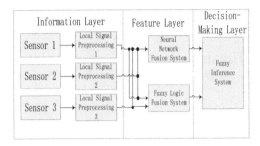

Figure 1. Working principle of data fusion system.

consisted of the information layer, feature layer and the decision-making layer, which are shown in figure 1. The acquisition and processing of the original data were mainly completed in information layer. Data fusion was completed in feature layer by using output signal of the information layer. All kinds of feature information output by feature layer were full used by decision-making layer to give out final result of the fire by adopting appropriate fusion technology and judgment rules [5–7].

3 REALIZATION OF DATA FUSION SYSTEM

The realization of highway tunnel fire detection system based on multi-information fusion technology was mainly the realization of the algorithm in information layer, feature layer and decision-making layer. Preprocessing of the selected information was completed in information layer. Feature fusion to preprocessed information was completed in feature layer by adopting neural network feature fusion machine and fuzzy logic feature fusion machine. After that, the tunnel fire was separated as flame fire, smoldering fire and non-fire. The final fire judgment result was gotten in decision-making layer by analyzing, judging and applying appropriate fuzzy logic inference.

3.1 *Realization of information layer*

3.1.1 *Parameters selection*
As the specialization of the fires in highway tunnel, the fire detection was completed by detecting changes of smoke concentration, temperature and flame signal.

3.1.2 *Signal preprocessing*
Fire detection sensors are based on analog signals. Since amplitude of output signal of sensors based on analog signals changes with the change of interference signal, the output signal of sensors cannot be used by the system directly. A series of processing, including shaping filter, A/D conversion and amplification, must be done to data selected by sensors. The output signals of sensors that after information preprocessing were sent to neural networks and fuzzy logic systems to get intelligent process of data fusion.

3.2 *Realization of feature layer*

3.2.1 *Structure of feature layer*
The parallel form of BP neural network fusion machine and fuzzy logic fusion machine was selected in feature layer to complete fusion of tunnel fire features. Three detected signals, respectively were temperature ($X1$), smoke concentration ($X2$), fire ($X3$), were sent to feature layer after preprocessed by local signal preprocessor, and finally be sent to BP neural network fusion machine. After these processes, the fire was identified as flame fire ($Y1$), smoldering fire ($Y2$) and non-fire ($Y3$). Then

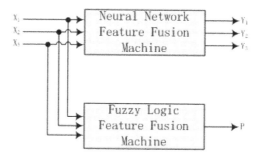

Figure 2. Structure of the fire detection fusion system in feature layer.

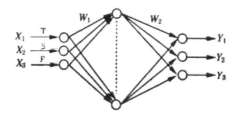

Figure 3. Structure of BP neural network feature fusion machine.

it was sent to fuzzy logic feature fusion machine and a corresponding fire probability P was gotten. The structure of feature fusion machine was shown in figure 2[8−10].

3.2.2 *Application of BP neural network*

(1) Designation of BP neural network feature fusion machine

BP neural network has three inputs, which are temperature (T) detection signal, smoke concentration (S) detection signal and flame (F) detection signal.

The output signal of BP neural network feature fusion machine was probability of flame fire (Y1), probability of smoldering fire (Y2), probability of non-fire (Y3).

In practical application, random numbers between −1 and 1 were taken as initial weights, and each random number cannot be equal.

The structure of BP neural network feature fusion machine was shown in figure 3. In the figure, the weight matrix of input layer and middle layer was Q_1, the weight matrix of middle layer and output layer was Q_2.

(2) Training of BP neural network

BP neural network was applied into the tunnel fire detection system as a feature fusion machine. First the known data was used by the network to do the learning and training. The it can be applied into practical fire detection. The weight vector was adjusted through network training.

3.2.3 *Application of fuzzy logic*

(1) Input and output of fuzzy logic fusion system

The fuzzy logic fusion system has three input signals, which are temperature (T) detection signal, smoke concentration (S) detection signal, and flame (F) detection signal. The corresponding output signal of it is probability of the fire (P).

(2) Domain and subordinate functions of fuzzy variables

The fuzzy subsets of three fuzzy variables above were {PB, PS}, which, PB means the probability of the fire was large and PS means the probability of the fire was small.

The conventional triangular subordinate function was adopted here as follow:

$$f(x) = \begin{cases} (x-a)/(b-a), a < x < b \\ (x-c)/(b-c), b < x < c \end{cases} \qquad (1)$$

(3) Rules of fuzzy language

The inference rules can be determined as follows:
if (T is PB and S is PS and F is PS)
then (the fire is PS)
else if(T is PS and S is PB and F is PS)
then(the fire is PS)
else if(T is PB and F is PB)
then(the fire is PB)
else if(S is PB and F is PB)
then(the fire is PB)

The fuzzy logic inference operation can be got by inference rules as follows:

$$f(P)=\max\{\min[T,S,F],\min[T,F],\min[S,F]\} \qquad (2)$$

where, P is output variable before defuzzification of the system, max{} is fuzzy logic "or", min [] is fuzzy logic "and", max or min is determined by maximum subordinate degree principle.

Finally, to fuzzy inference output P, the probability of the fire was output after defuzzification of the system completed by using gravity method.

3.3 *Realization of decision-making layer*

The main task of the decision making layer was to give out the final decision of the result by analyzing, judging and using appropriate fuzzy algorithm, based on output signal of the feature layer.

3.3.1 *Analyzing and judging of decision-making layer*
The analyzing and judging of decision-making layer was based on differences between output signal of neural network feature fusion machine and output of fuzzy logic feature fusion machine in feature layer, combined with experience.

The judgment rules were as follow:

(1). When $Y_1 > 0.5$ and $P > 0.7$, it was determined to be flame fire
(2). When $Y_1 > 0.5$ and $P < 0.7$, it was sent into flame fire fuzzy logic inference system to infer and judge.
(3). When $Y_2 > 0.5$ and $P > 0.7$, it was determined to be smoldering fire.
(4). When $Y_2 > 0.5$ and $P < 0.7$, it was sent into smoldering fire fuzzy logic inference system to infer and judge.
(5). When $Y_1 > 0.5$ and $P < 0.3$, it was determined to be non-fire.
(6). When $Y_3 > 0.5$ and $P > 0.3$, it was sent into non-fire fuzzy logic inference system to infer and judge.

3.3.2 *Realization of fuzzy logic inference in decision-making layer*
As the fire signal is a gradient signal and the interference noise signal only works a short time even it can cause big changes to output, taking duration of the fire signal as input variable of fuzzy inference system in decision-making layer can reduce falser alarm rate. The duration time of fire signal T is defined as follow:

$$T(n) = [T(n-1)+1]*u(P_i(x)-T_d) \qquad (3)$$

Figure 4.   Fire detection test in virtual tunnel platform.

Sample Values		Neural Network Parameters	
Input [21.801,20.600,20.001,18.601,17.500		Number of Intervals	100
Output [9.455, 11.895, 9.664, 9.652, 9.370.		Maximum Number of Cycles	8000
Neural Network Layers		Target Error	0.0001
		Learning Rate	0.02
Activation Function 1 tansig S			
Activation Function 2 purelin		Neurons in the Hidden Layer	5
		Train Neural Network	
Training Effect		Training Result	
Actual Training Times 985		Weights	Deviation
Final Error 9.96614e-005		-0.46015 -2.1153	3.1484
		1.2271 -1.6216	-1.9655
		-1.9393 -1.1736	0.091051
		-1.0456 -1.9055	-1.4696
		2.2803 0.68993	3.0628

Figure 5.   Neural Network Training Result in Matlab.

where, $u(x)$ is the step function, T_d is alarm threshold and the number of T_d is 0.5, $P_i(x)$ is fire probability P got by fuzzy logic feature fusion machine. When the output signal of fire probability P is bigger than alarm threshold T_d, the timer is started, otherwise not timing.

The output of fuzzy inference system in decision-making layer was the final fire probability P.

According to the existing fuzzy logic rules and certain other rules, the thought process of a proposition was inferred. The Mamdani method was taken to realize fuzzy logic inference. The fuzzy method of fire output probability P is the same as fuzzy rules in chapter 3.2.

Gravity method was selected as accurate method. After fire probability P was obtained, judging probability of flame fire and smoldering fire in [0, 0.5] as non-fire, otherwise as fire. Judging non-fire probability in [0, 0.5] as non-fire interrupt, otherwise as suspicious fire.

4   SIMULATION AND EXPERIMENT

Aiming at the characteristics of highway fire detection test system, the combination of tunnel test and simulation test in Matlab were adopted in fire detection test system based on data fusion technology. The virtual tunnel platform was built as shown in figure 4. The information when fire occurred was recorded in the system. Data fusion simulation system model was built in Matlab. Training the neural network in the model in off-line way and the trend of fire alarm thresholds was calculated based on existing conditions. The fire alarm decision algorithm was realized in the controller based on DSP, and the fire detection system was tested in the virtual tunnel platform.

BP neural network was adopted in data fusion system model built in Matlab. Five neurons and tansing S-type tangent function was adopted in hidden layer. Purilin linear function neural network was adopted in output layer. The training result was shown in figure 5.

Table 1. Experiment result of fire detection system in virtual tunnel.

Alarm Time (s)	Entrance	Tail
Fire detector	20	24
Smoke detector	18	26
Tem detector	24	33
System	18	24

Three detectors were installed in the entrance and tail of the virtual tunnel. The test result was shown in table 1. From the result it was found that the system can well detect the fire in the tunnel and the react speed was 18 s.

5 CONCLUSION AND PROSPECT

1) The data fusion method was taken into highway tunnel fire detection. Multiple sensor resources were full taken advantage of to obtain consistency interpretation and description of the tested object, which can overcome drawbacks of fixed threshold method taken by earlier fire detection algorithm.
2) The structure of highway tunnel fire automatic detection fusion system was built by combining the characteristics of high tunnel fire and fire alarm system. The intelligent algorithm of highway tunnel fire detection system was proposed.
3) The algorithm proposed can accurately determine the tunnel fire, and be able to identify interrupt signal that may easily cause false alarm, which can improve the performance of the fire detection system. The fire can be detected early by the system, so it can relief the damage of the tunnel fire and ensure the sage operation of the highway tunnel.

ACKNOWLEDGEMENT

This paper is funded by Research on Tunnel Traffic Fire Detection Based on Multi-information Fusion Technology (GN: 13C270).

REFERENCES

[1] Zhou, Z.B. (2010) Study on Highway Tunnel Fire Safety Guide System. Wuhan Institute University.
[2] Macii, D. Boni, A. & Tutorial14. (2008) Multisensor data fusion. *IEEE Instrumentation & Measurement Magazine*, 11(3), 24–33.
[3] Zhao, Y. & Chen, S.J. (2010) Fire early alarm system based on multi-sensor data fusion. *Modern Electronics Technique*, 33(24), 173–176.
[4] Aners, N. (2001) *Multi-sensor Technology and Its Application*. Beijing, National Defense Industry Press.
[5] Richard, Antony. (1995) *Principle of Data Fusion Automation*. London, Artech House.
[6] Li, H., Ye, L. & Wei, F. (2006) Data Fusion Method for Intelligent Measurement System. *Automation & Instrumentation*, 21(6), 76–79.
[7] Zhang, K., Lu, Y.H. & Lei, G. (2006) Application Research of Data Fusion Technology in Automatic Fire Detection System. *Computer Measurement & Control*, 14(5), 610–613.
[8] Jang, J.S.R. (1999) Adaptive Network Based Fuzzy Inference System. *IEEE Transactions on Systems, Man, and Cybernetics*, 23(3), 665–685.
[9] Gao, X.B. (2009) Highway Tunnel Fire Detection Technology. *The 6th Seminar on China Expressway Electromechanical Project*, 89–92.
[10] Du, Q.D., Xu, L.Y. & Zhao, H. (2002) D-S evidence theory applied to fault diagnosis of generator based on embedded sensors. *Proceedings of the 2nd International Conference on Information Fusion. France*, pp. 1126–1131.

Engineering Technology and Applications – Shao, Shu & Tian (Eds)
© 2014 Taylor & Francis Group, London, ISBN 978-1-138-02705-3

Adaptive beamforming technique based on neural network

Zehua Fan, Jackson Wilson & Alex Grey
Library of Huazhong University of Science & Technology, Wuhan, Hubei, China

ABSTRACT: Adaptive beamforming can achieve better SNR by varying the weights of each of the sensors used in the array. The traditional beamforming methods cannot achieve the optimal performance in beamforming because of the mismatch between the assumed array response and true array response. A radial basis function neural network algorithm has been proposed in this paper to solve this problem by turning the processing of calculating weighting of arrays to mapping processing. The simulation results indicate that the proposed method can adapt the weighting according to the direction of source signal automotive, and the SNR can be increased significantly with the DOA mismatch at 2 degrees.

Keywords: adaptive beamforming, radial basis function neural network, BP algorithm

1   INTRODUCTION

Adaptive beamforming technology through the method of inputting different weight values to sensor arrays can realize the maximization signal receiving in a particular direction and inhibit other signals with the same frequency in other directions. Therefore, adaptive beamforming technology in smart antenna can achieve the enhancement of expected source and restrain noise and interference signal. In order to achieve the best effect, weight value selection has become the key issue in the adaptive beamforming technology. Traditional adaptive beam forming method use iterative method under specified standards to calculate optimized weight, such as minimum mean square error (LMS), DMI, RLS algorithm[1−3], but these algorithms have disadvantages in the actual application that the source and the sensor does not match so as to reduce its performance[4−5]. In recent years, many scholars begin to study strong adaptive beamforming technology under the precondition of small mismatch, such as linear constrained minimum variance (LCMV) algorithm[6]. LCMV algorithm can in the desired source direction resist mismatch, but it return loses signal's interference freedom. Diagonal loading (DL) is also a kind of widely used of the strong adaptive beamforming algorithm[7], but in DL technology choice of Diagonal loading level is uncertain. This paper puts forward a new adaptive beamforming method based on radial basis (RBF) neural network algorithm. It will convert the problem of calculating adaptive antenna array weight to mapping problem, which can narrow the desired signal oriented vector error range, avoid beamforming algorithm performance problems caused by the excessive oriented vector error, and can save computing resources. Using Matlab based on RBF neural network algorithm to do simulation, the results show that the proposed method for signal oriented vector has very strong adaptability, can get nearly ideal Signal to interference plus noise ratio (SINR) in not ideal situation.

2   THE ADAPTIVE BEAMFORMING MATHEMATICAL MODEL

Consider a sensor array with M units, the observation vectors $X(t) = [x_1(t), x_2(t), \ldots, x_M(t)]$ can be represented by an M dimension vector as:

$$X(t) = A * S(t) + I(t) + N(t) \tag{1}$$

In the equation, A is oriented vector, S is received signal vector, I is interference signal, N is noise. The output of the narrowband beamforming is:

$$Y(t) = W^H X(t) \qquad (2)$$

$W = [w_1, w_2, \ldots, w_M]^T$ is the weight vector of the beamforming algorithm.

Define the SINR is:

$$SINR = \frac{W^H R_s W}{W^H R_{i+n} W} \qquad (3)$$

R_s is the expectation of the received signal vector, calculation as $R_s = E\{S(t)S^H(t)\}$, R_{i+n} is covariance matrix of the signal and noise. In order to get the best communication signal to noise ratio (SNR), need to select the optimal weight vector W to achieve the maximum SINR. Under the condition, solve the solution (2), get optimal weight vector W for:

$$W_{opt} = \frac{R_{i+n}^{-1} A}{A^H R_{i+n}^{-1} A} \qquad (4)$$

At this time the corresponding $SINR_{opt}$ is:

$$SINR_{opt} = P_s^2 A^H R_{i+n}^{-1} A \qquad (5)$$

3   BEAMFORMING BASED ON RBF NEURAL NETWORK

As transformation between source signal angle and antenna measurements is continuous, small angle change of source signal will result in a change of measurement signal, which requires the design of the neural network algorithm can get approximate continuous transformation function from fewer samples by discontinuous training. Specific to the design of the neural network algorithm, neural network had to deal with time delay as constant, and will not increase with the increase of the input. The second is the demand with the increase of input, neural network should have as fewer as layers. Under these requirements, RBF neural network has become the most effective tools in the beamforming, a three layer of RBF neural network can theoretically simulate any continuous function[8].

3.1   *RBF neural network structure*

Three layer of RBF neural network is adopted in this design, its structure as shown in figure 1. From figure 1, the input layer received pretreated data $S = (s_1, s_2, \ldots, s_M)$ from antenna, and broadcast

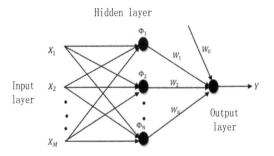

Figure 1.   RBF neural network structure.

it to each node of hidden layer. The role of the hidden layer is to transform the input space to the hidden space, usually hidden layer with high dimensions. Gaussian radial basis vector is adopted in this design, and the transformation relationship of each node in hidden layer is:

$$\phi_i = \exp(\sum_{k=1}^{N} \frac{(s_k - m_{ik})^2}{2\sigma_i^2})$$ (6)

In the equation, ϕ_i is the number i hidden layer node's output, m_i is gaussian center vector, σ_i is diffusion parameter. For each node, σ_i can be trained, we chose a fixed value in this design. Output layer will add ϕ_i together after weighting and make the result as output of the neural circuits.

3.2 *RBF neural network learning strategy*

According to the different choices of the central nervous basis of radial function, RBF neural network learning strategies can be divided into different ways. This paper chooses back propagation algorithm (BP algorithm) as a learning strategy[9−10]. BP algorithm has two processes including signal's forward transmission and signal error back propagation. Input the sample from the input layer and transfer it to output layer after processing in the hidden layer. The output layer will compare the actual output and desired output, if set conditions are not met then it backs to the process of signal error back propagation. Error back propagation is back transferring the output error in some form through the hidden layer to the input layer and allocating the error to all units of each layer, so as to acquire the error signal of each layer unit, and the error signal will be the basis of correcting the weights of each unit of. This signal weight adjustment process of the layers spreading between forward transmission and error back propagation is in cycles and is also the network learning training process. This process has been proceeding until the network output error reduced to an acceptable level, or to a predetermined number.

4 SIMULATION RESULTS

In order to verify the correctness of the design, we use MATLAB to do simulation experiment. In order to validate the expected signal adaptive enhancement and the suppression of interference signal of the beam forming algorithm based on neural network algorithm, we did simulation two different direction of the desired signal of 0 and 30 degrees, and choose the jamming signal of 60 degrees. In order to avoid signal overlap, chose space code as 0.5λ. Figure 2 and figure 3 show the

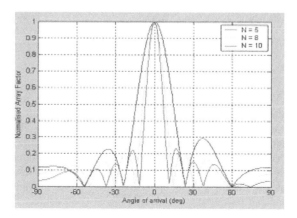

Figure 2.   Array factor at desired signal is 0 degrees, the jamming signal is 60 degree.

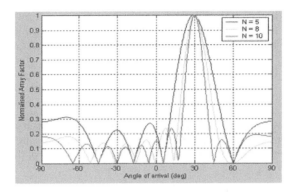

Figure 3. Array factor at desired signal is 30 degrees, the jamming signal is 60 degree.

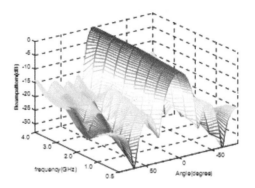

Figure 4. Beam in the 2 degree mismatch condition.

waveforms of beam forming array factor at 0 and 30 degrees of the desired signal AOA. Can be clearly seen from the graph, this method can be adaptive to maximize the expectations of different directions of the signal, and the suppress the jamming signal to 0.

At the same time, in order to validate the proposed method's adaptive processing capacity to the mismatch of input signal, we set up in our simulation DOA 2 degree of mismatch, and fixed its signal to noise ratio (SNR) of 10 db. In this case, the beam diagram as shown in figure 4. Can be seen from the graph it adjusts the direction of antenna to the signal adaptively and inhibit the interference signal of 60 degree to a great extent.

5 CONCLUSION

In the smart antenna, the traditional method of beam forming will reduce its performance because of the mismatch between the source and the sensor. This paper improved the traditional method, and used radial basis neural network algorithm to realize beam forming. The algorithm converts the problem of calculating the adaptive antenna array's weights to mapping problem, and avoids beam forming performance decline caused by the excessive oriented vector error. Using MATLAB to realize the algorithm, experimental results showed that when the direction of the desired signal changed from 0 to 30 degree, the array factor of beam forming gets its maximum value respectively at 0 and 30 degree, and can restrain the interference signal at 60 degree. In the case of DOA is 2 degree mismatch, the simulation results show that the algorithm can strengthen expected signal and restrain the interference signal.

REFERENCES

[1] Carl B. Dietrich, Warren L.Stutzman, Byung-Ki Kim, and Kai Dietze, "Smart Antennas in wireless communications: Base-station diversity and handset beamforming," IEEE Antennas and Propagation Magazine, 42(5): 512–519, 2000.

[2] A.H. EI Zooghby, C.G. Christodoulou, and M. Georgiopoulos, "Performance of radia basis function networks for direction of arrival estimation with antenna arrays," IEEE Antennas and Propagation Magazine, 45(3): 1611–1617, 1997.

[3] P.R. Chang, W.H. Yang, and K.K. Chan, "A neural network approach to MVDR beamforming problem", IEEE Antennas and Propagation Magazine, 40(8): 313–322, 1992.

[4] S. Bellofiore, C.A. Balanis, J. Foufz, and A.S. Spanias, "Smart-Antenna systems for mobile communication networks part I: Overview and antenna design," IEEE Antennas and Propagation Magazine, 44(3): 107–114, 2002.

[5] M.T. Islam, Z.A.A. Rashid, "MI-NLMS adaptive beamforming algorithm for smart antenna system applications," Journal of Zhejiang University Science A, Vol. 10, pp. 1709–1716, 2006.

[6] Li J., Stoica P. and Miller T.W., "On robust capon beamforming and diagonal loading," IEEE Trans. Signal Processing, vol. 51, pp. 1702–1715, 2003.

[7] Salvatore Bellofiore, Consfan fine A. Balanis, Jeffrey Foufz and Andreas S. Spanias, "Smart Antenna systems for mobile communication network," IEEE Antenna's and Propagation Maagazine, vol. 44(3): 145–151, 2002.

[8] H. Cox, R.M. Zeskind, and M.H. Owen, "Robust adaptive beamforming," IEEE Trans. Acoust., Speech, Signal Processing, vol. 35: 1365–1376, 1987.

Engineering Technology and Applications – Shao, Shu & Tian (Eds)
© 2014 Taylor & Francis Group, London, ISBN 978-1-138-02705-3

Design of real-time digital chaos system based on FPGA

Shiping Liu & Edmund Wayne
Air Force Early Warning Academy, Wuhan, Hubei, China

ABSTRACT: There has been considerable interest in digital chaotic communication over the past several years. In fact, researches in the application of chaos in communication have been greatly motivated. Therefore, the chaos theory has exceeded the stage of the laboratory simulations. The realization of chaos is based on Matlab/Simulink for algorithm design because of the complex algorithm and huge computation. But this method is not optimal since it uses the third party tools because as we approach the machine language as it is optimal and better. The results show that our method can generate chaos sequence correctly with a much convenient realization.

Keywords: chaos system, Runge-Kutta algorithm, FPGA, nonlinear circuits and systems

1   INTRODUCTION

Chaos theory is about nonlinear systems which show bifurcation under certain parameter conditions, cycle and non-cycle movement intertwined, leading to a non-cycle and orderly movement. In recent years, the chaotic system because of its advantages such as strong randomness and high security in communications, physical, and other fields has caused widely public concern. It leads to the study of chaotic systems from the algorithm simulation level to circuit realization. The circuit realization of chaotic systems can be divided into analog method and digital method[1]. Analog circuit performance depends on the internal parameters, and the optimization of these parameters often needs a lot of simulation and debugging, and analog components are very sensitive to environmental factors, such as temperature, aging components problems which will cause simulators have larger errors. And chaotic system is also sensitive to the initial state, the initial condition of tiny changes can also lead to the final state of the huge difference[2,3]. Digital method uses embedded systems (such as FPGA) to realize the chaotic system, which can solve the problems of analog circuits. Because the realization of chaotic systems need complexity algorithm and a large amount of computing resources, especially the use of multipliers, traditional digital method often uses FPGA based on DSP Builder to simplify the design process[4−6]. Because the DSP Builder needs a third party tool Matlab/Simulink for algorithm development, obviously this way cannot get optimal results, and the chaotic signal is not convenient for other chaotic system to call.

Take Rössler chaotic system as an example, this paper puts forward a complete real-time digital chaotic system method. This method uses hardware description language in FPGA with four order Runge-Kutta algorithm (R-K algorithm) algorithm to solve Rössler chaotic system. This method does not need to use DSP Builder, so as to solve the problems of using third-party tools and the optimization of FPGA embedded, and this method is easy to switch to any other kind of chaotic systems. Through the simulation results and the experimental results, it shows that this chaotic system can get accurate chaotic sequence. It is also convenient to be realized and occupies less resource. The study provides a new solution for the circuit realization of chaotic or super chaotic system.

2 RÖSSLER CHAOTIC SYSTEM MATHEMATICAL MODEL

O.E. Rössler proposed a chaotic system in 1979 and caused wide public concern. This chaotic system compared to the Lorenz and other commonly used chaotic system has only one nonlinear parameter. The mathematical model of Rössler chaotic system is as follows[7]

$$\begin{cases} \dfrac{dx}{dt} = -(y+x) \\[2mm] \dfrac{dy}{dt} = x + ay \\[2mm] \dfrac{dz}{dt} = b + z(x-c) \end{cases} \tag{1}$$

Runge-Kutta (R-K) algorithm[8] is a numerical method for solving differential equation with high precision, this paper uses 4 order R-K algorithm to solve Rössler chaotic system, and its numerical solution of general form is:

$$y_{n+1} = y_n + h \sum_{i=1}^{4} \lambda_i k_i \tag{2}$$

Solution equations are:

$$\begin{cases} y_{n+1} = y_n + h/6(k_1 + 2k_2 + 2k_3 + k_4) \\ k_1 = f(x_n, y_n) \\ k_2 = f(x_n + h/2, y_n + h/2 * k_1) \\ k_2 = f(x_n + h/2, y_n + h/2 * k_2) \\ k_4 = f(x_n + h, y_n + h * k_3) \end{cases} \tag{3}$$

This paper takes original classic Rössler chaotic system[9] as an example, the parameters are $a = 0.41$, $b = 1.15$, $c = 4.16$. Choose step 0.001 and use Matlab to solve equation (1), then get Rössler chaotic signals as shown in figure 1.

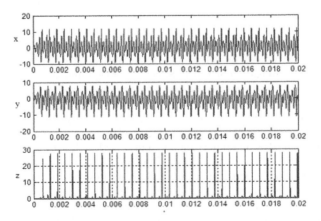

Figure 1.   Rössler chaotic system's corresponding chaotic signals.

3 THE REALIZATION OF DIGITAL CHAOS SYSTEM

Two main problems have to be solved in using hardware description language in FPGA to realize real-time completely chaotic system. One problem is how to decompose the continuous chaotic system so that the numerical method can be used in this system. The other one is how to use floating-point arithmetic in the FPGA. In this paper, it uses 4 order R-K algorithm to solve continuous chaotic system, and use the module reuse method to realize R-K algorithm, which can save the area of the FPGA resources and can reduce power consumption. Using the IEEE – 754 single precision standard floating-point format[10] in FPGA to do the decimal computing. Figure 2 shows the complete real-time digital chaos system block diagram of this paper. Four order R-K algorithm module is its core module. It is responsible for solving the numerical results of chaotic systems, and under the control of the control module it outputs the results with 32 array to analog-digital converter (DAC). The output of the DAC presents analog chaotic signal on the oscilloscope.

3.1 R-K algorithm module

Based on the four steps to achieve 4 order R-K algorithm, the system uses state machine to control the conversion of each step, the block diagram as shown in figure 3. Iterative equation calculation module and coefficient k generation module are the core of the algorithm model. These two modules are designed to reuse in 4 iterations, so that can save the area of the FPGA resources and power consumption. Its disadvantage is that it slows down the speed of data processing. Start the first iteration of R-K algorithm when the initial value $\{x_0, y_0, z_0\}$ is set to the register. Initial values are the input to two input terminal of the adders, the initial value generated from coefficient K module is 0. Iterative equation calculation module does the calculation in equation (3).

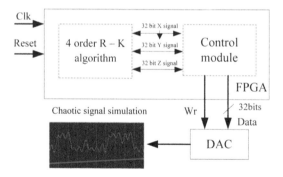

Figure 2.   Real-time digital chaotic system block diagram.

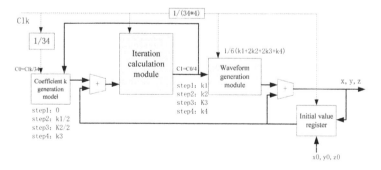

Figure 3.   R-K algorithm realization block diagram.

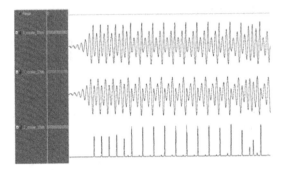

Figure 4.   Chaotic signal ModelSim simulation results.

The calculation results feedback to coefficient k generation module and generate coefficient k values. Waveform generation module calculates the formula $1/6(k_1 + 2k_2 + 2k_3 + k_4)$ after each iteration calculation, and sends the result to the second adder. Plus the result in the second adder and the initial value to get chaotic signal, chaotic signal is again sent into the initial value register for the next iteration until 4 iterations are finished and get the final chaotic signal $\{x, y, z\}$.

It needs at least 34 clock cycles for iterative equation calculation module calculates the results, so the coefficient k generation module uses the 34 frequency division of the system clock Clk. In addition, 4 order R-K algorithm needs 4 calls of coefficient k output module and iterative equation calculation module to generate a chaotic signal, so the clock C_1 generated from the waveform generation module is 4 frequency division of the clock C_0 generated from coefficient k generation module. Therefore, the implementation frequency of the whole system is: $f = Clk/(34 * 4)$. The design uses phase-locked loop (PLL) module provided in Altera Quartus IP library to synchronize these three clocks.

3.2   System design and simulation

In order to verify the correctness of the designed system, we used the Mentor ModelSim software to do time sequence simulation. Its development language is Verilog HDL. Figure 4 is the time sequence simulation diagram. It can be seen from Figure 4 the waveform of output chaotic signal $\{x, y, z\}$ is almost the same as the result of Matlab Simulation chaotic signals (Figure 1). Simulation results show the correctness of the design.

4   FULL REAL-TIME DIGITAL CHAOTIC SYSTEM RESULTS

We use Quartus to compile designed system, and choose Altera Cyclone II EP2C20F484C7. Compiling results show that the whole design uses 23% of the total logic units, 38% of the total number of multipliers and 41% of the total storage space. The results show that the resource occupation of this design can meet most practical situations. Use PLL changes system clock frequency to 200 MHz, and download the whole design to EP2C20F484C7 development board. The data grasped from FPGA with the tool of Signal Tap II tool in Quartus software is shown Figure 5. This figure displays all of the floating-point numbers as the IEEE754 standard, so as 16Q16 representation method. Save the data analysis can also get the waveform is as shown in figure 1. In this time, clock C1is about 1.5 MHz, that means the system processing speed is about 1.5 MHz.

The chaotic system's output signals pass through DAC, and the waveforms can be obtained on the oscilloscope shown in Figure 6. We can be seen from the diagram the results of real-time digital chaotic systems designed in this paper is almost the same as the simulation results. It verifies the correctness of the design.

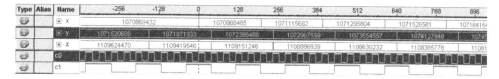

Figure 5. Signal Tap II grasped data's waveform.

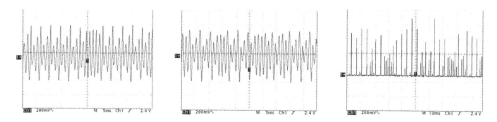

Figure 6. Output chaotic signal (the left is x signal, the middle is y signal, the right is z signal).

5 CONCLUSION

In order to solve the digital method's problems when it realizes chaotic system since it uses the third party tools and it has problems of the optimization of FPGA embedding, this paper puts forward a full real-time digital realization method of the chaotic systems. This method uses complete hardware description language to achieve 4 order R-K algorithm to solve chaotic system, thus can avoid using third party software and solve the system parameter sensitive issues of traditional analog circuit. This method can easily portable to other kinds of chaotic systems, such as the Lorenz chaotic system, Chen chaotic system, etc. Output is obtained by simulation experiment and field experiment and the output chaotic sequence can precisely restore its mathematic model sequence. The processing speed of the whole system is 1.5 MHz when the system clock is 200 MHz, and the whole design uses only about 4000 LE, 12 multipliers and 20 M4K storage blocks. The study provides a new solution of chaos or super system circuit.

REFERENCES

[1] Rossler, O.E. (1979) An equation for hyperchaos. *Phys Lett A*, 71(23), 155–157.
[2] C. Barbara. & C.Silvano. (2002) Hyperchaotic behaviour of two bi-directionally coupled Chua's circuits. *International Journal of Circuit Theory and Application*, 12(3), 625–629.
[3] Asseeri, M.A., Sobhi, M.I. & Lee, P. (2004) Lorenz chaotic model using field programmable gate array. *Midwest Symposium on Circuit and Systems*, 34(23), 64–69.
[4] Kvarda, P., (2002) Investigating the Rossler attractor using Lorenz plot and Lyapunov exponents, *Radioengineering*, 11(3), 145–148.
[5] M. Azzaz., C. Tanougast., S. Sadoudi. & A. Dandache. (2009) Real time FPGA implementation of Lorenz's chaotic generator for ciphering telecommunications. *Circuits and Systems and TAISA Conference*, 5(11), 1–4.
[6] T. Ueta. & G. Chen. (2000) Bifuracation analysis of chen's attractor. *Int. J. Bifurcation and Chaos*, 10(4), 1917–1931.
[7] N.E. Lorenz. (1963) Deterministic non-periodic flow. *Journal of Atmospheric Science*, 20(2), 130–141.
[8] K. Sivaranjani., J. Venkatesh. & P.A. Anakiraman. (2007) Realization of a digital differential analyzer using CPLDs. *International Journal of Modelling and Simulation*, 27(3), 223–228.
[9] B.R. Andrievskii. & A.L. Fradkov. (2003) Control of Chaos: methods and applications. *Automation and Remote Control*, 64(5), 673–713.
[10] O.E. Rossler. (1976) An equation for continuous Chaos. *Physics Letter*, 57(5), 397–398.

Engineering Technology and Applications – Shao, Shu & Tian (Eds)
© *2014 Taylor & Francis Group, London, ISBN 978-1-138-02705-3*

Impact analysis on the voltage waveforms of the low voltage grid with respect to electronic equipment

Quanmin Li, Ling Cui & Hongli Lv

School of Information & Electrical Engineering, Shandong Jianzhu University, Jinan, Shandong, China
Shandong Provincial Key Laboratory of Intelligent Buildings Technology, Jinan, Shandong, China

ABSTRACT: In real word application, electronic equipments have potential influence on the power grid voltage waveform, in order to help the engineer find the effective solutions to reduce the influence, this paper takes the fluorescent lamps in a teaching building as an example, analyzes the circuit structure of single-phase fluorescent lamps and power supply circuit, and establishes the circuit model. Then, according to the features of fluorescent lamp, the paper studies the voltage waveform of the end of power supply circuit in stages, and gives the mathematical analysis and mathematical pattern. The calculation results are verified by Matlab modeling simulation and real world experiment. The voltage waveform of the fluorescent lamp at the end of the power supply shows that the electronic equipments use the energy of power grid voltage peak time too much, which makes the grid voltage waveform distortion.

Keywords: fluorescent lamp; power grid voltage; electronic equipment; differential equation

1 INTRODUCTION

With the constant progress of technology, and the continuous improvement of living standards, in all kinds of buildings, a growing number of computer equipments, electronic ballasts, electronic energy-saving lamps and other electronic equipments and components are widely used. These electronic equipments will inevitably produce a large amount of harmonic current, if not timely treatment, which can make the voltage waveform of the low voltage power supply system distortion, and cause great influence on the power quality, and the safe and economic operation of power system [1, 2].

At present, there are two basic ideas to solve the harmonic pollution problems with respect to electronic equipment and other harmonic sources [3]: one kind is compensate harmonic by installing filters, active filter can compensate the harmonic current generated by the load accuracy, but its price is higher, and the compensation capacity is small [4]. Passive filter cannot be controlled, and the adaptability of the system impedance and frequency variation is poor. The other idea is to modify the electronic device itself, but the cost is high and the efficiency is low [5]. So it has realistic significance to study the voltage waveform at the end of the power electronic equipments, and find economic and practical scheme to reduce the influence of the end of power grid voltage waveform from electronic equipment.

As the current total harmonic distortion of fluorescent lamp can exceed 100% [6], fluorescent lamps become a significant harmonic source of power grid. To study the power grid voltage waveform of fluorescent lamps is a great significance to govern harmonics effectively and operate maintenance of the power grid safely. Many domestic and foreign scholars study the causes of harmonic form fluorescent lamp and the harmonic characters, literatures [1, 8] analyzed the cause of harmonic from fluorescent lamp and its current harmonic character, literature [7] analyzed that the influence of nonlinear load in power grid voltage was relate to the length of the grid lines theoretically, but it didn't have actual experiment validation, literature [9] collected the harmonic

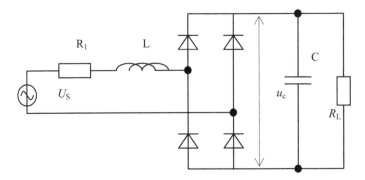

Figure 1.   Equivalent circuit for fluorescent lamp and line impedance.

current of different brand energy-saving lamps by oscilloscope, and developed a simulation model which could study the content of current harmonic, but these studies didn't analyze the influence of the end of the power supply voltage waveform based on field experimental data, through the calculation. In order to further study the influence, and put forward more targeted schemes to reduce harmonic, taking the fluorescent lamps in a school building as an example, the circuit model of fluorescent lamp and line impedance was established, the end of the power supply voltage waveform input by electronic equipment was studied by mathematical method in stages, and the voltage waveform diagram of mathematic analysis was given, finally, the results of theoretical analysis was verified by Matlab modeling simulation and actual voltage waveform from oscilloscope, which illustrated that the electronic devices lead to the power grid voltage waveform distortion.

2   CIRCUIT MODEL FOR FLUORESCENT LAMP AND LINE IMPEDANCE

Inductance-capacitance filter type and single phase bridge rectifier circuit is commonly used by harmonic analysis of electronic devices (fluorescent lamp, desktop computers, laptops etc.) [11, 12, 13], the difference is only the values of filter capacitors C and the equivalent load R, fluorescent lamp is one of the typical examples. Taking fluorescent lamps in a campus building for example, firstly, we assume that the capacity of the distribution station in the school is greater than the electricity consumption of teaching buildings, so distribution station is equal to the constant voltage source which resistance is zero. The cable from distribution to the building transformer room is YJV − 3 + 1 * 75 * 95, and its length is 200 m. The resistance and impedance of the cable per unit length are 0.217 Ω/km and 0.089 km/km [14]. Total (length of cable* 2) resistor R1 is about 0.087 Ω, inductance L is about 0.113 mH.

The main electricity load in the teaching building is 36 W T8 fluorescent lamp at ordinary time, there are about 2100 lights, which are distributed equally to the three-phase alternating current (ac), every phase ac is 700 lights. The power of every light bulb is 36 W, electronic ballast is about 4 W, the total power consumption of each fluorescent lamp is 40 W. The filter capacitor in electronic ballast is 22 uF, when the supply voltage is 220 V, the dc voltage measured of capacitance is 290 V. Thus the dc equivalent resistance of every fluorescent lamp (including ballast) is 2102 Ω. The equivalent resistance of every phase alternating current (ac) is $R_L = 2102 \div 700 = 3.00\,\Omega$. The equivalent capacitance is $C = 22\,\mu F \times 700 = 15400\,\mu F$. The equivalent circuit for every phase fluorescent lamp in this building is shown in figure 1.

In the discontinuous conduction mode, the circuit has charge and discharge two working conditions. When the voltage uc of the capacitance C is less than the end of the power supply voltage uc, rectifier diodes conduct, capacitor charges, there are ac current. When the voltage uc of the capacitance C is greater than the end of the power supply voltage uc, rectifier diodes are cut-off, capacitor is discharge, there are not ac current. Therefore, ac current is discontinuous pulse waveform, which contains rich harmonic, and make the low voltage power grid voltage waveform distortion.

3 MATHEMATICAL ANALYSIS FOR THE IMPACT OF POWER GRID VOLTAGE WAVEFORM FROM FLUORESCENT LAMP AND LINE IMPEDANCE

3.1 *Mathematical analysis for capacitor is discharge freely by RL*

There are two stages in this circuit state analysis: the freedom discharge stage of capacitor, which is the period from b to d as shown in figure 2; Capacitor charging stages, which is the time from a to b as shown in figure 2.

In the period when capacitor is discharge freely from b to d, point b is the alternating current peak, point c is the maximum voltage of capacitor could charge, due to the effect of inductive energy storage, there's difference at points b and c, but the difference is not big, we can ignore the difference between them, and think point c is point b. Because the speed of capacitor discharge is slow from point d to point b, capacitor voltage is higher than the ac voltage, rectifier diodes are cutoff, capacitor is discharge until point d through resistance RL. D point is chosen as the origin of coordinates, according to the equivalent circuit during charging time as shown in figure 3, the discharge rule of capacitor from point b to point d is follow the formula:

$$u_C = U_m e^{-\frac{t+0.005}{R_L C}} = U_m e^{-\frac{\omega t + 0.5\pi}{\omega R_L C}} \tag{1}$$

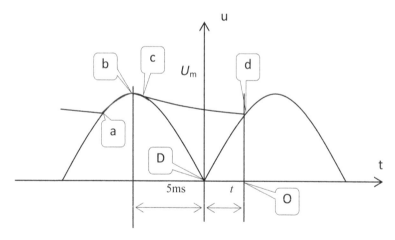

Figure 2.   Charge and discharge process.

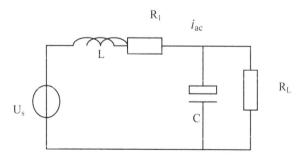

Figure 3.   Equivalent charging circuit.

227

where U_m is the peak of ac, $U_m = \sqrt{2} \times 220\,V = 311.1\,V$, ω is the angular frequency of power frequency alternating current. When uc lands to point d, the voltage of capacitor $u_c = U_0 = U_m \sin \omega t$, and U_0 is the initial value for charging voltage of capacitor the next time. They are as follows:

$$u_c = U_m e^{-\frac{t+0.005}{R_L C}} = U_m e^{-\frac{\omega t + 0.5\pi}{\omega R_L C}} = U_m \sin \omega t \tag{2}$$

From the above equation, $t = 3.16\,ms$, which is equivalent to 57° electrical point. Therefore, the voltage of point d is $U_0 = U_m \sin \omega t = 261\,V$.

3.2 Mathematical analysis of capacitor charging

According to the equivalent circuit, when the capacitor is discharge to point d, the capacitor voltage is lower than the ac voltage, rectifier diodes conduct, on the one hand, power supply charge for capacitor, on the other hand, provide power for resistance RL. Then, we ignore the diode resistance, the equivalent circuit converts from figure 1 to figure 3. According to the figure 2, O point is chosen as the new origin of coordinates, the voltage balance equation of equivalent circuit diagram 3 is as follows:

$$u_c + L\frac{di_{ac}}{dt} + R_1 i_{ac} = U_s = U_m \sin(\omega t + 57°) \tag{3}$$

And the current instantaneous value on the ac mains is shown in the following formula:

$$i_{ac} = C\frac{du_c}{dt} + \frac{u_c}{R_L} \tag{4}$$

So

$$\frac{di_{ac}}{dt} = C\frac{d^2 u_c}{dt^2} + \frac{1}{R_L}\frac{du_c}{dt} \tag{5}$$

When formula (4), (5) are substituted into formula (6), we can get

$$LC\frac{d^2 u_c}{dt^2} + \left(\frac{L}{R_L} + CR_1\right)\frac{du_c}{dt} + \left(1 + \frac{R_1}{R_L}\right)u_c$$
$$= U_m \sin(\omega t + 57°) \tag{6}$$

The above formula is second-order non-homogeneous linear differential equation.

According to the solution of second-order differential equation, the solution uc of formula (6) consists of two parts:

$$u_c = u_c' + u_c'' \tag{7}$$

where u_c' is the solution of zero-input response, u_c'' is the solution of zero-state response.

3.2.1 Solution of the second-order circuit zero-input response

From second-order inhomogeneous linear differential equation (6), we can get homogeneous equation shown as follows:

$$\frac{d^2 u_c}{dt^2} + \left(\frac{1}{CR_L} + \frac{R_1}{L}\right)\frac{du_c}{dt} + \frac{R_L + R_1}{LCR_L} u_c = 0 \tag{8}$$

228

The characteristic equation of formula (8) can be expressed as

$$p^2 + \left(\frac{1}{CR_L} + \frac{R_1}{L}\right)p + \frac{R_L + R_1}{LCR_L} = 0 \qquad (9)$$

We can solve

$$p = \frac{-\left(\frac{1}{CR_L} + \frac{R_1}{L}\right) \pm \sqrt{\left(\frac{1}{CR_L} + \frac{R_1}{L}\right)^2 - 4\frac{R_L + R_1}{LCR_L}}}{2}$$

$$= -\frac{1}{2}\left(\frac{1}{CR_L} + \frac{R_1}{L}\right) \pm \sqrt{\left(\frac{1}{2CR_L} - \frac{R_1}{2L}\right)^2 - \frac{1}{LC}}$$

Through calculating, we can know $\left(\dfrac{1}{2CR_L} - \dfrac{R_1}{2L}\right)^2 - \dfrac{1}{LC} < 0$, so

$$p = -\frac{1}{2}\left(\frac{1}{CR_L} + \frac{R_1}{L}\right) \pm j\sqrt{\frac{1}{LC} - \left(\frac{1}{2CR_L} - \frac{R_1}{2L}\right)^2}$$

$$p_1 = -\frac{1}{2}\left(\frac{1}{CR_L} + \frac{R_1}{L}\right) - j\sqrt{\frac{1}{LC} - \left(\frac{1}{CR_L} + \frac{R_1}{L}\right)^2}$$

$$p_2 = -\frac{1}{2}\left(\frac{1}{CR_L} + \frac{R_1}{L}\right) + j\sqrt{\frac{1}{LC} - \left(\frac{1}{CR_L} + \frac{R_1}{L}\right)^2}$$

We command $\delta = \dfrac{1}{2}\left(\dfrac{1}{CR_L} + \dfrac{R_1}{L}\right)$, $\omega_z = \sqrt{\dfrac{1}{LC} - \left(\dfrac{1}{CR_L} + \dfrac{R_1}{L}\right)^2}$, then,

$$P_1 = -\delta - j\omega_z = -393.83 - j659.22$$

$$P_2 = -\delta + j\omega_z = -393.83 + j659.22$$

The common form of second-order homogeneous differential equations is as follows:

$$u'_c = e^{-\delta t}\left(A_1 \cos\omega_z t + A_2 \sin\omega_z t\right) \qquad (10)$$

According to initial conditions of the circuit, we can get $\begin{cases} u'_c(0) = U_0 \\ \frac{du'_c}{dt}\big|_{t=0} = 0 \end{cases}$.

Through calculating, we can get $\begin{cases} A_1 = 260.93 \\ A_2 = 155.93 \end{cases}$

When A_1, A_2 are substituted into formula (10), we can get zero-input response of second-order circuit shown as follows

$$u'_c = e^{-393.83t}\left(260.93 \cos 659.22t + 155.93 \sin 659.22t\right) \qquad (11)$$

Namely

$$u'_c = 303.90e^{-393.83t} \sin\left(659.22t + 59.15°\right) \qquad (12)$$

3.2.2 Solution of the second-order circuit zero-state response

According to the solving principle of the second-order differential equation, the solution for zero state of second-order circuit has two parts, they are compulsory component of zero state u_{c1} and free component of zero state u_{c2}, respectively. u_{c1} is obtained by the phasor method:

$$\dot{U}_{c1} = \dot{U}_s \frac{Z_c /\!/ R_L}{Z_L + R_1 + Z_c /\!/ R_L} \tag{13}$$

Where u_{c1} is the phasor of u_{c1}, $Z_C = \frac{1}{j\omega C}$, $Z_L = j\omega L$, $\dot{U}_s = U_m\angle 57°$, the solution of formula (13) is

$$\dot{U}_{c1} = \dot{U}_s (0.93 - j0.47) = 1.042\dot{U}_s\angle -26.8° = 324.20\angle 30.2°$$

$$u_{c1} = 324.20\sin(314.16t + 30.2°)$$

The form of zero-state free component is as follows: $u_{c2} = e^{-\delta t}(B_1 \cos \omega_z t + B_2 \sin \omega_z t)$
Because $u''_c = u_{c1} + u_{c2}$

So
$$\begin{aligned} u''_c = &324.2\sin(314.16t + 30.2°) + \\ &e^{-393.83t}(B_1 \cos 659.22t + B_2 \sin 659.22t) \end{aligned} \tag{14}$$

In the zero-state response:
$$\begin{cases} u''_c(0) = 0 \\ \dfrac{du''_c}{dt}\Big|_{t=0} = 0 \end{cases} \tag{15}$$

When formula (14) are substituted into formula (15), we can get $\begin{cases} B_1 = 260.93 \\ B_2 = 155.93 \end{cases}$

So the zero-state response of the second-order circuit is

$$\begin{aligned} u''_c = &324.2\sin(314.16t + 30.2°) - \\ &e^{-393.83t}(163.08\cos 659.22t + 230.96\sin 659.22t) \end{aligned} \tag{16}$$

Namely

$$\begin{aligned} u''_c = &324.2\sin(314.16t + 30.2°) - \\ &282.75e^{-393.83t}\sin(659.22t + 35.23°) \end{aligned} \tag{17}$$

Finally, we can get the complete response of circuit model is

$$\begin{aligned} u_c = &u'_c + u''_c \\ = &324.2\sin(314.16t + 30.2°) + \\ &123.22e^{-393.83t}\sin(659.22t + 127.53°) \end{aligned} \tag{18}$$

3.3 Mathematical expressions for the end of power supply voltage u(t) waveform

According to the above analysis, the mathematical expressions of the ac voltage waveform can be obtained, the formulas (19) are mathematical expressions of a cycle voltage waveform. There are four periods in every cycle, which consist of two charging stages, they are respectively $0 \leq t \leq 3.60$ ms, $10.00 \leq t \leq 13.60$ ms; And two discharging stages, they are respectively

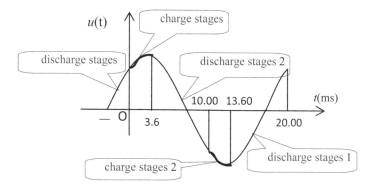

Figure 4. Voltage waveform for the end of the cable.

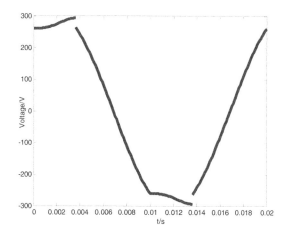

Figure 5. Mathematical graph for power supply voltage.

$3.60 \leq t \leq 10.00$ ms, $13.60 \leq t \leq 20.00$ ms. A cycle voltage waveform for the end of the cable is shown as figure 4.

$$u(t) = \begin{cases} 324.2\sin(314.16t + 30.2°) + 123.22e^{-393.83t}\sin(659.22t + 127.53°), & 0 \leq t \leq 3.60, \\ 311.1\sin(314.16t + 57°), & 3.60 \leq t \leq 10.00, \\ 324.2\sin(314.16t + 30.2°) + 123.22e^{-393.83t}\sin(659.22t + 127.53°), & 10.00 \leq t \leq 13.60, \\ 311.1\sin(314.16t + 57°), & 13.60 \leq t \leq 20.00. \end{cases} \quad (19)$$

4 MATLAB SIMULATION EXPERIMENT

According to the mathematical expression of the power supply voltage, the mathematical graph is made by plot function in Matlab, as shown in figure 5.

The system model is built by Simulink of Matlab simulation environment, which is for single-phase 700 fluorescent lights and power supply line impedance, and we get the power supply terminal voltage waveform of the analog circuit, as shown in figure 6.

Compare figure 5 and figure 6, in the charging stage, both figures appear a small platform firstly, and then, the voltage increase slowly, due to the charge and discharge characteristics of capacitor, when the voltage reaches maximum, supply terminal voltage jump to charging stage quickly, this stage is part of the standard sine wave. Based on the above analysis, the mathematical

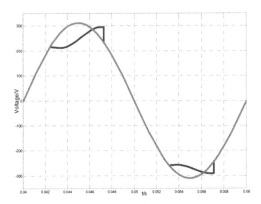

Figure 6.　Simulation voltage waveform and standard sine wave.

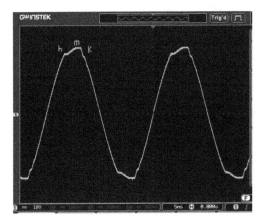

Figure 7.　The actual voltage waveform for fluorescent lamp.

analysis figure for voltage waveform of fluorescent lamp and the voltage waveform from modeling simulation are consistent, modeling and simulation is verified the validity of mathematical analysis.

5　EXPERIMENTAL VERIFICATION

In order to verify the correctness of the mathematical analysis and modeling simulation, grid is simulated by standard 220 V sine wave power supply, a single fluorescent lamp is supplied by RV-2 * 1.0 wire length of 200 m which analog to power lines, the voltage waveform for fluorescent lamp in the terminal power supply is observed by oscilloscope, as shown in figure 7. Points h, m, k in figure 7 are similar to points a, b, c in the waveform from Matlab simulation, which verify the correctness of the mathematical analysis and modeling simulation.

According to mathematical analysis, modeling simulation and practical experiment, there is a certain difference between voltage waveform of terminal power supply and the standard sine wave, the cause analysis is as follows: when the power supply voltage is higher than the capacitor voltage, namely point h in the figure 7, rectifier diodes conduct, but because the role of the inductor, current can't mutate, so there is a small platform between point h and m firstly, then, the current increases slow, and recharge capacitor. When the power supply voltage is lower than the capacitor voltage, due to the action of inductance self-induced electromotive force, the sum of the power supply voltage and inductance self-induced electromotive force is still greater than the capacitor voltage,

232

power supply will continue to recharge capacitance. When the sum of the power supply voltage and inductance self-induced electromotive force is equal to capacitor voltage, namely the point m in figure 7, rectifier diodes are cutoff. At this time, the voltage of the fluorescent lamp terminal power supply voltage transforms from the charging stage to the discharging stage, there is a jump between point m and k in the voltage waveform.

The maximum for actual voltage of the terminal power supply is smaller than the peak of standard sine wave, and the position of the maximum offsets, which is only due to the current for electronic equipment exists when it is near the peak voltage, the line impedance and load use the energy near the peak of sine wave power supply excessively, which lead to energy losses excessively near the peak of the power grid voltage wave, and contrarily the voltage near the peak place is not high.

6 CONCLUSION

Taking fluorescent lamps in a school building for example, circuit structure of fluorescent lamps and line impedance is analyzed, circuit model is established, voltage waveform of the fluorescent lamps at the end power supply is studied in stages, and the mathematical analysis and mathematic analytic figure are given, and the terminal power supply voltage waveform is simulated and modeled by Matlab. The actual voltage waveform collected by oscilloscope is consistent with the mathematic analytic figure and simulation voltage waveform, which verify the correctness of the theoretical analysis. Due to the structure of electronic devices is similar to that of fluorescent lamps, the voltage waveforms of electronic devices is similar to that of fluorescent lamps. The input voltage waveform of electronic devices at the end of power supply is proved that electronic equipments lead to grid voltage waveform distortion, which is due to line impedance and load use energy near the peak of sine wave excessively. Since the power of the fluorescent lamp is small relatively, the fluorescent lamps could use the energy either side of the peak of sine wave energy, and avoid using energy near the peak of sine wave, which can improve distribution of the energy in the power grid, and reduce the influence of grid voltage waveform distortion from fluorescent lamp.

ACKNOWLEDGEMENT

The national natural science fund project, project grant No.: 61074149, Project name: Research for Energy-saving Nash Equilibrium Control Strategy of Building Electrical Equipment based on Internet. The national natural science fund project, project grant No.: 61374187, Project name: Whole Dynamic Optimization and Scheduling for Intelligent Building Environment and Distributed Energy System.

REFERENCES

[1] Tu Qiang. Harmonic Source and Treatment Power Distribution System in Commercial Duilding [J]. Modern architecture electrc, 2012 (3): 1–4.
[2] Zhang Wenpeng. Introduction of harmonic source and treatment in building [J]. Building science, 2012, (27): 64.
[3] Song Daizhi. Research on harmonic suppression technology for Electronic devices [D]. Xi 'an university of electronic science and technology, 2006.
[4] Hua jun. Introduction of the harmonic problems of electronic devices inside building and measures of controlling the harmonics [J]. Construction and design for project, 2013, (8): 98–101.
[5] Li Jianfeng. Detecting Method and Suppression of Power System Harmonic [D]. Guangxi University, 2008.
[6] Jing Yong, Liang Chen. A Frequency-Domain Harmonic Model for Compact Fluorescent Lamps [J]. IEEE Transactions On Power Delivery, 2010, 25(2): 1182–1189.

[7] Hu Longjian Cao Man, Zhao Jincheng. Nonlinear load and its impact on quality of Electricity [J]. Movable Power Station & Vehicle, 2011 (4): 13–16.

[8] Yangguang. Discuss of harmonic problems for compact fluorescent lamp and lighting distribution [J]. Light source and lighting, 2008, (2): 22–24

[9] A.S. Koch, J.M.A. Myrzik. Harmonics and Resonances in the Low Voltage Grid Caused by Compact Fluorescent Lamps [C]. 2010 14th International Conference on Harmonics and Quality of Power, Bergamo, 2010: 1–6.

[10] Ding Tongren. Ordinary differential equations [M]. Higher education press, 2010.

[11] Yong Jing, Chen Liang, Chen Shuangyan. Frequency-Domain Harmonic Model and Atetenuation Characteristics of Desktop PC Loads [J]. Proceedings of the CSEE, 2010, 30(25): 122–129.

[12] Investigation on the Harmonic Attenuation Effect of Single-phase Nonlinear Loads in Low Voltage Distribution System [J]. Proceedings of the CSEE, 2011, 31(13): 55–62.

[13] Alexandre B. Nassif, Janak Acharya. An Investigation on the Harmonic Attenuation Effect of Modern Compact Fluorescent Lamps [C]. 2008. 13th International Conference on Harmonics and Quality of Power, Wollongong, NSW, 2008: 1–6.

[14] Design manual for industrial and civil powe distribution [M]. Beijing: China Electric power press, 2005.

Engineering Technology and Applications – Shao, Shu & Tian (Eds)
© 2014 Taylor & Francis Group, London, ISBN 978-1-138-02705-3

Theory of automatic transformation from Computation Independent Model to Platform Independent Model for MDA

Xue Chen, Dongdai Zhou & Liuguo Wu
Ideal Institute of Information and Technology, Northeast Normal University, Changchun, Jilin, China
Engineering and Research Center of E-learning, Northeast Normal University, Changchun Jilin, China

ABSTRACT: Because of problems such as the low degree of automation and the incompact relations' description in CIM-to-PIM (Computation Independent Model to Platform Independent Model) transforming, this paper proposes a relationships-oriented CIM-to-PIM transformation method. Regard use case specification as the input, combining with relationships-oriented thought and affinity analysis to automatically transform the use case specification into a class diagram based PIM. Using method above to do CIM-to-PIM transformation can guarantee the perfection of the model, reduce the risk of uncertainty caused by manual work, and then promote the automation process of MDA (Model Driven Architecture).

Keywords: MDA; CIM-to-PIM transformation; relationships-oriented; affinity analysis

1 INTRODUCTION

MDA defines four different abstract levels of models – CIM (Computation Independent Model), PIM (Platform Independent Model), PSM (Platform Specific Model) and Code, and drives software development through the mutual transformation between each of the four models.[1] Since MDA has developed for more than ten years, researches about MDA has always focused on the PIM, PSM and PIM-to-PSM transformation. The CIM-to-PIM transformation methods, as the higher abstract level of MDA, were ignored by people.[2] That has led to the cut off of MDA passage, not only affects MDA automated code generation, but also appears problems that seriously affect system quality such as information lost during the transformation process, the code cannot be executed, etc. So, how to build a complete and reliable CIM, and transform the CIM to a correct PIM is very important.

2 RELATED WORK

For the current study, the researches of CIM-to-PIM transformation are roughly divided into four kinds, namely, transformations based on the analysis of use case event flows[3~5], transformations based on features and components[6], transformations based on QVT[7~8], and transformations based on meta-model[9~11]. Although these transformations use different methods when modeling PIM, an important issue are commonly existed – the imperfect semantic of PIM because of the incomplete model transforming algorithm, either containing the dynamic model as sequence diagram and state diagram, or containing the static model as class diagram. Especially in the transformations based on the class diagram, these methods are short of supports of the relationship between class and class. The ultimate goal of researching model transformation is to realize the code's automatic generating, and it is the class diagram which is the most closely to the code. Therefore, this article focuses on the research of the transformation based on class diagram to solve the current problems in the insufficient support for relationships between classes.

Table 1. Class diagram based model transformation and its technology.

NO.	T	I	R	A
[12], [13]	ER model	RS	Y	N
[14]	Object identification	RS in English	N	N
[15]	Textual analysis	RS in UML	N	N
[16]	Textual analysis	RS in English	N	N
[17]	Object identification	RS	Y	SA
[18]	Object identification	Use case	N	N
[19]	Texual analysis	Use case	N	N
[20]	ER model	RS	Y	SA

Note: T – Technique; I – Input; R – Relationship; A – Automated; RS – requirement specification; SA – Semi-automatic.

Table 1 shows the techniques of transformations based on class diagram summarized through the literature[12]~[20].

There are also some issues existed in methods above.

1) The mainly method of transformation is still relying on experienced designers' handwork, only a few realize semi-automatic, the automatic degree is lower;
2) The generated class model is lack of the micromesh description of relationships such as the attribute, function and generalization between classes.

Focusing on the issues above, this article proposes a relations-oriented CIM-to-PIM transformation method. Use use case diagram to modeling CIM, regard the use case specification as the input, finally combine with relations-oriented thinking and affinity analysis[21] to generate the CM (the model based on class diagram). It will solve problems such as the low degree of automation and the incompact relations' description in CIM-to-PIM transforming.

3 THE ESTABLISHMENT OF A USE CASE SPECIFICATION

CIM is used for describing the system requirements. In this process, models like a use case diagram which can make a clear and direct expression to the system are needed. But only rely on a use case diagram that consist of roles, use cases and relationship between them can't describe the system in detail, there also needs to write a use case specification to assist them. The current use case specification is described by natural language. While the ambiguity of natural language is the biggest defect that leads to the use case specification cannot accurately describe the use case.[22] In order to solve the problems that exist in the natural language in use case specification, this section discusses how to realize the standardization of use case specification, it will help to accurately describe use cases and implement the automatic transformation from use case diagrams to class diagrams.

3.1 The use case specification template

The traditional use case specification template only focus on a single use case, overlooking the inheritances, dependencies between use cases. This causes the transformed model can't describe the system completely. Therefore, this article will improve the traditional use case specification template, adding the relationship between other use cases, called relational use case specification template, as shown in Table 2.

3.2 The use case specification language

The event flows in the use case specification in UML2.0 are generally described by natural language. But because of the ambiguity and random syntax, it's difficult to let a machine to

Table 2.　Relational use case description template.

Id	Use case Id
Name	The name of the use case. It usually starts with a verb.
Description	Summarizes the use case in a short paragraph.
Pre-condition	What should be true before the use case is executed.
Actor	Only one actor, because use case is small.
Dependency	Include, extend and dependent relationships to other use cases.
Generalization	Generalization relationships to other use cases.
Event Flow	All the steps of the event.

Table 3.　Description of English sentence structure in thirteen cases.

No.	Syntactic Structure
1	subject verb object
2	subject verb object (to) verb1 (object1)
3	subject verb object participle (object1)
4	subject verb object adjective
5	subject verb object object1
6	subject verb object conjunctive to verb1 (object1)
7	subject verb gerund (object)
8	subject verb object preposition object1
9	subject verb object object1
10	subject verb (for) complement
11	subject verb
12	subject be predicative
13	subject verb preposition object1

understand the semantics described by natural language. In order to realize the automatic transformation from use cases to the PIM, this article will use semi-formal language to describe the use case.

Liwu Li defined 13 kinds of simple sentences in literature[22] through summarizing the articles about the verb form of 25 kinds of English sentence patterns of A.S.Hornby et al., as shown in Table 3.

Keith Thomas compared the sentence patterns of CREWS, CP and Liwu Li in literature[23] and drew a conclusion that Liwu Li's thirteen kinds of sentence patterns can describe use cases more clearly and accurately. So this article will adopt these 13 kinds of English sentence patterns to describe use cases.

In addition to describing event flows in the sentence structure above, Li also stipulated eight language rules that were suitable for the standardization of the use case description[22].

1) Use simple sentences, each sentence only contains one subject and one predicate or activity, and ends with a semicolon.
2) Use active instead of passive sentence patterns.
3) Use the same verb when describing the same action in different sentence.
4) Use the name of a subject or a object instead of the pronouns which represent them.
5) Use If-clause to begin a branch structure, and use EndIf-cause to end it. Also, the Else-clause is optional.
6) Use StartConcurrency-clause to start a set of concurrent actions, and use endConcurrency-clause to end it. Introduce each of the concurrent actions, except the first one, with a concurrent-clause.

7) Use StartSynchronization-clause to start a set of synchronized actions, and use endSynchronization-clause to end it. Introduce each of the synchronized actions, except the first one, with a synchronized-clause.
8) Use While-clause to begin a loop structure, and use EndWhile-clause to end it.

4 THE TRANSFORMATION FROM CIM TO PIM

4.1 *The thought of CIM-to-PIM transformation*

CIM is in the requirements analysis stage of a system, it describes system in a computation independent view. PIM is in the design stage of a system, it describes system in a platform independent view. Although there are a lot of PIM modeling methods currently and various methods are not the same, when choosing PIM meta-model, most of them choose UML's class diagram. So this article thinks that implementing automatic transformation of CIM-to-PIM must be able to implement the automatic transformation from use case diagram to class diagram.

The class diagram based transformation methods mentioned in literature [7], [10], [19] and [22] almost base on use case driven method[24] when transforming use case diagram to class diagram. First, add attributes to initial class diagram through robust analysis, then add functions to class diagram by sequence diagram. But such method can only generate a separate class, relationships like inheritance, dependency etc. between classes cannot be automatically generated. The objective world is composed of relationships, all the objective existents have their worth because of relationships. Relationship is the most direct expression of the objective world, the lack of support for the relationships between classes, will reduce the integrity and consistency of model, increase the risk of the system. Therefore, this article puts forward a relationship-oriented model transformation method, abstracts the relationships between classes by extracting functions and data, and clustering them, makes the transformed model more complete.[25]

4.2 *Transformation form use case specification to class diagram based model*

4.2.1 *Transformation steps*
a) Obtain relationship matrix as Function-Data by extracting Function and Data according to rules in Table 4.
b) Cluster by affinity analysis to obtain entity class as Entity.
c) Replace Function-Data matrix by Entity to obtain Function- EntityData matrix.
d) Do affinity analysis to Function- EntityData matrix to obtain business class.
e) Refine the model.

4.2.2 *Transformation algorithm*
- Firstly, extract the Function-Data matrix (F-D).
  a) Read each event flow in the use case specification circularly.
  b) Do extraction for each event flow in addition to "[]".
    i. If the event flow begins with "[n]", then remove the "[n]", and get the value of n, obtain Function and Data (Argument) according to the rules in table 5.
    ii. If the event flow does not begin with "[n]", skip the event flow.
- Secondly, obtain the entity class and Function-DataEntity matrix (F-DE).
  a) Do affinity analysis between Data and Data
    i. Calculate affinity for each (Data, Data) pair as $A(D_i, D_j)$.

$$A(Di, Dj) = \frac{Number\ of\ function\ in\ common\ to\ Di\ and\ Dj}{Number\ of\ function\ to\ Di} \tag{1}$$

Table 4. The corresponding rules of thirteen English sentences and function-argument.

No.	Syntactic Structure	Function	Argument
1	Sub V Obj	V	Obj
2	Sub V Obj (to) V1 (Obj1)	V1(+Obj1)	Obj, (Obj1)
3	Sub V Obj P (Obj1)	P V (+Obj1)	Obj, (Obj1)
4	Sub V Obj Adj	be+Obj1	Obj
5	Sub V Obj Obj1	set+Obj1	Obj, (Obj1)
6	Sub V Obj Con to V1 (Obj1)	V	Sub, Obj, V1(+Obj1)
7	Sub V G (Obj)	V	Sub, G V(+Obj)
8	Sub V Obj Prep Obj1	V+Obj	Obj1, (Obj)
9	Sub V Obj Obj1	V	Obj, Obj1
10	Sub V (for) Com	V	Sub, Com
11	Sub V	V	Sub
12	Sub be Pred	be+Pred	Sub
13	Sub V Prep Obj1	V+Prep	Sub, Obj

Note: Sub – subject; V – verb; Obj – object; P – participle; Adj – adjective; Con – conjunctive; G – gerund; Prep – preposition; Com – complement; Pred – predicative

  ii. Sort the $A(Di,Dj)$ by descending order, let the two level of affinity approximate natural interrupt.
  iii. If both Data in a (Data, Data) pair are not in a group, then create a new group for them.
  iv. If Data A which is in the (Data A, Data B) pair is in a group, and its affinity level in this group is high enough (average 0.5, adjustable), then join the Data B in the same group.
  v. If both Data in a (Data, Data) pair are in different groups, and the level of this (Data, Data) pair achieves the merged level, then combine the two groups where each Data of the (Data, Data) pair is in.
  vi. Each group is an entity class, and naming the entity class use the most Data owner's name.
 b) Generate Function-DataEntity (F-DE) matrix
  i. Re-order the Data in the F-D matrix, arrange the Data which belongs to the same DataEntity attribute together.
  ii. Use DataEntity to replace Data of the same category, transfer the relationships in F-D to the new generated F-DE matrix.
• Thirdly, obtain the business class.
 a) Do affinity analysis between Function and Function
  i. Calculate affinity for each (Function, Function) pair as $A(Fi,Fj)$.

$$A(Fi,Fj) = \frac{Number\ of\ DataEntity\ in\ common\ to\ Fi\ and\ Fj}{Number\ of\ DataEntity\ to\ Fi} \qquad (2)$$

  ii. Sort the $A(Fi,Fj)$ by descending order, let the two level of affinity approximate natural interrupt.
  iii. If both Function in a (Function, Function) pair are not in a group, then create a new group for them;
  iv. If Function A which is in the (Function A, Function B) pair is in a group, and its affinity level in this group is high enough (average 0.5, adjustable), then join the Function B in the same group.
  v. If both Function in a (Function, Function) pair are in different groups, and the level of this (Function, Function) pair achieves the merged level, then combine the two groups where each Function of the (Function, Function) pair is in.
 b) Do affinity analysis between Function and DataEntity:
  i. Sort the F-DE matrix until appearing DataEntity logical order.
  ii. Circle the logical group of DataEntity and Function in the sorted F-DE matrix.

- Finally, refine the class model.
  a) Scan all the Functions in a business class.
    i. If multiple business classes contain many of the same Functions, then extract these same Functions as interfaces that the multiple business classes must abide.
    ii. The rationality of interfaces will be checked by handwork, if check is passed, then name these interfaces, else backtrack.
  b) Extract the parent class.
    i. Scan all entity classes, if an attribute or a function contains the entity class name they are belonged to, then delete the entity class name from the attribute or the function.
    ii. Scan all the entity classes, if multiple entity classes contain many of the same attributes and functions, then extract these attributes and functions into a public class, regard the extracted public class as these entity classes' parent class.
    iii. The rationality of extracted parent classes will be checked by handwork, if check is passed, then name these parent classes, else backtrack.
  c) Confirm the relationships between classes.
    i. Confirm the relationships between business classes and entity classes. If there is a DataEntity that contained in a business class, and the DataEntity can exactly represent an entity class, then the relationship between the business class and the entity class is association.
    ii. Confirm the relationships between entity classes and entity classes. If an entity class' attributes contain another entity class' instance, then the relationship of the two entity classes is association.
    iii. Confirm relationships according to matching the attribute and the class name. For example, if a Reader class contains the borrowed book, then the relationship between the Reader class and the Book class is association.

5 CASE STUDY

This section only describes Borrow Books use case of Library Management System due to space limitations, as shown in Table 5.

Extract Functions and Data for all the use case specification, then do affinity analysis with them and obtain entity classes, use entity classes to replace Data and obtain Function-DataEntity matrix, as shown in Fig. 1.

According to describing all the use cases, extracting Function-DataEntity relationship matrix, using the affinity analysis algorithm, finally obtain the class model, as shown in Fig. 2.

Compared with traditional methods, this article improves the use case template, adding the description of relationships between use cases, in this way, the CIM can express the system more completely. In the process of CIM-to-PIM transformation, this article realizes the automatic transformation based on the class diagram, and strengthens support for relationships between classes, can guarantee the transformed model more completely than before.

6 CONCLUSION

The relationship-oriented CIM-to-PIM transformation method this article proposed solves problems in transformation, such as the low degree of automation, the incompact relationship description, etc., fills the faults of MDA passage and improves the process of MDA. But this article only does class diagram based PIM transformation, only expresses the static semantics of PIM, the dynamic semantics of PIM are not covered. We will continue improving the model transformation method in the future research to ensuring the integrity of the PIM.

Table 5. Borrow book use case.

Id	02005
Use Case Name	Borrow Book
Brief Description	Reader borrow a book
Pre-condition	Ready book
Actor	Reader, Admin
Dependency	Search Book
Event Flow	[1]The Admin scans the reader's cardNumber;

[1]The Admin scans the reader's cardNumber;
If
   [12]The cardNumber is wrong;
   [1]The System throws errorMessage;
Else
   [1]The System checks the reader's borrowStatus;
   If
      [12]The borrowStatus is false;
      [1]System throws errorMessage;
   Else
      [1]The Admin scans the book's ID;
      [1]The System checks the book's bookBorrowStatus;
      If
         [12]The bookBorrowStatus is false;
         [1]The System throws errorMessage;
      Else
         [1]The Reader borrows the book;
      EndIf
   EndIf
EndIf

Function/DataEntity	admin	reader	book	(admin)
addReader	*	*		
viewReaderInfo	*	*		
deleteReader	*	*		
modifyReader	*	*		
viewBookInfo			*	*
modifyBookInfo			*	*
deleteBook			*	*
addBook			*	*
borrowBook		*	*	
renewBook		*	*	
returnBook		*	*	
scanReaderCard		*		
checkBorrowStatus		*		
modifyPersonalInfo		*		
viewPersonInfo		*		
readerLogin		*		
adminLogin	*			
scanBookID			*	
checkBookBorrowStatus			*	
searchBookByAuthor			*	
searchBookByTitle			*	
viewBookBrifeInfo			*	

Figure 1. Function-dataentity analysis diagram of library management system.

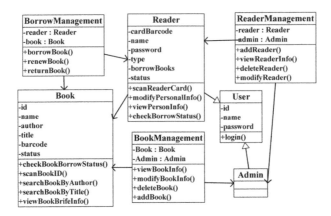

Figure 2.   Library management system class diagram.

ACKNOWLEDGEMENT

This paper is sponsored by National Science-technology Support Plan (GN: 2014BAH22F00).

REFERENCES

[1] Miller, J. & Mukerji, J. (2003) MDA Guide Version 1.0. 1. *Object Management Group*, 234, 51.

[2] Karow, M., Gehlert, A., Becker, J., et al. (2006) On the Transition from Computation Independent to Platform Independent Models, *AMCIS 2006*, pp. 469.

[3] Xin, Y.B. (2009) *Research on UML Model Transformation based on MDA*. Xi'an: Xidian University.

[4] Jiang, L. (2009) *MDA-based Transformation between UML Models*. Xi'an: Xidian University.

[5] Liu, T. (2010) *Research on Model Transformation from CIM to PIM in MDA*. Xi'an: Xidian University.

[6] Zhang, W., Mei, H., Zhao, H., et al. (2005) Transformation from CIM to PIM: A feature-oriented component-based approach. *Model Driven Engineering Languages and Systems*. Springer Berlin Heidelberg, pp. 248–263.

[7] Rodríguez, A., Fernández-Medina, E. & Piattini M. CIM to PIM transformation: A reality [M]//Research and Practical Issues of Enterprise Information Systems II. Springer US, 2008: 1239–1249.

[8] Rodríguez, A., Fernández-Medina, E. & Piattini, M. Towards CIM to PIM transformation: from secure business processes defined in BPMN to use-cases [M]//Business Process Management. Springer Berlin Heidelberg, 2007: 408–415.

[9] Kherraf, S., Lefebvre, É. & Suryn, W. Transformation from CIM to PIM using patterns and archetypes [C]// Software Engineering, 2008. ASWEC 2008. 19th Australian Conference on. IEEE, 2008: 338–346.

[10] CAO XiaoXia, MIAO HuaiKou & SUN JunMei. An Approach to Transforming from CIM to PIM Using Pattern [J]. Computer Science, 2007, 34(6): 265–269.

[11] Sharifi, H.R., Mohsenzadeh, M. & Hashemi, S.M. CIM to PIM Transformation: An Analytical Survey [J]. Int. J. Computer Technology & Applications, 2012, 3(2): 791–796.

[12] Chen, P. Entity-relationship modeling: historical events, future trends, and lessons learned [M]//Software pioneers. Springer Berlin Heidelberg, 2002: 296–310.

[13] Chen, P.P.S. English sentence structure and entity-relationship diagrams [J]. Information Sciences, 1983, 29(2): 127–149.

[14] Beck, K. & Cunningham, W. A laboratory for teaching object oriented thinking [C]//ACM Sigplan Notices. ACM, 1989, 24(10): 1–6.

[15] Wahono, R.S. & Far, B.H. OOExpert: distributed expert system for automatic object-oriented software design [C]//Proceedings of the 13th Annual Conference of Japanese Society for Artificial Intelligence. 1999: 456–457.

[16] Mich, L. & Garigliano, R. NL-OOPS: A requirements analysis tool based on natural language processing [C]// Proceedings of Third International Conference on Data Mining Methods and Databases for Engineering, Bologna, Italy. 2002.

[17] Harmain, H.M. & Gaizauskas, R. Cm-builder: A natural language-based case tool for object-oriented analysis [J]. Automated Software Engineering, 2003, 10(2): 157–181.

[18] Giganto, R. Generating class models through controlled requirements [C]//NZCSRSC-08, New Zealand Computer Science Research Student Conference Christchurch, New Zealand. 2008.

[19] Dennis, A., Wixom, B.H. & Tegarden, D. Systems analysis and design with UML version 2.0 [M]. Wiley, 2005.

[20] Montes, A., Pacheco, H., Estrada, H., et al. Conceptual model generation from requirements model: A natural language processing approach [M]//Natural Language and Information Systems. Springer Berlin Heidelberg, 2008: 325–326.

[21] Gutierrez, N. Demystifying Market Basket Analysis [J]. DM Review Special Report, 2006.

[22] Li, L. Translating use cases to sequence diagrams [C]//Automated Software Engineering, 2000. Proceedings ASE 2000. The Fifteenth IEEE International Conference on. IEEE, 2000: 293–296.

[23] Keith Thomas Phalp, Jonathan Vincent & Karl Cox. Improving the quality of use case descriptioins: empirical assessment of writing guidelines [J]. Springer Science+Business media, LLC, June 2007, 15: 383–399.

[24] Rosenberg, D. & Stephens, M. Use Case Driven Object Modeling with UML:Theory and Practice [M]. Apress, 2007.

[25] Wu LiuGuo. Theory of an MDA_based Method on Automatic Modeling and Model Transformation [D]. Jilin: Northeast Normal University, 2013.

Engineering Technology and Applications – Shao, Shu & Tian (Eds)
© 2014 Taylor & Francis Group, London, ISBN 978-1-138-02705-3

The research on how to improve the reliability of county-level distribution grids using the TD-LTE process

Junliu Zhang, Lijuan Xiong, Jun Yuan, Xiaohong Duan & Bin Wu
Taiyuan Power Supply Company, State Grid Shanxi Electric Power Company, Taiyuan Shanxi, China

ABSTRACT: Utility reliability is an indicator that measures the ability of a utility system to supply customers with electric power and energy with the acceptable quality and in the required quantity without interruption. The service reliability of a county-level power distribution grid is one of the factors that decide the production value of the customers. To improve the reliability of distribution grids, potential ways of improving the reliability level are investigated from the perspectives of reliability constraints, and the effect of TD-LTE penetration on smart distribution grids is examined to provide theoretical basis for improving the intelligence and reliability of county-level distribution grids. First, the average service availability of urban distribution grids in China is reviewed, and a method and an assessment index system for rating the reliability level of a distribution grid is proposed. Next, the TD-LET process is described, with focus on its superiority and application. Finally, after examining the LTE-based access layer communication system framework of distribution grids, a cooperative wireless resource management framework based on quarantine service is proposed. Our results indicate that a TD-LTE-based country-level distribution grid can expect much higher service reliability.

Keywords: distribution grid service; reliability; index system; TD-LET process; communication system framework

1 INTRODUCTION

The service reliability of a county-level distribution grid is the ability of the utility to supply power to the customers without interruption. As any interruption during the use of power could cause unnecessary damage to the customer, higher service reliability county-level distribution grids have always been the goal pursued by both power customers and suppliers. As science advances and intelligence technology expands, a TD-LET-based connection method for county-level distribution grids is provided with a view to improving the power quality for both customers and suppliers.

Many researchers have focused their effort on the TD-LET process and the service reliability of distribution grids. Their outputs have provided theoretical reference for power utilities and contributed to the improved customer power reliability in China. Examples include Tang L.R. et al (2013) who, for the purpose of improving the service quality of intelligent distribution communication, established a heterogeneous network model oriented to smart distribution grid, and analyzed the service quality requirements of different types of communications in light of the objective of smart distribution automation systems, as a contribution to the improved reliability level of distribution grids[1]; Qu L.B. (2014), who outlined the TD-LET process, described the implementation of the TD-LET wireless network program and then analyzed the TD-LET performance as support for the commercial application of TD-LET[2]; Qiu H.J. et al (2013), who studied the reliability index calculation of distribution grids, proposed reliability assessment methods of distribution grids, mainly including hierarchical assessment, and their realization strategy, and verified the reliability of the study subject with practical examples[3].

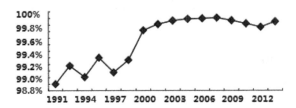

Figure 1. Year-to-year service reliability variation of urban power distribution grid in China.

In this paper, the application of TD-LET to county-level distribution grids for the purpose of improving service reliability is discussed by combining previous findings with the service reliability improvement target of distribution grids and the application of TD-LET to smart distribution grids.

2 CONSTRAINTS ON THE RELIABILITY OF COUNTY-LEVEL POWER DISTRIBUTION GRID

Lu Y. et al (2009) noted that, as the society develops and people are more and more dependent on electric power, any power interruption could bring about heavy damage to customers as well as great cost to the power suppliers themselves. One of the most important indicators that measure the ability of a supplier to supply power to its customers is its service reliability, which reflects how well the power industry meets the energy demand of the national economy, and how mature a power supplier has grown in terms of planning, design, construction, operation and management[4].

Fig. 1 shows the year-to-year service reliability of urban power distribution grids in China.

In this section, reliability constraints on county-level distribution grids are analyzed in two subsections, the first of which introduces the method of analyzing the reliability of a distribution grid, and the second contains a service reliability index system, with a view to providing directional guidance on incorporating the state-of-the-art technology.

2.1 *Methods used to analyze the reliability of county-level power distribution grid*

Service reliability is an integral indicator that measures the grid construction, equipment health and service management of a distribution grid. To improve the service reliability of a distribution grid, reliability constraints have to be identified. In this subsection, a formula method and a statistics method are introduced for the analysis of reliability constraints.

In our study, the utility customer reliability is defined as the percent of the available service hours of customer to the total hours during the calculation period. Assuming that $\overline{T}$ stands for the customer average interruption hours, t stands for interruption hours per time per customer; n the total customers, τ the interruption duration per time, m the customers experiencing interruption per time, then the relation of these parameters is indicated by expression (1) below.

$$\overline{T} = \sum \frac{t}{n} \times \sum \frac{\tau \times m}{n} \tag{1}$$

If η is the service reliability, T is the time of the statistics period, then their relation is indicated by expression (2).

$$\eta = \left(1 - \frac{\overline{T}}{T}\right) \times 100\% \tag{2}$$

Table 1. Statistics service parameter *t* of the power distribution grid during 2010~2012.

Type	2010	2011	2012	3-year avg
Planned interruption	92.8%	94.9%	98.2%	95.3%
Failure interruption	1.8%	2.04%	1.54%	1.79%
Unplanned interruption	5.4%	3.06%	0.26%	2.91%

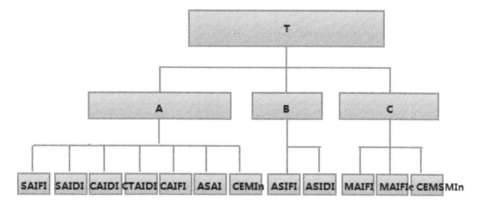

Figure 2. Hierarchy of the service reliability assessment index system of distribution grid.

As can be seen from expressions (1) and (2), the greatest constraint on the service reliability of a distribution grid is $\overline{T}$ while the greatest constraint on $\overline{T}$ is *t*.

Hence when investigating the reliability of a county-level distribution grid, more attention will have to be paid on the interruption time. One thing we are sure about is that the longer interruption time will mean the lower reliability, and vice versa. By nature, power interruption can be divided into planned, failure and unplanned interruption. In this subsection, the application of statistics method to reliability analysis is described using the annual interruption hours per time per user statistics of a county-level distribution grid during 2010~2012 as listed in Table 1.

As can be seen from the figures in Table 1, the absolute majority of the interruption events were planned ones, with the average percent to total interruption hours of 95.3%; failure and unplanned interruption made up a very small proportion and together accounted for only 4.7%, suggesting that planned interruption formed the principal contributor to the deteriorated service reliability. Of the three types of interruption, however, failure interruption is more adverse on customer value and should be avoided to the largest extent.

As analyzed above, the main constraint on customer service reliability of a distribution grid is the interruption hours per time per customer, mostly contributed by planned interruption though failure interruption is the most adverse on customer value.

2.2 *Reliability assessment index system of power distribution grid*

Song Y.T. et al (2008) suggested that the service reliability indexes of a distribution grid include A – sustained interruption index, B – load-based index and C – momentary interruption index which, when put together, can reflect the service reliability of a distribution grid. In our study, the index hierarchy structure method of the AHP (Analytical Hierarchy Process) is used to derive a service reliability assessment index system for distribution grids as shown in Fig. 2.

Table 2 defines the symbols used in Fig. 2.

Table 2. Definitions of symbols in Fig. 2.

Symbol	Definition	Symbol	Definition
T	Integrated Service Reliability Assessment	CAIFI	Customer Average Interruption Frequency Index
A	Sustained Interruption Index	ASAI	Average Service Availability Index
B	Load Chain-based index	CEMIn	Customer Experiencing Multiple Interruptions
C	Transient Interruption Index	ASIFI	Average System Interruption Frequency Index
SAIFI	System Average Interruption Frequency Index	ASIDI	Average System Interruption Duration Index
SAIDI	System Average Interruption Duration Index	MAIFI	Momentary Average Interruption Frequency Index
CAIDI	Customer Average Interruption Duration Index	MAIFIe	Momentary Average Interruption Event Frequency Index
CTAIDI	Customer Total Average Interruption Duration Index	CEMSMIn	Customers Experiencing Multiple Sustained and Momentary Interruptions

When expressed with symbols shown below, we can get a three-level index calculation formula indicated by expression (3):

x: customers experiencing each interruption; n: total customers; t: customer interruption duration

y: total customers experiencing interruption; z: interruption of customer affected; w: available hours of customer

g: service hours of customer needed; a: customers experiencing more than k sustained interruption; p: load interrupted

P: total service load; b: momentary interruption events of customer affected; c: total customers served

d: customers experiencing more than k interruptions

$$
\begin{cases}
\text{SAIFI} = \dfrac{\sum x}{n}; \text{SAIDI} = \dfrac{\sum t}{n}; \text{CAIDI} = \dfrac{\sum t}{y}; \text{CTAIDI} = \dfrac{\sum t}{y} \\[3mm]
\text{CAIFI} = \dfrac{\sum z}{y}; \text{ASAI} = \dfrac{w}{g}; \text{CEMIn} = \dfrac{a}{n}; \text{ASAI} = \dfrac{\sum p}{P} \\[3mm]
\text{ASIDI} = \dfrac{\sum p}{P}; \text{MAIFI} = \dfrac{\sum b}{c}; \text{MAIFIe} = \dfrac{\sum b}{c}; \text{SCEMSMIn} = \dfrac{\sum d}{c}
\end{cases}
\tag{3}
$$

3 BRIEF INTRODUCTION OF TD-LTE

Xu X.H. (2009) believes that distribution grids are an important part of an intelligent grid that connects the main grid and supplies power to customers as well as a critical part deciding the normal operation of the entire power system. As China's national economy grows, the state is starting to construct and reform urban and rural power grids, the key of which is the intelligence and information-oriented reform of distribution grids. In this context, information detection, analysis, processing and decision have to be transferred via an efficient, reliable communication platform so as to shorten the response time to system failure and minimize potential damage as a result of distribution grid failures[6]. Hence, intelligent service of distribution grids is a trend for improving service reliability. Only by improving the intelligence service level of the distribution grid can its reliability be guaranteed.

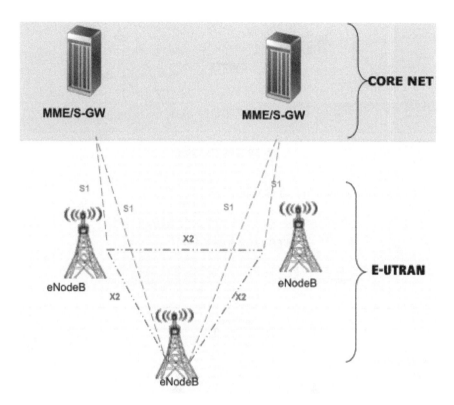

Figure 3.   Main structures at LTE wireless side.

Zhong Y.F. et al (2011) noted that conventional access layer communication technologies of distribution grids include data broadcasting station, communication cable and power line carrier. These technologies, however, are yet to meet the network intelligence requirements in terms of the transmission speed, reliability, timeliness and easy maintenance. New mobile communication technologies, on the other side, are believed to possess advantages incomparable for wire communication[7].

LTE is an acronym for Long-Term Evolution, a 4G wireless communications standard developed by the 3rd Generation Partnership Project (3GPP) from 2004 for the next generation of wireless communication. This technology is designed with three "high", two "low" and one "flat". Three "high" refers to high peak rate, high spectrum efficiency, high mobility; two "low" refers to low time delay, low cost; one "flat" means that the entire of the system is a flat structure based on packet switch. Fig. 3 shows the main interfaces at the LTE wireless side.

In the TD-LTE frame structure, the frame is 10 ms long and contains 10 subframes. Fig. 4 shows the length configuration of all parts constituting a special subframe. The total length of time is a constant 1 ms.

Here, DWPTS stands for Downlink Pilot Time Slot; GP stands for Guard Period; UpPTS stands for Uplink Pilot Time Slot. Given that a wireless frame length is 10 ms, the calculation formula is $Tf = 307200 \times Ts = 10$ ms.

4   ACCESS FRAMEWORK OF DISTRIBUTION GRID BASED ON TD-LTE

The system in the power supply of county-level distribution grids to which TD-LTD is applied is a hierarchically distributed framework consisting of a service system and a broadband (BB) wireless

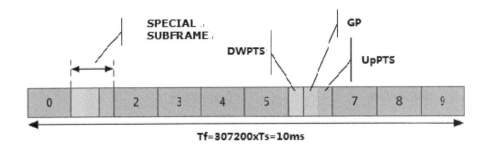

Figure 4.   Length configuration of special subframe.

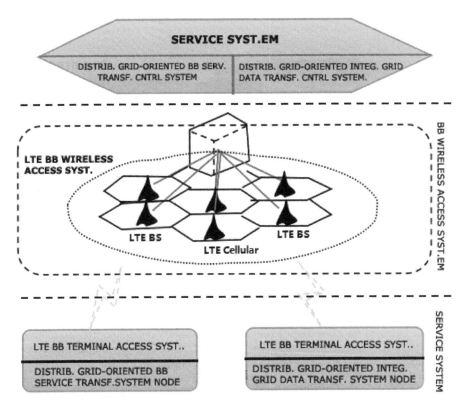

Figure 5.   Communication system framework to improve the service of reliability of distribution grid based on LTE.

access system. Fig. 5 shows the LTE-based access layer communication system framework of a distribution grid

Wu W. (2012) suggested that a broadband wireless access system supports the data transfer to distribution grids mainly based on LTE. The communication system of the distribution grid requires that the system is designed with timeliness and a high bandwidth transfer guarantee mechanism. The flat structure of the LTE system framework makes the networking framework even simpler, reducing the delay in data transfer. The incorporation of new physical layer technologies like OFDM also provides very high system transfer bandwidth[8]. Wu W. (2012) also noted that services undertaken by a LTE-based access layer communication platform of a distribution grid

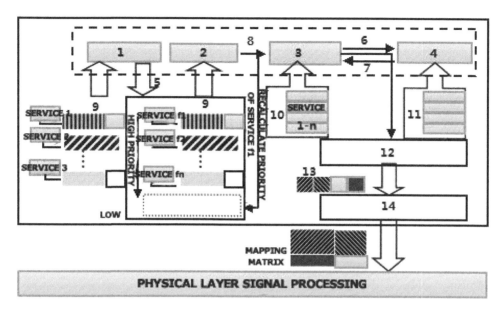

Figure 6. Cooperative wireless resource management framework based on quarantine service.

Table 3. Definitions of symbols in Fig. 6.

Symbol	Definition	Symbol	Definition
1	Service priority calculation	8	SERVICE 1
2	Service quarantine hierarchical scheduling	9	Queue of services to be scheduled
3	Multidimensional wireless resource allocation	10	Service parameter
4	Distribution grid-oriented optimization	11	Channel quality
5	Rearrange	12	Data processor
6	Service link 1	13	Scheduled data
7	Modulation and coding mode	14	Resource quarantine mapper

include broadband service transfer and integrated grid data transfer. The former includes two-way voice/video conversation and video monitoring. The latter includes centralized metering, load control and other automation services[8].

The emergency communication system of an intelligent distribution grid is characterized by large data transfer volume, many collecting devices, high timeliness requirement for data receiving and monitoring and distribution grid information transfer. According to these requirements, link-level and system-level cooperative wireless resource management strategies can be employed to distinguish and quarantine the communication service at the access side.

At the system level, the services are arranged by priority level according to the customer QoS requirement. At the link level, a resource scheduler controls the time-frequency resource allocation dynamically by selecting the time-frequency two-dimensional resources having the best channel quality through adaptive modulation and coding (AMC) before the customer services are mapped onto the physical resource block by a resource quarantine mapper for data delivery.

As described above, a TD-LTE based intelligent distribution grid is flexible, reliable and efficient, and can significantly improve the service reliability of county-level distribution grids. Intelligent distribution grids are what are needed for the power use of people. Fig. 6 shows the cooperative wireless resource management framework based on quarantine service.

Table 3 defines the symbols 1~14 used in Fig. 6.

5 CONCLUSIONS

This paper consists of two parts. One analyzes the reliability constraints on distribution grids, the other discusses the possibility of applying TD-LTE to intelligent distribution systems. The purpose of the first part is to find out ways to improve the service reliability of county-level distribution grids while that of the second part is to describe the application of TD-LTE to the power supply of intelligent distribution grids, demonstrating that this technology is also an effective means of improving the service reliability of distribution grids.

First, the average service reliability level of distribution grids in China is reviewed to provide reference for identifying the data objectives. To find out ways of improving the service reliability of distribution grids, reliability constraints on distribution grids are analyzed before an analysis method and an assessment index system are provided. Next, the TD-LTE process is outlined with focus on its superiority and application, providing technical support for developing an intelligent distribution grid discussed herein. Finally, after examining the access layer communication system framework of a distribution grid based on LTE, a cooperative wireless resource management framework based on quarantine service is proposed, laying foundation for the incorporation of the TD-LTE process into intelligent distribution grids. From the characteristics of this technology, our intelligent distribution system can significantly improve the service reliability of county-level distribution grids.

REFERENCES

[1] Tang L.R. et al. Dynamic load balancing in heterogeneous integrated communication networks oriented to smart distribution grid [J]. Proceedings of the CSEE. 2013.33(1): 39–49.
[2] Qu L.B. TD-LTE radio network planning and performance analysis [J]. New Focus. 2014.
[3] Qiu H.J. et al. Reliability assessment of distribution grid – method and application [J]. Power Supply Technologies and Applications. 2013(6): 239–240.
[4] Lu Y. et al. The countermeasures of improving the power supply trustiness rate for the county level distribution network [J]. Ningxia Electric Power. 2009(1): 14–16.
[5] Song Y.T. et al. Comparison and analysis on power supply reliability of urban power distribution network at home and abroad [J]. Power System Technology. 2008. 32(23): 13–18.
[6] Xu X.H. Intelligent Grid Introduction [M]. Beijing: China Electric Power Press. 2009.
[7] Zhong Y.F. et al. ZigBee wireless sensor network [M]. Beijing: Beijing University of Posts and Telecommunications Press. 2011.
[8] Qu W. et al. The application research of LTE in smart distribution network [J]. Communications for Electric Power System. 2012. 33(234): 80–84.

Engineering Technology and Applications – Shao, Shu & Tian (Eds)
© *2014 Taylor & Francis Group, London, ISBN 978-1-138-02705-3*

Service reliability evaluation of regional distribution networks using the AHP method

Lijuan Xiong, Huibin Zhao, Haiyan Shang, Huifei Wu & Jun Yuan
Taiyuan Power Supply Company, State Grid Shanxi Electric Power Company, Taiyuan Shanxi, China

ABSTRACT: The service reliability of a power distribution network more or less decides the production and living of its users. Any reliability problem in connection with power failures could greatly damage people's interest. Hence, a complete service reliability evaluation index system for distribution networks can serve a critical guide for reliability improvements. In this paper, an evaluation framework consisting of four indices concerning the service liability of a distribution network is built: network structure level, load supply capacity, equipment and technical level and operation and management level, with a view to find out a scientific evaluation algorithm as guide for improving the service reliability of regional distribution networks. First, a fish-bone diagram is used to introduce the indices evaluating the service reliability of a distribution network, on which bases an index hierarchy for reliability evaluation is built to enable subsequent AHP analysis. Then, after describing how the evaluation indices are weighted, outlined the evaluation process of the AHP method as theoretical basis for the application of a scientific algorithm. Finally, an empirical study was carried out on the service reliability of the distribution networks with base data of 8 distribution networks A~H to verify how well this algorithm works.

Keywords: evaluation system; AHP analysis; weight; service reliability; curve fitting tool

1 INTRODUCTION

As urban economy grows, users are calling for higher levels of service security of distribution networks. Technically, improvement of the service reliability of a distribution network will necessitate technical renovation and equipment updating. Administratively, this is to ensure healthy operation of all functional mechanisms of a network. Either way, directional guide will have to be provided if we are to upgrade the service reliability of our distribution networks. In this paper, a service reliability evaluation index system combining the technical and administrative aspects of a distribution network, together with the associated evaluation process, is proposed.

Many researchers have worked on how to evaluate the service reliability of distribution networks. It is their work that has provided the direction toward which our power supply systems should develop and made contribution to the improved service reliability in China. Qiu H.J. et al (2013) noted that the service reliability indices are indications of the ability of a supplier to supply power to its users, its equipment capabilities and management level; improving the service reliability level for users will play an important role in providing better quality service in future[1]. Fang H.H. et al (2013) examined the distribution network evaluation indices from safety, economy, flexibility, reliability and coordination perspectives. They evaluated distribution networks on a comprehensive basis, with focus on the ability of a distribution network to continue to supply power when one single transformer fails at a substation and to transfer load when the line fails[2]. Zhang J. et al (2013) introduced an improved algorithm for evaluating the service reliability of an MV distribution network based on fault influence traverse, which incorporates the influence of fuse faults and that of the load branch due to the fuse failure into the feeder where this load branch is located, and verified that this algorithm is correct and effective[3].

After reviewing these findings, we established a service reliability evaluation system for distribution networks and tried to find out an index system evaluation algorithm as a directional guide for reliability improvements.

2 SERVICE RELIABILITY EVALUATION SYSTEM FOR POWER DISTRIBUTION NETWORK

Dai X.H. et al (2012) noted that, as a distribution network involves a complicated structure, tremendous quantities of equipment and frequent updates, many factors are included is any comprehensive evaluation is to be performed. So far, distribution networks are evaluated generally from singular characteristic indices like service reliability, economy, safety and service quality, which measure the technical level of a distribution network from different angles[4]. However, we believe that the service reliability of a distribution network can be evaluated as an integral index. In other words, the service reliability evaluation of a distribution network can be performed as an integrated evaluation of this network.

A fish-bone diagram can be used to identify the service reliability factors of a distribution network as shown in Fig. 1.

Table 1 defines the symbols shown in Fig. 1 above.

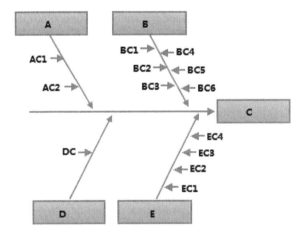

Figure 1.   Fish-bone of service reliability factors for distribution network.

Table 1.   Definition of symbols shown in Fig. 1.

Symbol	Definition	Symbol	Definition	Symbol	Definition
A	Load supply capacity	AC2	% MV line available for power wheeling	BC6	% single transformer substations
B	Network structure level	BC1	% interconnected MV line	EC1	% cable line
C	Service reliability	BC2	% typical connection networks	EC2	% insulated line
D	Operation & management level	BC3	% acceptable main line cross-section	EC3	% oil-free MV switches
E	Equipment & technical level	BC4	% inter-station connected MV line	EC4	% automation of distribution network
AC1	Main transformer N-1 criterion	BC5	% single power line substation	DC	Lift cycle index

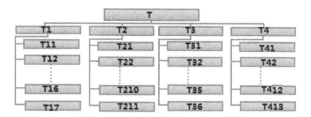

Figure 2. Service reliability evaluation index system for distribution network.

2.1 *Establish a reliability evaluation index system*

Feng X.L. et al (2013) suggested that two kinds of indices reflect the development of a distribution network: performance indices, which are indications of the service result of a distribution work as well as the most concerned indices for the stakeholders of the network; property indices, namely what kind of a distribution network is needed, which are clear indications of the details and characteristics of a distribution network and directly decide the performance[5]. In this paper, the service reliability of distribution networks is evaluated using four first-level indices – network structure level, load supply ability, equipment and technical level, and operation and management level, and second-level indices under these first-level ones. Fig. 2 shows our service reliability evaluation index system for distribution networks.

In this diagram, T is the target layer; T1–T4 are the first-level indices, respectively network structure level, load supply ability, equipment and technical level, and operation and management level. To evaluate the operability of the first-level indices, second-level indices are defined under each of these first-level indices, with the purpose of providing basis for the integrated evaluation of the service reliability of distribution networks.

37 second-level indices are defined, including 7 for T1, 11 for T2, 6 for T3 and 13 for T4. Fig. 2 shows our service reliability evaluation index system for distribution networks. To highlight the essentials of this algorithm, an example is given on network structure level T1.

2.2 *Define the weight of the hierarchical evaluation indices*

In this section, the 7 second-level indices under network structure level T1 are weighted, with the purpose of presenting how second-level indices are weighted. T11 stands for the % single power line substations; T12 is the % single transformer substations; T13 is the % interconnected MV line; T14 the % inter-station connected MV line; T15 is the % typical connection networks; T16 is the % overlong main lines; T17 is the % acceptable main line cross-section.

Li K.G. et al (200&) noticed that the consistency of a judgment matrix can be easily satisfied when there are more than nine indices within a layer, in which case the indices can be weighted using the Delphi method. In practical application, both methods are based on recommendations of many specialists so that the result will provide more objective basis, and the result of the weighting is approximated so that the calculation is more direct and convenient[6].

A judgment matrix is the single ranking of factors of each layer relative to a certain factor of the upper layer, which can be simplified as the judgment of a number of pair-wise factors. In our study, Saaty (1–9 ratio scale) is introduced and written into a matrix as shown in Table 2, Saaty (1–9 ratio scale). The resulting judgment matrix is shown in expression (1) below.

$$
\begin{array}{c}
\begin{array}{ccccc} A_k & B_1 & B_2 & \cdots & B_n \end{array} \\
\begin{array}{c} B_1 \\ B_2 \\ \vdots \\ B_n \end{array}
\begin{bmatrix}
b_{11} & b_{12} & \cdots & b_{1n} \\
b_{21} & b_{22} & \cdots & b_{2n} \\
\vdots & \vdots & \vdots & \vdots \\
b_{n1} & b_{n2} & \cdots & b_{nn}
\end{bmatrix}
\end{array}
\tag{1}
$$

Table 2. Saaty (1–9 ratio scale).

Rating of score	Pair-wise relative importance	Definition
1	Equal importance	Both are equally important
3	Moderate importance	One is moderately more important than the other
5	Strong importance	One is strongly more important than the other
7	Demonstrated importance	One has very strongly more important than the other
9	Extreme importance	One is extremely more important than the other
2, 4, 6, 8	Between adjacent important levels	The relative importance is between any two above

Table 3. Weight of 2nd-level indices under the network structure.

2nd-level index	Weight	2nd-level index	Weight	2nd-level index	Weight
T11	0.15	T14	0.15	T17	0.10
T12	0.15	T15	0.10		
T13	0.20	T16	0.15		

Note: T11–T17 are symbols representing the 7 second-level indices under the network structure level.

Then, the initialized judgment matrix is converted into an integrated judgment matrix, which is derived from the final matrix using a calculation method indicated in expression (2).

$$\begin{cases} A(S) = \left| a(S)_{ij} \right|_{n \times n} \\ a_{ij} = k^* \sqrt[k^*]{\prod_{S=1}^{k^*} a(S)_{ij}} \end{cases}, S = 1, 2, \cdots, k; i, j = 1, 2, \cdots, n \tag{2}$$

Table 3 lists the weights of the second-level indices of the network structure level derived from the said method.

3 AHP EVALUATION PROCESS

Again, let us see how the service reliability of a distribution network is evaluated using the same 7 second-level indices under the network structure level. First, the rating criteria are defined as presented in Table 4. Each of the 7 second-level indices is assigned fixed values. The correlation between the physical values and their ratings is given in this table.

To allow quick calculation of the scores of an index of any value, the rating criteria can be processed with a curve fitting tool so as to derive the functional relation between the index values and their scores. Expression (3) shows the functional relation between the index value of T16 and its score. Here, y_{16} is the score of the % overlong main line, and x_{16} is the index value of the % overlong main line.

$$y_{16} = \begin{cases} 100 - 0.8 x_{16} & 0 \leq x_{16} \leq 75 \\ 160 - 1.6 x_{16} & 75 \leq x_{16} \leq 100 \end{cases} \tag{3}$$

When all the second-level indices have been scored, their scores are calculated layer by layer on a bottom up basis according to expression (4). Symbols in the expression are defined below:

1) $s^{(k+1)}$ is the score of attribute $A^{(k+1)}$ in layer $k+1$.
2) n is the number of subattributes of attribute $A^{(k+1)}$ in layer k.
3) $s_j^{(k)}$ is the score if subattributes of attribute $A^{(k+1)}$ in layer k.

Table 4. Comparison scores of the 7 2nd-level indices under the network structure level.

Second-level index	Score interval					
	0 score	40 scores	60 scores	80 scores	90 scores	100 scores
T11	100	35	20	10	5	0
T12	100	35	20	10	5	0
T13	0	55	70	85	95	100
T14	0	20	35	55	65	100
T15	0	20	50	75	87.5	100
T16	100	75	50	25	12.5	0
T17	0	40	60	80	90	100

Note: T11 is the % single power line at substation; T12 is the % single transformer substations; T13 is the % interconnected MV line; T14 is the % inter-station connected MV line; T15 is the % typical connection networks; T16 is the % overlong main line; T17 is the % acceptable main line cross-section.

Table 5. 1–9 average random consistency indices of the 1-9 scale matrix[7].

1	2	3	4	5	6	7	8	9
0.00	0.00	0.58	0.90	1.12	1.24	1.32	1.41	1.45

4) $w_j^{(k)}$ is the weight of subattribute j.

$$s^{(k+1)} = \sum_{j=1}^{n} s_j^{(k)} w_j^{(k)} \tag{4}$$

The indices are weighted in two steps as described below:

Step 1: Calculate the characteristic vector corresponding to the maximum characteristic root of the matrix. Then formalize this vector to derive the weight of each index. First, formalize or normalize each column of the judgment matrix. Then add each row of the resulting judgment matrix and build the summation of each line into a column vector. Next, normalize the resulting column vector, which is then used as the characteristic vector corresponding to the maximum characteristic root. Finally, derive the maximum characteristic root.

Step 2: Check the consistency of the integrated judgment matrix. Expression (5) shows the calculation method of the consistency index CI.

$$CI = \frac{\lambda_{max} - n}{n - 1} \tag{5}$$

When the integrated judgment matrix is fully consistent, CI = 0. A larger CI will mean a lower consistency. To assign the satisfaction of CI, the average random consistency index RI of the 1–9 order matrix is incorporated as presented in Table 5.

When the order of an integrated judgment matrix is larger than 2, the ratio of the consistency index CI to the same-order random consistency index RI is called the random consistency ratio of this matrix. When CR < 0.10, the judgment matrix has a satisfactory consistency level. Otherwise adjustment will have to be made.

4 EMPIRICAL ANALYSIS

The service reliability of 8 distribution networks A∼H is evaluated. The 8 distribution networks in the area are all 10 kV systems. Table 6 lists the base data of these networks.

Table 6. Base data of a 10 kV distribution network.

Class of net	Supplied zone	P1	P2	P3	P4	P5
A	C, E	1.658	115	761.98	868	330.4
B	D, E, F	0.273	64	1963.07	1124	111.3
C	D, F	0.817	86	1583.86	1031	145.5
D	D, E, F	0.662	53	832.12	713	146.5
E	E, F	0.825	98	2517.95	1356	107.1
F	D, E, F	0.465	48	994.31	760	90.60
G	D, F	0.490	104	2497.39	2116	214.7
H	D, F	0.680	74	1950.81	1764	242.7

Note: P1 is the power supplied in TWh; P2 is the number of lines; P3 is the length of the line in km; P4 is the number of distributing transformers; P5 is the capacity of the distributing transformer in MVA.

Table 7. Scores of the 7 second-level indices under the network structure level of 8 distribution networks.

Class of net	T11	T12	T13	T14	T15	T16	T17
A	78.69	82.89	64.62	81.68	82.61	98.61	87.83
B	4.65	79.65	20.93	40.44	100	85.00	90.00
C	69.62	28.56	34.23	48.01	92.06	96.28	0.00
D	81.39	67.32	65.5	58.32	95.47	96.98	100
E	46.54	51.25	16.03	23.91	100	96.74	76.57
F	79.65	36.75	49.87	79.18	95.00	85.00	100
G	27.23	82.56	21.61	77.15	98.16	61.60	29.63
H	17.41	35.24	33.87	66.72	94.59	88.11	83.15

Note: T11 is the % single power line substations; T12 is the % single transformer substations; T13 is the % interconnected MV line; T14 is the % inter-station connected MV line; T15 is the % typical connection networks; T16 is the % overlong main line; T17 is the % acceptable main line cross-section.

Table 8. Resulting service reliability by distribution network.

Class	T	T1	T2	T3	T4	Class	T	T1	T2	T3	T4
A	76.06	81.25	86.16	52.83	76.38	E	61.07	53.63	80.28	23.22	81.24
B	70.43	54.65	86.63	52.29	87.93	F	64.28	71.56	72.93	40.30	64.34
C	63.94	52.42	80.88	37.69	82.07	G	53.70	54.41	79.71	24.46	42.85
D	72.88	78.25	87.03	29.46	87.05	H	57.54	55.67	76.73	16.39	72.71

Note: T1 is the network structure level; T2 is the load supply ability; T3 is the equipment and technical level; T4 is the operation and management level; T is the integrated distributing service reliability index.

Table 7 lists the resulting rating of the 7 second-level indices of the 8 distribution networks under the network structure level.

Table 8 lists the service reliability rating of distribution networks after integrating the four first-level indices.

Fig. 3 shows a histogram of the resulting service reliability rating by distribution network presented in Table 8.

As can be seen from Fig. 3, network A has the highest service reliability among the 8 counterparts. Of the four first-level indices, if the equipment and technical level T3 is higher, the service reliability of the distribution work will be higher too. The T3 values of all the other networks are quite low, with that of network H being the lowest, though the difference is not significant.

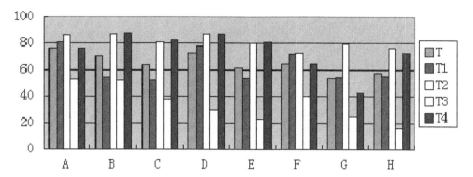

Figure 3. Statistics overall reliability rating.

Networks B and D have the highest service reliability levels after network A, but the former has a low T1 value and the latter has a low T3 value.

5 CONCLUSIONS

The purpose of our study is to try to find out a scientific algorithm for evaluating the service reliability of distribution networks. Two non-negligible aspects are an evaluation index system and an evaluation method. In our study, first, a service reliability evaluation index system for distribution networks is established. Then, the method of evaluating these indices is described. Finally, using the base data of 8 distribution networks, the service reliability of these networks is evaluated with a view to providing direction for further service reliability improvements of different distribution networks.

REFERENCES

[1] Qiu, H.J., et al. (2013) Method and realization of reliability evaluation for power distribution network. *Power Supply Technologies and Applications*, (6), 239–240.
[2] Fang, H.H., et al. (2013) Indices system of distribution network planning evaluation. *Proceedings of the CSU-EPSA*, 25(6), 106–111.
[3] Zhang, J., et al. (2013) The research on the reliability evaluation algorithm based on fault influence traverse [J]. Journal of Guangzhou University, 12(6), 64–68.
[4] Dai, X.H., et al. (2012) Power distribution network operation safety evaluation methods based on game theory and grey relational grades. *Shaaxi Electric Power*, 40(4), 30–33.
[5] Feng, X.L., et al. (2013) Comprehensive evaluation index system of distribution network and evaluation method. *Guangdong Electric Power*, 26(11), 20–26.
[6] Li, K.G., et al. (2007) Application of AHP and fish-bone diagram in the diagnosis of logistics enterprise's problems. *Logistics Technology*, 26(11), 212–215.
[7] He, Z.X. (1985) *Fuzzy Mathematics and its Application*. Tianjin: Tianjin Science & Technology Press.

Engineering Technology and Applications – Shao, Shu & Tian (Eds)
© 2014 Taylor & Francis Group, London, ISBN 978-1-138-02705-3

Reliability evaluation of distribution networks with DG connection using statistics simulation

Chengxiao Zhang, Huibin Zhao, Liang Guo, Xiaoyan Wang & Yating Zhang
Taiyuan Power Supply Company, State Grid Shanxi Electric Power Company, Taiyuan Shanxi, China

ABSTRACT: This paper describes a service reliability evaluation process for distribution networks with DG connection using the statistics simulation algorithm. First, the measuring method of voltage quality, frequency deviation and voltage waveform distortion, and the calculation algorithm of the five indices for eventual service reliability evaluation are given as basis for further content quantification of service reliability evaluation indices. Then, the power output model of wind turbine is characterized as a cut-in point for presenting the statistics simulation method in question as well as providing source material for designing a service reliability evaluation algorithm for distribution networks without DG connection, with normal source connection, and with DG connection, in hopes of offering theoretical reference for further improving the service reliability evaluation system of distribution networks.

Keywords: distributed generation; technical index and service reliability index; output power model; statistics simulation algorithm

1 INTRODUCTION

Growing science and technology has deepened the extensive penetration of distributed generation (DG) into distribution networks. While this is fully attributable to the superiority of DG over any of its peers, a challenge has also been brought to the service reliability evaluation of distribution networks. In order to find out a way to address this problem, a service reliability evaluation algorithm based on statistics simulation is introduced with a view to providing useful reference for the service reliability of our distribution networks.

Many researchers have worked on the reliability evaluation of distribution networks with DG connection. It is right their work that has laid theoretical foundation for China's power supply systems. Zhang Y.X. et al (2014) investigated the reliability evaluation of distribution networks with DG, and built a DG reliability model after examining the effect of DG connection on the reliability evaluation of distribution networks and characterizing the output stochasticity of photovoltaic (PV) and wind power generation[1]. Li H.Y. (2013) analyzed the technical indices of DG and those of distribution networks with DG connection, and structured a technical index system of distribution networks with DG connection, offering reference for the future robust development of DG[2]. Su A.X. et al (2013) introduced a new algorithm for evaluating the reliability of distribution networks considering wind power effect, which is based on segment-network simplification and pseudo sequential Monte Carlo simulation, and allows quick and correct calculation of system reliability. They compared the reliability level of the IEEE RBTS system under different conditions and verified that this algorithm is both effective and useful[3].

On the basis of previous findings, this paper analyzes the reliability evaluation indices of distribution networks with DG connection, and designs a reliability evaluation algorithm based on statistics simulation, with a view to offering useful recommendations for further improving China's power supply systems.

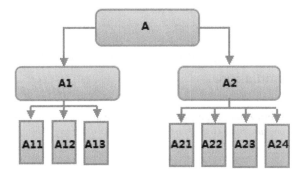

Figure 1.   Technical index structure system of distributed networks with DG.

2   TECHNICAL INDICES OF DISTRIBUTION NETWORKS WITH DG CONNECTION

Distributed generation (DG) is a minor power generation system that is:

1) installed in a distributed manner nearby where the users live;
2) designed with power output ranging from kilowatts to tens of megawatts; and
3) economical, environment-friendly, small-sized, flexible and efficient.

Because of these features, distributed generation is widely connected to distribution networks. Connection of this power source, however, is also accompanied by a change in the power flow distribution that is already affecting the protection of distribution networks and threatens their energy quality and service reliability.

To evaluate the service reliability of distribution networks with DG connection, the technical index system of distribution networks with DG will have to be discussed from the perspectives of both DG and distribution networks with DG. Fig. 1 shows the technical index system of distribution networks with DG.

Here, A stands for the technical index system of distribution networks with DG connection; A1 stands for the technical indices of DG itself; A2 the technical indices of distribution network with DG connection; A11 the resource conditions; A12 the connection address; A13 the technical adapatibility; A21 the voltage quality; A22 the frequency deviation; A23 the voltage waveform distortion; A24 the service reliability. In this section, the technical indices and service reliability indices are discussed in two separate subsections.

2.1   *Content of technical indices*

For the purpose of this paper, technical indices include voltage quality, frequency deviation and voltage waveform distortion. The first index, voltage quality, can be investigated from user-side voltage increase, voltage deviation, voltage fluctuation and flicker, and three-phase imbalance.

When DG is connected to a distribution network, it provides part of the active and reactive power needed by the load, thereby reducing the line current and increasing the user-end voltage. By incorporating the *VPII* variable that represents the user-side voltage increase, the calculation formula is indicated by expression (1).

$$\begin{cases} VPII = \dfrac{VP_{W/DG}}{VP_{WO/DG}} \\ VP = \sum_{i=1}^{N} V_i L_i K_i \\ \sum_{i=1}^{N} K_i = 1 \end{cases} \tag{1}$$

Here V_i is the voltage amplitude of line i; L_i is the load of line i; K_i is the weight factor of load line i; and N is the total number of load lines.

The voltage deviation can be indicated by expression (2).

$$\Delta U\% = \frac{U - U_e}{U_e} \times 100\% \tag{2}$$

Here U is the measured voltage at the detection point of a distribution network with DG connection, and U_e is the measured voltage at the detection point of the distribution network without DG connection. The voltage fluctuation and flicker indices refer to sudden voltage changes the degree of which can be indicated by expression (3).

$$\delta U = \frac{(U_{max} - U_{min})}{U_e} \tag{3}$$

Here U_e is the rated voltage; U_{max}, U_{min} are the maximum and minimum voltage fluctuations within a time period. The three-phase imbalance can be indicated by expression (4).

$$\varepsilon_u = \frac{U_2}{U_1} \times 100\% \tag{4}$$

Here U_1 is the amplitude of the positive sequence component of three-phase voltage after decomposed with the symmetrical component, and U_2 is the amplitude of the negative sequence component of three-phase voltage after decomposed with the symmetrical component.

The degree of deviation of the supply power frequency is indicated by expression (5), where f is the actual supply frequency, and f_e is the rated frequency of the supply system.

$$\Delta f\% = \frac{(f - f_e)}{f_e} \times 100\% \tag{5}$$

The voltage harmonic distortion can be calculated by expression (6), where U_n is the usable voltage of harmonic n, and U_1 is the usable voltage of the fundamental wave.

$$k = \sqrt{\frac{1}{U_1^2} \sum_{n=2}^{\infty} (U_n)^2} \times 100\% \tag{6}$$

2.2 Content of service reliability indices

Five indices, i.e. customer average interruption duration ($CAID$), customer average interruption frequency ($CAIF$), average service availability index ($ASAI$), customer prescheduled average interruption duration ($CPAID$), and customer prescheduled average interruption frequency ($CPAIF$), are used to evaluate the service reliability of a distribution network with DG connection.

The calculation method for the these five indices is indicated by expression (7), where N_i is the number of users affected by each interruption; U_i is the duration of the interruption; N_0 is the total number of the designated users; N_k is the number of users affected by each prescheduled interruption; and U_k is the duration of the prescheduled interruption.

$$\begin{cases} CAID = \dfrac{\sum_i N_i U_i}{N_0}; CAIF = \dfrac{\sum_i N_i}{N_0}; CPAID = \dfrac{\sum_k N_k U_k}{N_0} \\ ASAI = \left(1 - \dfrac{CAID}{8750}\right) \times 100\%; CPAIF = \dfrac{\sum_k N_k}{N_0} \end{cases} \tag{7}$$

3 SERVICE RELIABILITY EVALUATION ALGORITHM FOR DISTRIBUTION NETWORKS WITH DG CONNECTION BASED ON STATISTICS SIMULATION

Liu C.Q. (2008) noted that conventional service reliability evaluation of a distribution network generally assumes that the feeder is supplied by a single source on a Caoyong radiation basis so that any element failure in a feeder could result in outage of all the loads behind this feeder. When DG is connected to a distribution network, the power system becomes a multi-source networke connected to the users so that any element failure could result in isolated island operation with DG in the distribution network, and consequently changes in the reliability evaluation modeling and algorithm[4].

Distribution systems with PV or wind generation are different from conventional ones with backup sources. To allow better investigation, a distribution network with wind generation is used as an example to present a possible service reliability evaluation algorithm for distribution networks with DG connection. Zhang S. et al (2010) suggested that power system risks are mainly assessed by analytical method and Monte Carlo simulation, the latter of which is more suitable for risk analysis of major generation and transmission systems with wind power generation due to its simple principle and easy realization[15].

Hence, in this section, the wind output power of wind generation and the statistics simulation algorithm are discussed in two separate subsections, in hopes of providing theoretical references for evaluating the service reliability of distribution networks with DG connection.

3.1 Power output model of wind turbine

If a wind turbine generation unit is examined from the energy transformation perspective, it can be looked at as a wind turbine and a power generator. Luo W. (2011) noted that wind speed measurements have demonstrated the subjection of regional wind speed to the two-parameter Weibull distribution. If we use a random variable V to represent the wind speed, the distribution function can be indicated by expression (8)[6].

$$F(V) = P(v \leq V) = 1 - \exp\left[-\left(\frac{V}{C_W}\right)^{k_w}\right] \tag{8}$$

Here k_W is the shape parameter, and C_W is the dimensional parameter. Fig. 2 shows a Jensen model that well simulates the wake situation from flat terrain.

Symbols shown in this diagram is defined below:

R: turbine impeller radius; R_W: wake radius; V: average wind speed; V_T: Wind speed through the impeller; V_x: wind speed affected by wake; X: distance between turbines

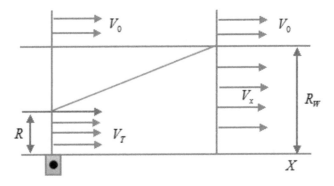

Figure 2. Jensen wake effect model.

With this, we can calculate the wind speed affected by wake as indicated by expression (9).

$$V_x = V_0 \left[1 - \left(1 - \sqrt{1 - C_T}\right)\left(\frac{R}{R + kX}\right)^2 \right] \tag{9}$$

Aerodynamics confirms that the electric power from a wind generator is proportional to the wind speed. As such, the relation of the output power of a wind generator and the wind speed can be indicated by expression (10).

$$P_W(v) = \begin{cases} 0 & v < v_{ci} \cup v > v_{co} \\ \dfrac{P_{WR}}{v_r^3 - v_{ci}^3} & v_{ci} \leq v \leq v_r \\ P_{WR} & v_r < v < v_{co} \end{cases} \tag{10}$$

Here v_{ci} is the cut-in wind speed; v_{co} is the cut-out wind speed; v_r is the rated wind speed; P_{WR} is the wind generation active power under the rated wind speed; v is the actual wind speed of the turbine; $P_W(v)$ is the active power of the turbine under wind speed v.

If the forced outage probability q_{WT} of the wind turbine is considered, we can derive the probability density function $f(P_{WT})$ of the turbine output power P_{WT} as indicated by expression (11).

$$f(P_{WT}) = \begin{cases} (1 - q_{WT})[1 - F(v_\infty) - F(v_{ci})] + q_{WT} & P_{WT} = 0 \\ (1 - q_{WT})\dfrac{n}{k_1 m}\left(\dfrac{P_{WT} - k_2}{k_1 m}\right)^{n-1} \exp\left[-\left(\dfrac{P_{WT} - k_2}{k_1 m}\right)^n\right] & 0 < P_{WT} < n_{WT} \cdot P_{WT} \\ (1 - q_{WT})[F(v_\infty) - F(v_r)] & P_{WT} = n_{WT} \cdot P_{WT} \end{cases} \tag{11}$$

Here, $k_1 = \dfrac{n_{WT} \cdot P_{WT_r}}{v_r - v_{ci}}, k_2 = -k_1 v_{ci}$.

3.2 Statistics simulation evaluation algorithm

In a distribution network with DG connection, the state of each element can be decided by the probability of this state, and the combination of the states of all elements constitutes the state of the entire system. If a system element is represented by m, the outage probability of this element is represented by P_m, and the random number of interval [0,1] from a sampling process is represented by U_m, the state S_m of this element can be indicated by expression (12).

$$S_m = \begin{cases} 0 & U_m > P_m \\ 1 & 0 \leq U_m \leq P_m \end{cases} \tag{12}$$

Then all the elements of the distribution network system is sampled according to expression (13). The sampling state of a system containing M elements can be indicated by expression (13).

$$\mathbf{S} = \begin{pmatrix} S_1 & S_2 & \cdots & S_M \end{pmatrix} \tag{13}$$

If the sampling quantity is massive, the sampling frequency of the system state $\mathbf{S}$ can be used as the unbiased estimation of its probability. If the probability of easy system state is estimated by sampling, the mathematical expectation of this index can be indicated by expression (14).

$$\begin{cases} \hat{P}(\mathbf{S}) = \dfrac{n(\mathbf{S})}{N_S} \\ \hat{E}(F) = \displaystyle\sum_{\mathbf{S} \in Fault} F(\mathbf{S})\hat{P}(\mathbf{S}) \end{cases} \tag{14}$$

Here N_s is the number of the sampling processes; $n(\mathbf{S})$ is the number of occurrences of state $\mathbf{S}$; $F(\mathbf{S})$ is the index function under state $\mathbf{S}$; $\hat{P}(\mathbf{S})$ is the probability of state $\mathbf{S}$. To estimate the uncertainty of the system indices with expression (14), the mean variance of the samples can be used. In such case, the accuracy level of statistics simulation can be expressed with variance coefficient η as indicated by expression (15).

$$
\begin{cases}
V\!\left(\hat{E}(F)\right) = \dfrac{1}{N_s(N_s-1)}\displaystyle\sum_{k=1}^{N_s}\left(F_k - \hat{E}(F)\right)^2 \\
\eta = \dfrac{1}{\hat{E}(F)}\sqrt{V\!\left(\hat{E}(F)\right)}
\end{cases}
\tag{15}
$$

To evaluate the service reliability of a distribution network with DG connection, we can make the following assumptions provided that the accuracy is not compromised:

1) all elements are repairable;
2) permanent failure instead of transient ones are considered. In other words, the system will not assume operation after an element fails until this element is repaired;
3) malfunction of the circuit breaker or other switching gear is not considered; and
4) only a single fault is expected.

The calculation is carried out in the following steps.

Step 1. Sample the normal service time T_{TF} and failure time T_{TR} of all the elements with expression (16), and arrange the sampling result into an operation state duration sequence of each element during the total simulation time in an one-off manner. Here λ_i and μ_i are the outage rate and repair rate of element i, and u is the random number between $(0,1)$ subject to uniform distribution.

$$
\begin{cases}
T_{TF} = -\dfrac{1}{\lambda_i}\cdot \ln u \\
T_{TR} = -\dfrac{1}{\mu_i}\cdot \ln u
\end{cases}
\tag{16}
$$

Step 2. Integrate the operation state duration sequences of all the elements and identify all failure events of the system in a given simulation time.

Step 3. Analyze each failure duration, identify the loads affected by the failures and group them as unrepairable loads, repairable loads and loads in isolated island.

Step 4. For a repairable load, accumulate this outage to the total interruption time T_{TR} and the total number of interruptions N_1 at this load point. For an unrepairable load, regard it as the outage time of the load.

Step 5. For a load within isolated island, the following steps are taken:

Check that the power generation system is functional during the isolated island operation according to the system operation state sequence derived from Step 1. If yes, go back to 2). If no, proceed to Step 6;

Switch according to formula (1). If $S_m = 0$, the switching is regarded as a failure. If 1, the switching is successful;

Calculate the service time of the DG during the isolated island operation.

Step 6. According to the total interruption time T_{TR} and total number of interruptions N_1 at each load point, calculate the N_i, U_i, N_0, N_k and U_k.

Step 7. Calculate the customer average interruption duration $CAID$, customer average interruption frequency $CAIF$, average service availability index $ASAI$, customer prescheduled average interruption duration $CPAID$ and customer prescheduled average interruption frequency $CPAIF$ according to Step 6.

4 ALGORITHM VALIDATION

Next, our algorithm is validated with the example system shown in Fig. 3. Here, nodes 11, 17, 21 and 26 are the points at which wind generation is to be connected; the rated capacity of the

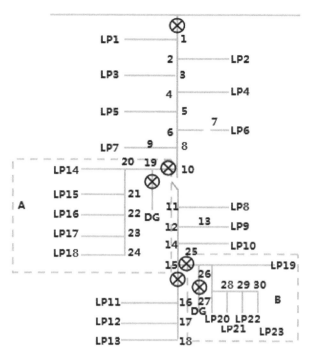

Figure 3. Example system.

Table 1. Statistics λ, u parameters at four load points under three different connection modes.

Connection Mode	Parameter name	Point LP1	Point LP9	Point LP13	Point LP16	Point LP20
Zero connection	Parameter λ	1.193	1.193	1.577	1.834	1.724
	Parameter u	6.154	10.177	13.136	11.348	15.017
Connection at node 19	Parameter λ	1.193	1.193	1.577	0.652	1.724
	Parameter u	6.154	10.177	13.136	5.909	15.017
Connection at nodes 19	Parameter λ	1.193	1.193	1.577	0.652	0.633
and 25	Parameter u	6.154	10.177	13.136	5.909	5.653

Table 2. Comparison resulting reliability levels by distribution network.

Connection position	ASAI	CPAID	CPAIF
None	99.9363%	6.452 (h/user)	1.9235 (Times/user)
Node 19	99.9373%	6.367 (h/user)	1.9235 (Times/user)
Node 19, 25	99.9530%	5.110 (h/user)	1.9235 (Times/user)

turbine is 1.1 MW; the cut-in, cut-out and rated wind speeds are 3 m/s, 25 m/s and 14 m/s, respectively. The reliability parameters and load parameters of the elements are quoted from reference [7]. First, the λ, u parameters are collected at load points LP1, LP9, LP13, LP16 and LP20 without DG connection, with DG connection at node 19, and with DG connection at nodes 19 and 25. The result is presented in Table 1. Then, the service reliability is evaluated with the average service availability index ASAI, the customer prescheduled average interruption duration CPAID and the customer prescheduled average interruption frequency CPAIF under these three different conditions. The result is presented in Table 2.

267

5 CONCLUSIONS

The purpose of our study is to find out a possible service reliability evaluation algorithm for distribution networks with DG connection, which involves the investigation of the service reliability evaluation technical indices of distribution networks with DG connection and the output model of the power output of DG before a workable evaluation algorithm can be worked out.

First, the measuring method of voltage quality, frequency deviation and voltage waveform distortion, and the calculation algorithm of the five indices for eventual service reliability evaluation are presented. Then, to find out a possible way for reliability evaluation, the power output model of wind turbine is investigated, and the reliability evaluation algorithm steps for distribution networks with DG connection based on statistics simulation are given. Finally, by using the example system quoted from reference [7] and our own statistics simulation algorithm, the reliability evaluation index simulation results of distribution networks without DG connection, with normal source connection and with DG connection are yielded, and the operability of the algorithm validated.

REFERENCES

[1] Liu, Y.X., et al. (2014) Reliability evaluation of distribution system with distributed generators. *Electric Switches*, (01), 4–7.
[2] Li, H.Y. (2013) Technology index system of distributed generation integrating distribution grid. *Popular Science & Technology*, 15(169), 74–76.
[3] Su, A.X. (2013) Reliability evaluation of distribution system considering wind power effect. *Power System Protection and Control*, 41(1), 90–95.
[4] Liu, C.Q. (2008) *Involves the Distribution Generation of the Distribution System Reliability Evaluation*. Shanghai, Shanghai Jiaotong University.
[5] Zhang, S., et al. (2010) Reliability assessment of generation and transmission systems integrated with wind farms. *Proceeding of the CSEE*, 30(7), 8–14.
[6] Luo, W., et al. (2011) Short-term wind speed forecasting for wind farm. *Transactions of China Electrotechnical Society*, 26(7), 68–74.
[7] Roy, B., et al. (1996) A test system for teaching overall power system reliability assessment. *IEEE Transaction on Power System*, 11(4), 1670–1676.

Engineering Technology and Applications – Shao, Shu & Tian (Eds)
© 2014 Taylor & Francis Group, London, ISBN 978-1-138-02705-3

Maintenance strategy optimization of main transmission and transformation equipment in distribution networks based on integrated risk assessment

Kangning Wang, Tianzheng Wang & Shan Lu

Electric Power Research Institute, State Grid Shanxi Electric Power Company, Taiyuan, Shanxi, China

ABSTRACT: This study is performed to find out a main equipment maintenance strategy for distribution networks having the minimum risk assessment function as their objective function. In order to relate proper maintenance to the risk assessment function, first, the conflict between scheduled maintenance and condition maintenance is described to provide basis for further determination of the maintenance risk assessment function. Then, the function expression for risk assessment of distribution networks is presented, and constraints on the objective function designed using the safety-resource-maintenance relation to improve the optimization model. Finally, a PSO-based solution to the optimization model is presented. Empirical data are used to validate the operability of the algorithm, and the results derived under three different scenarios confirm that our integrated risk assessment model is highly superior over others and can provide theoretical reference for the scientific maintenance strategy optimization of China's distribution networks.

Keywords: condition maintenance; risk assessment; probability function; optimization model; PSO algorithm; integrated assessment

1   INTRODUCTION

Ultrahigh pressure and large capacity power systems are a trend amid the ever-growing human economy, bringing higher demand for the transmission and transformation equipment in distribution networks. However, equipment failures and their maintenance will also result in some damage to the users. That explains why maintenance strategies concerning main transmission and transformation equipment in distribution networks have been put on the agenda. In this paper, a maintenance strategy optimization model based on integrated risk assessment is presented in hopes of offering a course of protection for the power supply and use in China.

Many researchers have worked on maintenance strategies concerning the transmission and transformation equipment in distribution networks. As maintenance strategies become more scientific, the nation is witnessing balanced increases in its power supply reliability. La Y. et al (2010) introduced a condition-based maintenance concept as a substitute for conventional regular maintenance for transmission and transformation equipment to offset problems intrinsic to the conventional technique. They compared three different condition-based maintenance techniques: maintenance based on life cycle management, on reliability, and on condition and risk assessment, and discovered outstanding merits in the condition-based maintenance technique based on condition and risk assessment[1]. Guo H.B. et al (2013) designed an assistance maintenance decision system model for the condition-based maintenance of transmission and transformation equipment in order to allow better management of condition-based maintenance activities, and used this model to assess the initial conditions of pilot equipment. The resulting equipment maintenance turned out to be significantly upgraded in terms of the relevance and effectiveness[2]. Zhang Z.H. et al (2014) presented an assessment-based maintenance strategy to address the maintenance management of power transformers, and rated the risks according to the assessment result using the ALARP principle.

Example calculations confirmed high practicality of the risk assessment-based power transformer maintenance strategy [3].

On the basis of previous findings, this paper relates the maintenance of main transmission and transformation equipment in distribution networks and the related risks, and establishes an optimization model that minimizes the maintenance assessment risk function for substation maintenance as an example. The purpose is to find out a scientific maintenance strategy that can provide theoretical basis for optimizing the maintenance strategies of transmission and transformation equipment in China's distribution networks.

2 MAINTENANCE MODEL OF MAIN EQUIPMENT IN DISTRIBUTION NETWORK

Zhang H.Y. et al (2009) noted that, as one of the key techniques that support a robust smart power supply system, condition-based maintenance has demonstrated marked success in offsetting the over- and under-maintenance frequently found in conventional scheduled maintenance programs[4]. Present researches on condition-based maintenance are focused on the equipment condition assessment and maintenance decision. The former looks at condition monitoring, fault prediction and condition assessment, while the latter examines the equipment itself, the power system operation and the combination of the power system and its equipment.

As our purpose is to provide a high quality maintenance strategy for main substation equipment which needs to take into account maintenance related risks, it has to be designed to minimize risk potential or, in other words, to identify the minimum function under constraints using the risk assessment function as the objective function.

Within the technical constraint where the risk assessment function is the objective function, the equipment failure probability is included for in considering possible consequences of maintenance and failures. However, as equipment failures are non-scheduled ones, we can use the equipment failure probability to represent the health condition of the equipment.

The opposite of condition-based maintenance is scheduled maintenance, which is frequently based on an average failure rate curve derived from historical failure records. Given that we already know the equipment parameters as shown below, we can arrive at the equipment failure rate $\lambda(t)$ indicated by expression (1).

t_{act} : length of equipment service; f_E: environment index; f_L load index; t_{eq}: equivalent age of service; a: shape parameter of Weibull curve; η: scale parameter of Weibull curve.

$$\begin{cases} t_{eq} = \dfrac{t_{act}}{f_E f_L} \\ \lambda(t) = \dfrac{a}{\eta}\left(\dfrac{t}{\eta}\right)^{a-1} \end{cases} \tag{1}$$

The basis on which condition-maintenance is performed is usually a health index δ_{HI} resulted from complete and advanced condition assessment. The equipment failure-health index and the formula of the health index are indicated by expression (2).

$$\begin{cases} \lambda = \exp(\delta_{HI} C)^K \\ \delta_{HI} = \exp(B\Delta T)^{\delta_{HI0}} \end{cases} \tag{2}$$

Here K is the scale factor; C is the curvature factor; δ_{HI0} is the initial health index of the equipment; B is the aging factor; ΔT is the time interval from the initial to the final time. As such, given that the failure rate of the equipment within a certain period is known, we can derive the equipment reliability $R(t)$ and failure probability $F(t)$ within any period, as indicated by expression (3).

$$\begin{cases} R(t) = \exp\left[-\int_{t_1}^{t_2} \lambda(\tau)d\tau\right] \\ F(t) = 1 - R(t) \end{cases} \tag{3}$$

3 MAINTENANCE STRATEGY OPTIMIZATION MODEL FOR DISTRIBUTION NETWORKS BASED ON RISK ASSESSMENT

Li M. (2011) noted that, as far as maintenance is concerned, the general risk facing the operation of a power system is defined as the power system operational risks R_0, consisting of power system maintenance risks R_M and power system failure risks R_F that conflict each other. To solve this conflict, we have to find a balance point by coordinating them so that the operational risks of the power system after maintenance are the minimal. This is the time when a maintenance program is deemed to be the most proper, since the power system is neither exposed neither to high maintenance risks in connection with over-maintenance nor to high failure risks in connection with under-maintenance[5].

In our study, scheduled maintenance and condition-based maintenance are investigated using the equivalent service age and service age reduction factor α to represent the result of scheduled maintenance, and the health index and health recovery factor β to represent the result of condition-based maintenance. With expressions (1) and (2), we can get the expression of the maintenance result indicated by expression (4), in which t_{act0} is the present length of service of the equipment; t_1 is the time when equipment maintenance starts; T_1 is the time lasted for equipment maintenance; Δt is the time elapsed since the last assessment.

$$
\begin{cases}
t_{eq} = \begin{cases} \dfrac{t_{act0}+t-1}{f_L f_E} \\ (1-\alpha)t_{ed}(t_1)+\dfrac{t-t_1-T_1+1}{f_L f_E} \end{cases} & \delta_{HI}(t)=\begin{cases}\delta_{HI0}\exp[B(\Delta t+t-1)] \\ \beta\delta_{HI}(t_1)\exp[B(t+T_1-t_1+1)]\end{cases} & \begin{array}{l} t \le t \\ \\ t > t_1 \end{array} \\
\lambda(t)=\lambda(t_{ed}(t)) & \lambda(t)=\lambda(\delta_{HI}(t))
\end{cases} \tag{4}
$$

This section comprises two subsections that respectively discuss distribution network risks and optimization modeling, with a view to providing basis for a scientific algorithm design.

3.1 *Expression of distribution network risks*

As discussed above, distribution network risks can be divided into maintenance risks and failure risks. As any main equipment to be maintained has to be taken out of service, this will lead to direct loss of part of the load of the network. This part of the load is called scheduled load loss. Besides, there are occasions when failures in other equipment result in greater risks of load loss to the distribution network. This is called non-scheduled load loss. Scheduled and non-scheduled load losses are calculated as the formulas below, in which symbols are defined as follows:

$R_M(t)$ is the distribution network maintenance risk in time t; R_{M1} is the scheduled load loss; R_{M2} is the non-scheduled load loss; M_t is the maintenance mode set in time t; $F(t)$ is the failure mode set under maintenance mode m in time t; $P_{M1,m}, P_{M2,m}$ are the scheduled and non-scheduled load losses under maintenance mode m in time t; $P_{M2,0}$ is the non-scheduled load loss under non-equipment maintenance mode; P_f is the probability of failure mode f in time t; N_F, N_R are the failure and non-equipment sets under failure mode f; c_1, c_2 are the unit power prices of scheduled and non-scheduled load losses; T is the number of times within a period.

$$
\begin{cases}
R_M = \displaystyle\sum_{t=1}^{T} R_M(t) = \sum_{t=1}^{T}\sum_{m\in M_t}\left(R_{M1,m}+R_{M2,m}\right) \\
P_f = \displaystyle\prod_{i\in N_F}F_i(t)\prod_{j\in N_R}R_j(t) \\
R_{M1,m} = P_{M1,m}T_m c_1 \\
R_{M2,m} = \displaystyle\sum_{f=F_t}P_f\left(P_{M1,m}-P_{M2,0}\right)T_m c_2
\end{cases} \tag{5}
$$

271

As any equipment in a distribution network is exposed to failures, scheduled and individual load losses are resulted. These load losses are calculated by expression (6), in which R_{F1}, R_{F2} are the scheduled and the individual load losses; μ is the recovery probability; C_{Mi} is the failure recovery price of equipment i; C_{Ri} is the replacement price of equipment i; T_f is the power interruption time caused by failure f.

$$\begin{cases} R_F = \sum_{t=1}^{T} R_F(t) = \sum_{t=1}^{T} \sum_{f \in F_t} \left[R_{F1,f} + R_{F2,f} \right] \\ R_{F1,f} = P_f P_{F1,f} T_f c_2 \\ R_{F_2} = P_f \sum_{i \in N_F} \left[\mu C_{Mi} + (1 - \mu) C_{Ri} \right] \end{cases} \qquad (6)$$

Here $R_F(t)$ is the power system failure risk in time t; $R_{F1,f}, R_{F2,f}$ are the non-scheduled and self-load losses caused by failure mode f in time t.

3.2 Optimization model

As analyzed above, the objective function of our optimization model is the minimal risk, as indicated by expression (7).

$$F = \min(R_O) = \min(R_M + R_N) \qquad (7)$$

The technical constraints of the objective function shown in expression (7) can be based on safety, maintenance relationship and maintenance resources. Safety constraints on distribution networks can be subdivided into node voltage constraints and line power flow constraints. If $U_i, U_{i\max}, U_{i\min}$ are the voltage, the upper voltage limit and the lower voltage limit at point i; $S_j, S_{j\max}$ are the power flow and the permissible ultimate transmission power flow of line j, the safety constraint can be indicated by expression (8).

$$\begin{cases} U_{i\min} < U_i < U_{i\max} \\ S_j < S_{j\max} \end{cases} \qquad (8)$$

Any substation is equipped with a given set of resources, the number of equipments permitted to be simultaneously maintained if normally used to indicate the maintenance resource constraint. If u_{it} is the maintenance state variable, 1 represents that the equipment needs maintenance, 0 represents that the equipment does not need maintenance, m_i is the resource needed to maintain equipment i, and S_t is the upper limit of maintenance resource in time t, then the maintenance resource constraint can be indicated by expression (9).

$$\sum_{i=1}^{N} u_{it} m_i \leq S_t \qquad (9)$$

The exact maintenance constraints, identified according to particular conditions to be addressed, can generally be divided as equated constraints and excluded constraints.

4 ALGORITHM DESIGN AND VALIDATION

4.1 Algorithm design

The PSO, a bionic algorithm that provides quick solution to optimizations, is used to imagine each optimizing problem as a particle, and all these particles are in one D-dimension space. All of them are assigned a fitness level by a fitness function to see how well they are positioned. Each particle is assigned memory capacity to remember their best position identified for them. Each particle is also given a speed to decide their flying distance and direction, and this speed is normally updated from time to time according to its own flying experience and that of its partners. The PSO algorithm designed to solve our optimization model is presented below.

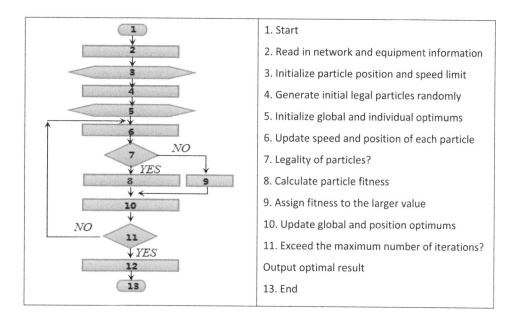

1. Start

2. Read in network and equipment information

3. Initialize particle position and speed limit

4. Generate initial legal particles randomly

5. Initialize global and individual optimums

6. Update speed and position of each particle

7. Legality of particles?

8. Calculate particle fitness

9. Assign fitness to the larger value

10. Update global and position optimums

11. Exceed the maximum number of iterations?

Output optimal result

13. End

4.2 *Algorithm validation*

Strategy optimization is provided for scheduled maintenance alone, condition-based maintenance alone and for the combination of scheduled and condition-based maintenance, to validate the applicability of our algorithm and the operability of our schematic design when applied to a distribution system as shown in Fig. 1.

Equipment information of the distribution system shown in Fig. 1 is quoted from Table 1 of reference[6]. With this distribution system and its maintenance program, the maintenance strategy is optimized using the POS algorithm. The optimization result is presented in Table 1.

Maintenance of the main equipment in a distribution network as arranged in Table 1 will result in RMB5.84 million risks under scheduled maintenance alone, RMB5.23 million risks under condition-based maintenance alone and RMB4.97 million risks under integrated scheduled and condition-based maintenance. This confirms that our optimization model is highly superior over others.

5 CONCLUSIONS

In this paper, first, the maintenance model of main equipment in distribution networks is analyzed, and equipment failure rates in connection with scheduled and condition-based maintenance are discussed with a view to providing basis for building a maintenance strategy optimization model for distribution network based on risk assessment.

Then, after examining the risk assessment function of distribution networks, a maintenance strategy optimization model is designed which, different from conventional scheduled maintenance,

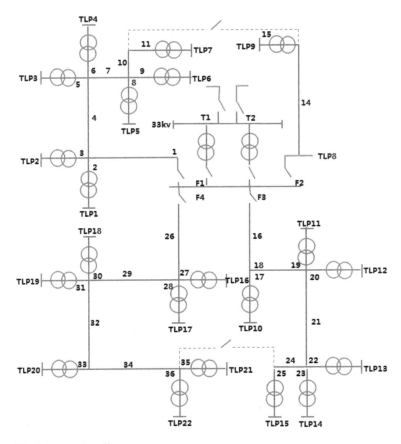

Figure 1.   Distribution system diagram.

Table 1.   Optimized time schedule.

Equipment name	Weeks after maintenance starts			Equipment name	Weeks after maintenance starts		
	Scheduled	Non-scheduled	Comprehensive		Scheduled	Non-scheduled	Comprehensive
T2	37	15	15	L32	29	4	12
1	33	15	14	TLP20	33	16	20
TLP1	39	39	32	L35	15	27	14
3	16	17	38	L36	19	30	10
4	35	39	27	TLP22	19	30	10
5	41	34	39	L16	15	36	35
TLP3	41	34	39	TLP10	32	5	20
TLP4	36	2	11	L18	35	35	16
7	12	1	8	L19	14	40	35
TLP6	24	26	8	TLP11	14	40	35
TLP7	37	8	9	TLP12	18	18	37
12	9	19	2	L22	32	8	8
13	17	16	14	L23	41	8	4
14	16	26	20	L24	24	28	13
36	22	42	1	L25	22	9	4
TLP17	40	5	21	TLP15	22	9	4
TLP18	15	3	21	Figures in this table are weeks after maintenance. 58 weeks are involved.			

is a combination of scheduled and condition-based maintenance, and takes into account constraints from power use safety, maintenance relationship and maintenance resources. This provides reference for developing a solution algorithm of our optimization model.

Finally, a PSO-based bionic algorithm is presented. After giving the realization steps of the algorithm, the optimization strategies under three scenarios are verified with the distribution system quoted from reference[6] as an example. The results confirm high superiority of our optimization model in the form of value risks.

REFERENCES

[1] La, Y., et al. (2010) The study of transmission and transformation equipment condition maintenance based on condition and risk assessment. *Guangdong Electric Power*, 23(10), 36–40.
[2] Guo, H.B., et al. (2013) Development and application of transmission and transformation equipment state overhaul assistant decision system. *Inner Mongolia Electric Power*, 31(4), 1–6.
[3] Zhang, Z.H., et al. (2014) Power transformer maintenance strategy research based on risk assessment. *China Rural Water and Hydropower*, (4), 1–6.
[4] Zhang, H.Y., et al. (2009) Research and implementation of condition-based maintenance technology system for power transmission and distribution equipments. *Power System Technology*, 33(13), 70–73.
[5] Li, M., et al. (2011) Basic concept and theoretical study of condition-based maintenance for power transmission system. *Proceedings of the CSEE*, 31(34), 43–52.
[6] Su, R., et al. (2013) Maintenance decision making optimization based on risk assessment for distribution system. *Electric Power Automation Equipment*. 33(11), 1–8.
[7] Li, M., et al. (2012) Decision-making model and solution of condition-based maintenance for substation. *Proceedings of the CSEE*, 32(25), 1–8.

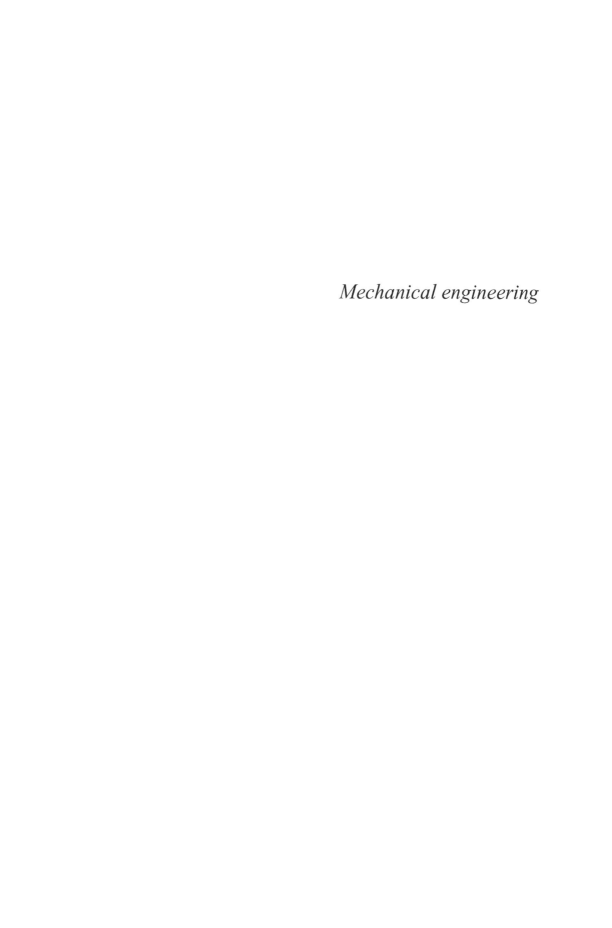

Mechanical engineering

Engineering Technology and Applications – Shao, Shu & Tian (Eds)
© 2014 Taylor & Francis Group, London, ISBN 978-1-138-02705-3

Research on vibrating cockpit systems about helicopter training simulator

Yongjun Gai
Military Simulation Technology Institute Aviation University of Air Force, Changchun, Jilin, China

Jingquan Wang
Economic Information Center of Jilin Province, Changchun, Jilin, China

Youyi Li
Military Simulation Technology Institute Aviation University of Air Force, Changchun, Jilin, China

ABSTRACT: We had researched and developed a new vibrating cockpit system. It realistically simulated the high frequency buffeting dynamic sense of helicopter training simulator. This study established a model of vibrating cockpit, according to vibration force model of engine giving and revolve winged. This paper described the system compositions and work principle, then gave hardware and software system. It proved that the design was reliable. Now it has applied production process successfully.

Keywords: helicopter training simulator, vibrating cockpit, vibration force, hardware, software

1 INTRODUCTION

There was different about helicopter and airplane with fixed wing, the helicopter always go with high frequency vibration, and it was changing along with change of flight state and condition, then immediately send pilots by seats, floor, controlled helm and footstool. Pilot feel state of bodywork by feeling of vibration characteristic change, such as ice over of revolve winged, lamina shatter, leaf lose speed. When flight simulator is training simulator of helicopter need afford to vibration for provide movement feeling to pilot, and he is personally on the scene.

Now vibrations of my country simulator usually adopt vibration of hydraulic flat, it is complexity and maintenance is not convenience and high cost. Using six-DOF hydraulic flat simulated vibrations did not realize vibration of high frequency low breadth because quality and inertia of movement flat; at the same time worked harm to other system such as projection tool of display system, thus need especially to design vibration system. This paper gives electro motion vibration means that has structure suppleness, maintenance convenience, high fidelity etc.

2 SIMULATE MODEL OF VIBRATING COCKPIT

Vibration fountain of helicopter was much such as engine drive system revolves winged and scull, they would produce change load when they worked, and become helicopter vibration. Furthermore there are random vibration forces of landing strike and collision, throw in missile and besides hang etc. So simulator system of vibration simulated cycle vibration force engine, scull and revolve winged. Vibration force of empennage is sameness together revolves winged, but size and direction is not sameness. Because under side mostly will analyze vibration force of engine and revolve winged.

2.1 Vibration force model of engine giving

Vibration force of turbine engine mostly comes from rotor mass imbalance. Rotor will bring levity inertia centrifugal forces, and then can bring vibration forces. Centrifugal force that was circum-rotate vector of uprightness to genie circumrotate axes, and then break up two positive vibration force f_x, f_y of phase discrepancy $90°$. Namely:

$$f_x = A\sin(2\pi\Omega t) \tag{1}$$

$$f_y = A\sin(2\pi\Omega t + \pi/2) \tag{2}$$

Formula: Ω–revolve winged angle speed.

It was direct ratio between vibration force frequency of turbine engine and engine rotate speed. Usually genie circumrotate axes is uprightness with seat floor, so f_x and f_y did not direction of portrait and landscape orientation.

2.2 Vibration force model of revolve winged

It is mostly reason about body vibration from vibrate rotor blade to vibration force of helicopter, this force may break up there directional force and moment: $X(t)$, $Y(t)$, $Z(t)$, $M_x(t)$, $M_y(t)$, $M_z(t)$. From relation to vibration force frequency and humorous, we known: blade is z, body angle speed is ω, rotor humorous germinations $2z\omega$ and $3z\omega$ give body and arose body vibration that rotor humorous $z\omega$ and combination rotor humorous $(z-1)\omega$. Underside gives vibration force model of upright axes (Y axes) directional.

$$Y = zP_0 + zP_{az}\cos z\omega t + zP_{bz}\sin z\omega t + zP_{a(2z)}\cos 2z\omega t + zP_{b(2z)}\sin 2z\omega t + \ldots \tag{3}$$

$$M = zh[Q_0 + Q_{az}\cos z\omega t + Q_{bz}\sin z\omega t + Q_{a(2z)}\cos 2z\omega t + Q_{b(2z)}\sin 2z\omega t + \ldots] \tag{4}$$

Formula: P_k–parallel to rotor axes force;
$\qquad$ Q_k–upright of radial force and parallel to rotor axes force.

3 SYSTEM REALIZATION

3.1 Constitute and theory of systems

Composing of vibration seat simulate system is fig. 1, hard core is electro motion humorous vibration of our development, it and seat and shock absorber make mass –spring –damp system. Vibration controlled computer bases flight state parameter that main controlled computer give it, through vibration force model produce there controlled signal, pass D/A transform zoom out drive there vibration, they produce vibration force of size mezzo, them choose damp radio and spring stiffness each other matching. For ensure vibration simulate fidelity, adapt to vibration signal closed loop controlled.

3.2 Design of hardware

In nature, vibration of helicopter was very complicated and arose non-linear moment by compounds vibration. To this characteristic, make use of computer controlled and PWM frequency conversion timing technology that designed electro motion humorous force vibration. System vibration frequency was $10\sim60$Hzó£vibration breadth was $0.1\sim2$ mm.

For realization to a living vibration simulates system and did not disturb other system work, we designed mass-spring-damp system and vibration moment limited cabin. (Refer with: fig. 2) Three vibrations set bracket of cabin according to X, Y, Z direction, bracket of cabin and vibration and base were connected and fit moment flat. So this makes mass-spring-damp system. Underside give simulate validate of single vibration and single direction mass-spring-damp system.

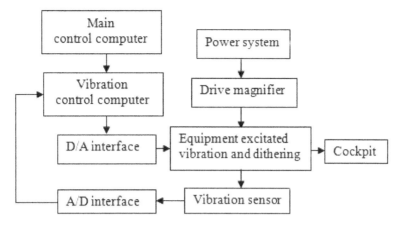

Figure 1. Simplified illustration of vibration cockpit system.

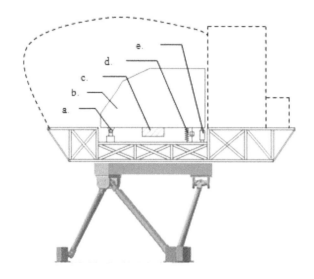

Note: a. axletree   b. vibration cockpit   c. excitated vibration equipment
d. mass-spring-damp system   e. dithering equipment

Figure 2. Digital vibration simulate system show.

3.3 *Design of software*

Vibration controlled computer counted vibration force from main controlled computer effect to parameter of vibration force, and then break up and progress F transform, so educed to vibration force of body. Humorous of frequency under 10 Hz go along simulation get across moment flat, Humorous of frequency hyper-60 Hz, owing to swing less, we did not feeling so can ignore and get rid of it. Humorous vibration force that was pitch on transformed input vibration humorous, sent to vibration implement. Frequency of humorous and frequency of vibration was homology, vibration breadth selected by experience. Then it would amend through acceleration true measure of helicopter vibration. Cabin produced corresponding vibration in humorous force; sensor measure vibration sign, by A/D transform make Fourier transform, and then educe practical vibration

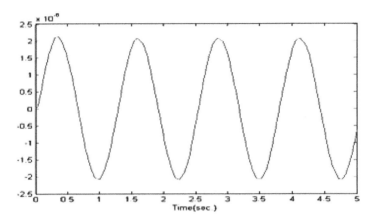

Figure 3.   Output curve of vibration simulate system.

humorous, and then compared about true vibration humorous sign and input vibration humorous sign. Then difference value would feedback output of vibration humorous, and make corrects value so that assured vibration living.

Refer to parameter of helicopter simulation vibration cabin system, suppose outside force of electro moment vibration was sine function, due to fig. 3 we known that deferent displacement approximate was steady sine curve. This system not only accurately produced vibration signal by vibration math model but also can disturb other system worked.

4   CONCLUSION

Applying this ways, we developed flight simulator of helicopter-simulated vibration of living helicopter, and this technology key was performance of humorous force vibration implement. Mass–spring-damp system and vibration force that was building of vibration simulate model and cabin seat and shock absorber were best matching.

REFERENCES

[1] Jingfeng He, Junwei Han. Research on six freedom degrees rotator control [J]. Machine Tool & Hydraulics, 2002, No. 15:98–100.
[2] Zhen Huang, Lingfu Kong, Yuefa Fang. Agency theory and control of parallel robot, Beijing Machinery Industry Press, 1997:303–306.
[3] Chunping Pan, Yongjun Gai. Parallel robot response amplitude non-linear revises [J]. Journal of Changchun University of Science and Technology, 2007, 30(1):121–124.
[4] Feng Guo. Motion simulator and platform system development and application prospects [J]. Mechanical design and manufacturing, 2008, 6:230–232.
[5] Xiaoxia Jiang, Ronglong Zhong. Magnetostrictive displacement sensors apply [J]. Sensor Technology, 2003, 22(1):50–53.
[6] Chunnan Li, Yun Lu. Magnetostrictive displacement sensor research [J]. Experimental Science and Technology, 2008, 2:10–12.
[7] Yongliang Wang, Lijun Qi. Displacement sensor applies in six freedom degrees [J]. Sensor Technology, 2003, 22(2):49–50.
[8] Qingshan Li, Rimin Pan. Magnetostrictive displacement sensor measurement and Implementation [J]. Instrument Journal, 2005, 26(8):50–56.

Engineering Technology and Applications – Shao, Shu & Tian (Eds)
© 2014 Taylor & Francis Group, London, ISBN 978-1-138-02705-3

A high current recorder for marine controlled source electromagnetic method

Zhongliang Wu & Xianhu Luo
Guangzhou Marine Geological Survey, Guangzhou, Guangdong, China

Ming Deng, Yuan Li, Kai Chen & Meng Wang
China University of Geosciences (Beijing), Beijing, China

ABSTRACT: Marine controlled source electromagnetic method is one of the hottest points of geophysics. The higher the quality of the original data, the better the result of the MCSEM method will be especially for the current data, which plays an important role of the inversion of the final data. So some higher reliability and better function current recorder systems are urgent demands. This paper introduced a high current recorder in MCSEM including GPS, DTCXO, MCU, FPGA and Hall sensor. This high current recorder is aim for marine control source EM, for its high current. The two important advantage of the heavy current recorder is that it can record current with max range of ± 250 A and the time accuracy of 40 us/hr. After several field surveys and marine testing, the current data shows the effectiveness of the current recorder, taking into account of time sync.

Keywords: current recorder; MCSEM; time sync

1   INTRODUCTION

With the fast development of the world economic, the high need of the resource is deeply demand to match the need of human, while the limited land resource is decrease. Therefore we have taken on consider of the ocean resource which account for 71 percent of the earth [1].

Taking account of the difficulty of the ocean resource development, the detection technology is still in the first step until now. Some developed country like The USA and England, paid great attention to the research of the marine electro magnetic technology.

Geophysics is based on the outdoor surveys and indoor tests, so as MT method. Some high reliability and better functional observe system is urgent demand. MT survey system hardware includes collection host computer, electronic sensing device, magnetic sensing device, and cable.

Magneto telluric method is a hot point of geophysics, mainly used for deep earth exploration, like metal mine, oil, hydrocarbon and geothermal energy resource [2]. As we all know, magneto telluric method depends on the electronic diversity of underground medium.

According to the natural field source MT signal containing Ex, Ey, Hx, Hy, Hz, which are intercrossed [3], then estimates impedance resistance and phase, finally the principle of the magneto telluric method take the result into two-dimension or three-dimension resistance conversion in order to get the underground electronic imager [11].

EM application attracts a growing interest of our marine geophysical scientists. In recent years, with the development of detecting instruments [10], data processing and retrieval, interpretation techniques, the reliability and usefulness of this method have been improved a lot. And it provides series basic techniques for Chinese seabed detection.

The technology of detection instruments dominated the marine geophysical development. During the beginning stage of Chinese marine geophysical science, the detection instruments' development plays an important role. And some of the pivotal technologies are monopolized by foreign companies. So we still need to improve the technology of detection instrument in the future.

2   THE IMPORTANCE OF MCSEM CURRENT

This integrated instrument with high parameter index such as intelligent control, low noise, and large dynamic range, is working out some solutions of some problems just like the record and save of seafloor environmental parameters in real time mode; the process of sensor making; the technology of the sealing and compressive stress bearing; the elimination of influence by the flowing water [14].

In recent process, the core of our research tend to invent the combined source electromagnetic current recorder with high level of sensitivity, low noise, low waste, and high reliability, a as well as high intelligent control, which used to collect the four level (Ex, Ey, Hx, Hy) and vertical (Ez) static electromagnetic data from the seafloor, meanwhile many environmental parameters such as time, sharing circular, orientation, obliquity, and temperature have to be recorded together.

Until now, as there are all kinds of current recorders for different detection method. But there are less controlled source electromagnetic current recorders, especially for marine detective [13]. On the other hand, the current record is a big problem in our monitor of marine detection [8].

Besides, our Chinese marine detection instrument is limited by the foreigner, especially in our current recorder [9]. We can only see some principle resource or some simple data, so we may have difficulty in doing our own marine instrument. The deeper research of our marine detection of the ocean, the more urge need of our marine instruments is.

Both wide frequency signal and the controlled source electromagnetic current signal are needed to be collected by electromagnetic marine combined source electromagnetic receiver [4]. Because of this, during the design of the instrument, some new technology should be considered such as more dynamic range and wider frequency.

Besides, the current collecting data of marine controlled source electro magnetic is much more complex than natural source land ones, as well as the data collection and process [5]. We can figure the MVO curve through the current data we get. After we record, the sync current data to do some further processing, and then we can get the result detection [6].

3   HARDWARE DESIGN

All the current recorders are made of the collection unit, the clock part. They are designed to be adapting to each other well.

3.1   Collection hardware unit

In our collection unit, we first use the hall sensor to make the analogy signal to PGA, then, the analogy signal which was amplified by PGA reach the ADC unit, so the analog signal will change to some digital signals. All the signals are sent to MCU to be reset or reorganized.

As the current data in MCSEM is a huge group quality, MCU then put the large amount of data in SD recorder in case of the data missing. At the same time, the RTC is a part unit connected to the clock board in order to make the data with the right time information. All the collection including the collection unit and the clock unit are been supplied by +5 V power to each unit.

Hall sensor is a magnetic sensor based on the hall efficiency. Hall efficiency is a little branch of the electro magnetic efficiency. This was found out in 1879 when Hall (A.H. Hall, 1855–1938) was studying the metal electoral features. Now the hall component depended on this Hall efficiency are used on the industry automation and the data processing.

As we all know, a high precision amplifiers can operate with inputs that exceed the supply rails, low quiescent current, reverse to supply protection and reliable. PGA we choose LT6910 to change the current signal. It can work between −40°C to 100°C. This is good for our ocean work because of the low temperature at the bottom of the ocean.

The analog to digital converter needs an integrated, low-noise program, so we choose ADS1282, which is suitable for the demanding needs of energy exploration and seismic monitoring

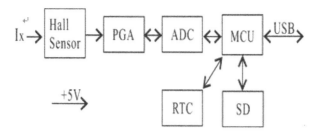

Figure 1.   The principle frame of collection unit.

environments. The flexible input MUX provides an additional external input for measurement, as well as internal connections.

Electromagnetic method is a branch of geophysics like seismic, so the ADC adapt to the seismic is suitable for our electromagnetic. It connects with PGA, and the PGA features outstanding low noise and high input impedance, allowing easy interfacing to MCU over a wide range of gains.

We choose MCU F320 to be our controlled core, for it is a new kind of USB microcontroller that designed by Cygnal Inc. Because of its greater instruction through put and larger capacity memory than usual, and its ISP function, it was considered as an ideal choice in the USB interface design. It is combined with 2304 byte RAM and 16 k flash risk, so it is a little SoC without other component when it works.

MCU plays an important link role between the software and the hard ware. It can executive the USB odder, the data communication, reset the mode of the output of signals, and the parameters of the three kinds of signals, which includes GPS unit parameter, the frequency of the DDS signals, and the amplitude value of PGA.

If we want to make the time accuracy to be 40 uS/hr, so we have to try to make the the high time accuracy clock unit, so the RTC is an important one to choose. After analysis about the marine controlled source electromagnetic, we finally choose DS1390. That provides hundredths of a second, seconds, minutes, hours, day, date, month, and year information.

The clock operates in either the 24-hour or 12-hour format with an AM/PM indicator. One programmable time-of-day alarm is provided. The date at the end of the month is automatically adjusted for months with fewer than 31 days, including corrections for leap year.

A temperature compensated voltage reference monitors the status of VCC and automatically switches to the backup supply if a power failure is detected. On this DS1390, a single open drain output provides a CPU interrupt or a square wave at one of four selectable frequencies. The DS1390 is programmed serially through an SPI-compatible, bidirectional bus.

In our collection unit, we choose the hall sensor of HALL sensor CHF 400F, taking account into our marine detective need the controlled source current to be 200 A, so we choose the open loop hall sensor. There are some indexes of the hall sensor shown in TAB 1.

3.2   *Clock hardware unit*

The time part contain GPS unit, DTCXO high stability clock, MCU controller, FPGA, band pass circuit (acceptor circuit), inner lithium battery and power translation circuit. As can be shown in the fig. 2, the high precision GPS unit can provide PPS and time information to the system. On the other hand, DTCXO connected to FPGA, provides high stability clock, which the permanence typical frequency is 50 ppb. Out of the system, this unit can communicate with the upper software though the USB joint in MCU, which can set different kinds output signals. In the contract, under the control of MCU, FPGA can generate CLK output PPS output, which also under control of MCU.

The new system has a perfect advantage that it makes the adaption of the PPS and DTCXO, in order to generate new PPS. Based on this FPGA, the new system also can get high precise time second pulse signal even though the signal invalid after the capture of GPS [12].

Table 1. The index of hall sensor CHF-400F.

Principle	Open loop Hall effect principle
Input current	50··· 600 A (DC, AC, Pulsed current)
Output voltage	±4 V
Frequency	DC~20 kHz
Accuracy	1.0%
Response time	<10 μS
Linearity	1.0%
Supply voltage	±12 V...15 V(±5%)
Consumption	25 mA
Isolation voltage	Between primary and secondary circuit: 3 kV RMS/50 Hz/1 min.
Operating temp.	−10°C··· +85°C

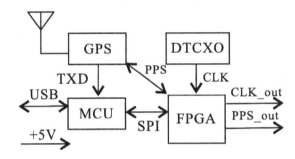

Figure 2. The principle frame of the clock unit.

GPS catches the time with high time accuracy. Because the marine detection need the high time accuracy with the current, in order to process and inverse the data. We choose Ublox-6T, which has five channel u-blox energy, and more than 100 million relations. It sets up or affiliate sets up less than 1 second for the first locating. The accuracy of the super sense is −160 Bm. So this u-blox can suitable to our need of time.

Passing from the MCU, the latest hardware platform for developing simple and low-end systems based on Alter MAX II devices. Our FPGA choose the MAX II devices, which includes the EPM240 device with 240 logic elements (LEs), 8,192 bits of user flash memory (UFM), and 25 uA standby power, in a 6 mm × 6 mm package.

This board also supports the capability to add a EPM240T100Cx device, it contains 64 general-purpose I/Os, two cap sense buttons, two push-button switches, user interface connector, external battery interface connector , and on-board power supply. So it controls the CLK signal out and the PPS out.

SOME IMPORTANT TECHNICAL INDEX

Max range: ±250 A
Accuracy: 0.05%
Time accuracy: 40 us/hr
Power dissipation: 2000 mW
Memory: 1 GB
Sample rate: 150 Hz
Band width: DC–50 Hz
Power voltage: 5 V

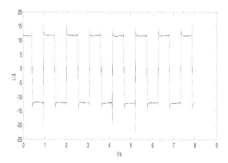

Figure 3.   The test result of field work.

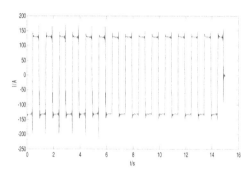

Figure 4.   The result of marine testing.

4   FIELD SURVEY RESULTS

We then connect the current recorder into the MT test, whose pole distance is 100 m, and make the test last 14 hr, in order to check the current recorder in field survey. Then we can get the apparent current data (in fig. 3) [7].

As can be shown in fig. 3, we can see the current recorder and the observation system are under good condition because of the continuation and smooth of the curve. During each square waves, the high point is the max current, our field work is a series work to make sure the precise current data without missing and the time sync. According to several tests, we can know the missing of the data is less compared before.

Marine field testing, we had tested this high current recorder in Dongsha sea in China, and the result is in fig 4. We can see the max current is reach 150A when we testing in the sea, this result clarify the efficiency of the recorder, means it can suffer the normal max range. After our testing in the ocean, the current recorder can contain the low temperature environment and can reach the max current data.

5   CONCLUSIONS

According to the test indoors and in the field, we can know that the current recorder used in MCSEM can record the current data timely and accuracy.

After the test field work and marine testing, we check the index of the current recorder, including the gain, band width, dynamic range, common mode rejection rate and so on.

The clock signals which generated by GPS and DTCXO, are combined with time information, checked the time sync of the system.

The test result of the field work and the marine data clarify the instrumentation is reasonable and effective.

ACKNOWLEDGMENT

Thanks go to Hai Yang Liu Hao first, which is a comprehensive geological and geophysical vessel with the finest, and foremost for introducing the author to the field of marine EM and making his knowledge and instrumentation available for future use and development. This generosity has had a truly significant impact on both the commercial and academic marine EM communities. Many thanks are also due to the numerous engineers and technicians who have helped design, develop, build and use the equipment described in this paper.

REFERENCES

[1] James Aird and Joseph A. Loux, 18th century regimental quick steps: from James Aird's A selection of Scotch, English, Irish & foreign airs, vol. II: for S. recorder, viol, or violin. 2006, Hannacroix, N.Y.: Loux Music Co. 7 p. of music.

[2] Nancy Beals and Marcella Lesher, Managing Electronic Resource Statistics. Serials Librarian, 2010. 58(1–4): pp. 219–223.

[3] Petr Beckmann and André Spizzichino, The scattering of electromagnetic waves from rough surfaces. International series of monographs on electromagnetic waves. 1963, goford, New York: Pergamon Press; distributed in the Western Hemisphere by Macmillan. viii, 503 p.

[4] J. van Bladel, Singular electromagnetic fields and sources. IEEE Press series on electromagnetic wave theory. 2000, New York: IEEE Press. xiii, 237 p.

[5] J. van Bladel and IEEE Antennas and Propagation Society, Electromagnetic fields. 2nd ed. IEEE Press series on electromagnetic wave theory. 2010, Hoboken, N.J.: IEEE/Wiley-Interscience. xiv, 1155 p.

[6] Henry G. Booker, Energy in electromagnetism. IEE electromagnetic waves series. 1982, London; New York: Peter Peregrinus on behalf of the Institution of Electrical Engineers. xiv, 360 p.

[7] Edwin L. Bronaugh and William S. Lambdin, Electromagnetic interference test methodology and procedures. A Handbook series on electromagnetic interference and compatibility. 1988, Gainesville, Va.: Interference Control Technologies.

[8] Christos Christopoulos, The transmission-line modeling method: TLM. IEEE/OUP series on electromagnetic wave theory. 1995, New YorkOxford: Institute of Electrical and Electronics Engineers; Oxford University Press. xi, 220 p.

[9] Donald G. Dudley and IEEE Antennas and Propagation Society, Mathematical foundations for electromagnetic theory. IEEE Press series on electromagnetic waves. 1994, New York: IEEE Press. xiii, 250 p.

[10] V. A. Fok, Electromagnetic diffraction and propagation problems. [1st ed. International series of monographs on electromagnetic waves, 1965, Oxford, New York: Pergamon Press. ix, 414 p.

[11] Linda Geisler, Esther Simpson, and Jan Mayo, Electronic Serials Cataloging Workshop (SCCTP). Serials Librarian, 2010. 58(1–4): pp. 14–19.

[12] Ameer H. Morad, GPS Talking For Blind People. Journal of Emerging Technologies in Web Intelligence, 2010. 2(3): pp. 239–243.

[13] Fethi M. Ulgen, Fault current recorder. 1945, University of Pittsburgh.

[14] V. V. Varadan, A. Lakhtakia, and V. K. Varadan, Field representations and introduction to scattering. Acoustic, electromagnetic, and elastic wave scattering. 1991, Amsterdam; New York; Sole distributors for the USA and Canada, Elsevier Science Pub. Co. xv, 355 p.

Engineering Technology and Applications – Shao, Shu & Tian (Eds)
© 2014 Taylor & Francis Group, London, ISBN 978-1-138-02705-3

A fusion edge detection method based on improved Prewitt operator and wavelet transform

Kunlin Yu & Zhiyu Xie

Changsha Aeronautical Vocational and Technical College, Changsha, Hunan, China

ABSTRACT: The traditional Prewitt operator only has two detection templates of horizontal and vertical direction, edge detection accuracy is not high. In order to improve the edge detecting precision and computing speed, the traditional Prewitt operator is improved in this paper, the edge detecting operator is increased from two to four. But Prewitt operators are sensitive to noise, pseudo edge prone to detection, and the detecting method based on wavelet transform edge not only has good anti-noise, but also can preserve the edge details, this paper puts forward a kind of edge detecting algorithm based on wavelet transform and improved Prewitt operator edge image fusion: first the source image is processed by using median filter method, then the filtered image is enhanced by histogram equalization method, the enhanced image is respectively detected by wavelet transform edge detection method and improved Prewitt operator edge detection method, finally the two edge detecting image is done image fusion by using wavelet image fusion algorithm based on adaptive wavelet image fusion algorithm, Experiments show that the fused image combines the advantages of two kinds of detection algorithm, the fusion detection method can suppress noise effectively, but also can improve the accuracy of edge detection. It is an ideal method of image edge detection.

Keywords: improved Prewitt operator edge detection; wavelet transform; adaptive wavelet image fusion; blade crack detection

1 INTRODUCTION

The traditional edge detection method is rely on edge detection operator to detect, such as Roberts operator, Sobel operator, Prewitt operator, LOG operator, Canny operator, which has the advantages of simple operation and strong real-time, but these operators have disadvantage of pseudo edge, while wavelet transform [1] can well remove the noise and can effectively reflect the image gray level change, In order to effectively to image edge detection, operator of the improvements and a variety of detection algorithm fusion has become a hot spot of image edge detection. This paper proposes a kind of edge image fusion detection Method based on the improved Prewitt operator edge detection and wavelet transform edge detection. The fusion algorithm combines the advantages of the two methods, and achieved better fusion effect.

2 PREWITT EDGE DETECTION OPERATOR

Prewitt edge detection operator is an edge template operator, which uses gray level difference of upper and lower, left and right neighboring points to detect edges.

2.1 *The traditional Prewitt edge detection operator*

The principle of the traditional Prewitt edge detection is in the image space using the template in two directions and image neighborhood convolution, the two direction templates respectively detects horizontal edges and vertical edges. This operator is usually expressed by the following formula:

$$G(x)=f(x+1,y-1)-f(x-1,y-1)+f(x+1,y)-f(x-1,y)+f(x+1,y+1)-f(x-1,y+1) \qquad (1)$$

$$G(y)=f(x-1,y+1)-f(x-1,y-1)+f(x,y+1)-f(x,y-1)+f(x+1,y+1)-f(x+1,y-1) \qquad (2)$$

$$P(x,y) = \max[G(x),G(y)] \text{ or } P(x,y) = G(x)+G(y) \qquad (3)$$

The Prewitt Operator template as follows:

$$G_x = \begin{pmatrix} 1 & 1 & 1 \\ 0 & 0 & 0 \\ -1 & -1 & -1 \end{pmatrix} \qquad (4)$$

$$G_y = \begin{pmatrix} -1 & 0 & 1 \\ -1 & 0 & 1 \\ -1 & 0 & 1 \end{pmatrix} \qquad (5)$$

Prewitt Operator think [2]: all that new pixel gray value greater than or equal to the threshold are edge points, because a lot of noise gray value is large, so the noise caused by misjudgment for edge point, but also for the smaller amplitude of edge point, Its edges have lost.

2.2 *Improved Prewitt edge detection operator*

The traditional Prewitt edge detection operator has only horizontal and vertical direction of two templates, Only the gray gradient changes on the two horizontal and vertical sensitivity, analysis of the two templates, G x templates within the first row is positive, so the image can obtain better edge detection [3], Sensitive image edge gradient is 180° direction, G y template within the third column is always positive, so the image on the right can get better edge detection, edge gradient is sensitive to 90° direction and the other direction, such as 0°, 45°, 135°, 225°, 270°, 315° these directions are not better edge detection. In order to increase the accuracy of operator in a pixel edge detection, templates can be increased from two to eight [4], Prewitt edge detection operator to increase from two to eight, the detection precision is greatly increased, the detection effect is increased, but the speed of operation is greatly reduced, in order to simultaneously determine the accuracy and speed, considering only increased two operator templates. Analysis of the original two operators that G x template: inside the first row is positive, so the image is get a good edge detection, template G y within the third column is always positive, so the image on the right can get better edge detection, and image below and left is not outstanding, not good edge detection, and so consider the operator template increases the two directions, so that the image below and left can be better edge detection operator template, increased operator templates as follows:

$$G_{XB} = \begin{pmatrix} -1 & -1 & -1 \\ 0 & 0 & 0 \\ 1 & 1 & 1 \end{pmatrix} \qquad (6)$$

$$G_{YL} = \begin{pmatrix} 1 & 0 & -1 \\ 1 & 0 & -1 \\ 1 & 0 & -1 \end{pmatrix} \qquad (7)$$

3   EDGE DETECTION ALGORITHM BASED ON WAVELET TRANSFORMATION

Suppose $\theta(x, y)$ is a two-dimensional smoothing function, and meet

$$\int_R \int_R \theta(x, y)dxdy = 1, \qquad \lim_{x^2+y^2\to\infty} \theta(x, y) = 0 \tag{8}$$

If we order

$$\theta_s(x, y) = \frac{1}{s^2}\theta(\frac{x}{s}, \frac{y}{s}) \tag{9}$$

Two dimensional signal f(x, y) smoothing is accomplished at different scales of s and θ s(x, y) do convolution operation to achieve, the first order derivative type along X and Y direction of two as the two basic wavelets [5] is shown as follows:

$$\phi_s^1(x, y) = \frac{\partial \theta_s(x, y)}{\partial x} = \frac{1}{s^2}\phi^1\left(\frac{x}{s}, \frac{y}{s}\right) \tag{10}$$

$$\phi_s^2(x, y) = \frac{\partial \theta_s(x, y)}{\partial x} = \frac{1}{s}\phi^2\left(\frac{x}{s}, \frac{y}{s}\right) \tag{11}$$

Suppose $f(x, y) \in L^2(R^2)$, Two dimensional wavelet transform in the scale of s includes two parts

$$WT_s^1 f(x, y) = f(x, y) * \phi_s^1(x, y) \tag{12}$$

$$WT_s^2 f(x, y) = f(x, y) * \phi_s^2(x, y) \tag{13}$$

Type as a vector in the form of

$$\begin{bmatrix} WT_{2^j}^1 f(x, y) \\ WT_{2^j}^2 f(x, y) \end{bmatrix} = \begin{bmatrix} WT_{2^j} f(x, y) \end{bmatrix} \tag{14}$$

In the formula, f s (x, y) is θ s (x, y) Images of the income smoothing, usually s take for 2 j (j $\in$ z), vector

$$\begin{bmatrix} WT_{2^j}^1 f(x, y) \\ WT_{2^j}^2 f(x, y) \end{bmatrix} = \begin{bmatrix} WT_{2^j} f(x, y) \end{bmatrix} \tag{15}$$

Be known as f (x, y) Binary wavelet, The $WT_{2^j}^1 f(x,y)$, $WT_{2^j}^2 f(x,y)$, model of local maximum points corresponding to smooth image in the corresponding position of the prominent point [6]. Its size reflects the position of the gray intensity, Therefore, the image edge points can be obtained as long as we detect wavelet transform modulus local maxima along the gradient direction.

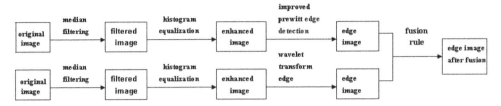

Figure 1.   Fusion improved Prewitt operator and wavelet transform edge detection method flow chart.

4   EDGE DETECTION METHODS BASED ON IMPROVED PREWITT OPERATOR AND WAVELET TRANSFORM IMAGE FUSION

4.1   *Image edge detection pretreatment*

4.1.1   *Image filtering*

Preprocessing of image Edge detection includes image filtering and image enhancement, edge detection algorithm is one order and two order derivative calculation based on image intensity, but the calculation of derivatives is very sensitive to noise, easily affected by noise, therefore, we must use the filter to reduce the noise, but in fact filter to reduce noise at the same time also led to the edge loss of strength. There are many methods of image smoothing filter, such as the mean filter, median filtering and so on, the image can reduce the noise by using the mean filter, but the filtered image edge become blurred, and the median filter to filter the image can reduce the noise (especially the salt and pepper noise), and can make the image edge is clear, especially for the salt and pepper noise, filtering effect is good, so here the median filter to preprocess the image edge detection.

4.1.2   *Image enhancement*

Image enhancement is to neighborhood (or local) intensity values have highlighted some significant changes, Histogram equalization [7] through some gray transform of the original image, so that the transformed image histogram can be evenly distributed, so as to be able to have similar gray area gray and possession of a large number of pixels in the original image to broadening, the tiny gray transform a large area of the display, the image is more clear, in order to gray difference expand the target and background and enhance the intensity contrast between them, in order to more easily detect the image edge details, the image histogram equalization enhancement method to preprocess the image edge detection.

4.2   *Detection methods of improved Prewitt operator and wavelet transform image fusion based on edge detection*

Fusion improved Prewitt operator and wavelet transform edge detection method process is shown in Fig. 1.

5   EXPERIMENTAL RESULTS AND ANALYSIS

Fig. 2 is the experiment of Lena image with Salt and pepper noise, Fig. 2(a) is the original Lena image, Fig. 2(b) is on the original image onto the salt and pepper noise, Fig. 2(c) is the Lena image after median filtering, filter will filter out salt and pepper noise in large, but also retains many important details of the image, the image is relatively clear, but in the denoising process will inevitably lose some detail edge. Fig. 2(d) is the image histogram equalization after median filtering, the purpose [8] is to expand the gray difference of target and background and enhance the intensity contrast between them, In order to more easily detect the image edge details, After image

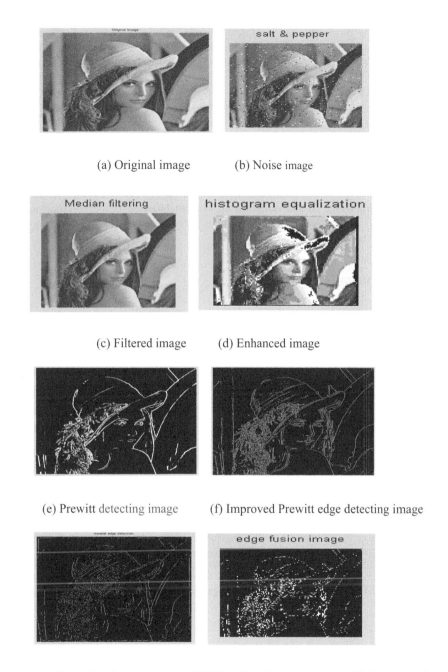

(a) Original image   (b) Noise image

(c) Filtered image   (d) Enhanced image

(e) Prewitt detecting image   (f) Improved Prewitt edge detecting image

(g) Wavelet transform edge detecting image (h) The edge detecting image of fusion algorithm

Figure 2.　The experimental results of different methods of image edge detection.

histogram equalization, image was enhanced, and the details of the image more clear. Fig. 2(e) is a Prewitt operator edge detection based on image enhancement, from Fig. 2(e) can be seen in Lena image edge details, many are not detected, lost a lot of edge detection details, The improved Prewitt operator edge detection result is much better than Fig. 2(e), the edge is clearly, as shown in Fig. 2(f), but still lost some gray value changes slowly edge detail, Fig. 2(g) is the image by means of image edge detection algorithm based on wavelet transformation, As can be seen from the image, edge contains the image detail information richer and more continuous, but the edge is rough. image edge detect method Based on improved Prewitt operator and wavelet transform edge image fusion edge is combined with the advantages of the two algorithms, which preserve the edge details, and remove the noise, no pseudo edge appears, as shown in Fig. 2(h) shows, it obtains the satisfactory effect of edge detection.

6 CONCLUSION

This paper presents a fusion method of edge detection of improved Prewitt edge operator and wavelet transform image, by analysis of the experimental results, the algorithm can suppress noise effectively, and can improve the accuracy of edge detection, detection effect is obviously superior to the single image edge detection.

ACKNOWLEDGEMENT

This work is supported by the scientific research foundation for Hunan Provincial Department of Education of China.

REFERENCES

[1] Abdullah, A.J., Mohammed, G.A. & Syed, A.A. (2013) Denoising of an image using discrete stationary wavelet transform and various thresholding techniques. *Journal of Signal and Information Processing*, 04(01), 33–41.
[2] Mark, S.N. & Alberto, S.A. (2007) *Feature Extraction & Image Processing*. 2nd edition. Academic Press, pp. 99–126.
[3] Zou, B.X. (2013) Improved detection method of Prewitt image edge. *Microelectronics and Computer*, 30(5), 23–25.
[4] Wang, K.F. (2011) Edge Detection of inner crack defects based on improved Sobel operator and clustering algorithm. *Applied Mechanics and Materials*, 1245(55), 467–471.
[5] Rafael, C.G., Richard, E.W. & Steven L.E. (2003) *Digital Image Processing Using MATLAB*. Prentice Hall, pp. 242–280.
[6] Tian, T., Zheng, Z. & Wang, H.X. (2012) Edge detection based on wavelet transform and fusion. *Applied Mechanics and Materials*, 1498(121), 3904–3908.
[7] Vorkle, S. (2013) Low-latency histogram equalization for infrared image sequences: a hardware implementation. *Journal of Real-Time Image Processing*, 8(2), 193–206.
[8] Turgay, C. (2012) Two-dimensional histogram equalization and contrast enhancement. *Pattern Recognition*, 45(10), 3810–3824.

Engineering Technology and Applications – Shao, Shu & Tian (Eds)
© 2014 Taylor & Francis Group, London, ISBN 978-1-138-02705-3

Using Directed-tree to control the production of remote sensing products

Bing Zhou, Shuang Xu & Xinxin Yang
School of Computer and Information Engineering, Henan University, Kaifeng, Henan, China

ABSTRACT: The Directed-tree is used to control the production of remote sensing products. The solutions of three key problems in the production process of remote sensing products are proposed, which are (1) how to produce the multi-period products; (2) how to improve the production speed with the use of some existing products; (3) the parallelization of the production process.

Keywords: remote sensing; multi-period product; single-period product; Directed-tree; data flow model

1 INTRODUCTION

In order to develop new algorithms for the production or image processing of remote sensing products more conveniently, one basic idea is to assemble a new one by making use of some existing processing algorithms, just as piling up the building blocks[1-3]. The dependency among algorithms (the calling sequence) can be described through the workflow or the algorithm flow diagram model[4-9]. But because the number of data sources used by single-period and multi-period products varies, the actual production process of remote sensing products is totally different even with the same algorithm calling sequence. A visual data flow model has been put forward by the author in [10] to describe the production process of remote sensing products, which includes not only the calling sequence among algorithms, but also the links between algorithms and the data being used. The visual data flow model is a conceptual model. In this paper, the Directed-tree structure is used to control the production of remote sensing products. The algorithms mentioned in this paper can be programmed with certain computer language with slightly changing.

There are three key problems[10] in the production of remote sensing products, which are (1) how to produce the Multi-period products; (2) how to improve the production speed with the use of some existing products; and (3) the parallelization of the production process. The problems mentioned above can be described and solved effectively by the visual data flow model in document [10]. In this paper, the control algorithms based on the Directed-tree could solve these problems also.

2 THE ALGORITHMS TO CONSTRUCT THE DIRECTED-TREE

Any arbitrary calling sequence of algorithms can be expressed by AOV-network (the Directed digraph) in which vertexes and arcs stands for algorithms and the calling sequence respectively. In an AOV-network, there should not be any directed circle and can be only one terminal point. Figure 1 illustrates an example of an algorithm calling sequence.

Suppose that A, B, C and D are single-period products, then with only one group of original data, the production process of product D can be illustrated by Figure 2[7]. If A, B, C are still single-period products except D, which is multi-period product, then with N groups of original data, the production of D can be illustrated by Figure 3[7].

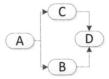

Figure 1. An example of the algorithm calling sequence.

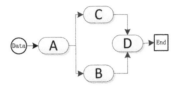

Figure 2. An example of the production process for single-period products.

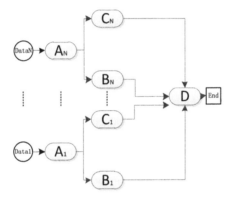

Figure 3. An example of the production process for multi-period products.

It is obviously that because of the different number of data sources used by single-period and multi-period products, the actual production process of remote sensing products is totally different even with the same algorithm calling sequence.

The production process of single-period products is consistent with the calling sequence of the algorithms. For the multi-period products, the algorithms of single-period products would be called many times to process different groups of data while the algorithms of multi-period products are called only once. The whole process of the production is tree structure.

Because most algorithm flow designers are used to describe the algorithm calling sequence with workflow or flow chart (in which not including any data information), in the first step of the production of remote sensing products, the original data and the algorithm calling sequence should be combined to determine the actual process of the production, based on whether they are single-period or multi-period products. In this paper, the production process will be described by the Directed-tree. Next the basic algorithms for constructing Directed-tree of single-period products, the modification for constructing Directed-tree of multi-period products, and how to prune the Directed-tree when existing products be used are introduced respectively.

2.1 *The basic algorithms to construct the Directed-tree*

The basic Directed-tree construction algorithm is used when all algorithms are for single-period products. The input requirements are one original data and a directed-graph that describes the algorithm calling sequence. The directed-graph is expressed by two-tuples. As for the example in

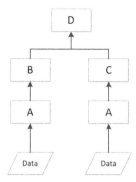

Figure 4. The Direct-tree for the production of single-period products mentioned in Figure 1.

Figure 1, the corresponding directed-graph is expressed as {(B, D), (C, D), (A, B), (A, C), (data, A)}. Two-tuple (B, D) meanings that product D will be produced after product B.

The basic idea to construct a Directed-tree is to expand the directed-graph by making copies of the same node. Figure 4 illustrates the Directed-tree for the production of single-period products mentioned in Figure 1. In Figure 4, although there are two nodes of A, they are same essentially, so as the two data nodes. The duplication of these nodes is just for the purpose of constructing the tree structure. While programming, node A and data node will be created only once, and their reference (the pointers) are actually used in Directed-tree.

The specific description of the algorithm is as follows:

(1) Create an array named Array as a global variable to store all the nodes. Determine and create the root node (node D in this example), and insert its reference into Array. Take the root node as the parent node and execute step (2).

(2) Search though the set of two-tuples and find out those tuples that their second element are the parent node (the parent node in this example is node D and the corresponding two-tuples are (B, D) and (C, D)). And if there is any, execute step (3).

(3) Take turns to carry out the operations below on each two-tuple that is found in step (2):

    a) *Fetch out the first element in the two-tuple and see if there is any array element in Array which has the same name with it. If there exists, execute (b). Otherwise, execute (c).*

    b) *Add a child node to the current parent node. Assign the reference of the element found in step (a) in Array to the new child node to make sure that the new child node points to it. Then continue step (d).*

    c) *Create a new element and insert its reference into Array. Then add a new child node to the current parent node and make the new node point to the element.*

    d) *Treat the first element which is fetched out from the two-tuple as the parent node and continue step (2).*

2.2 *Modified algorithms for multi-period products*

From the example in Figure 3, it can be observed that in the production of multi-period products, with the increasing of data sources, the running time of the single-period products algorithms will increase, meanwhile the algorithms of multi-period products will be executed only once. In terms of one data source, the calling sequence of the algorithms has not changed during the whole process. Thus to construct the production process for multi-period products, the first step is to construct the Directed-tree for single-period products with only one data source. Then make copies of the Directed-tree constructed above for the other data sources. During the duplication, new algorithm node stands for single-period products should be created while the algorithm node stands for multi-period products remains unique. In addition, as different data will be processed by

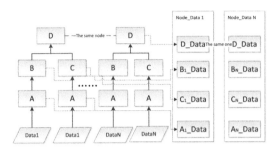

Figure 5.   The Directed-tree for the production of multi-period products mentioned in Figure 1.

different algorithms of single-period products, it is necessary to introduce the data structure named Node_Data to store the corresponding data information related to every algorithm node. Algorithm nodes are used to reflect the algorithm calling sequence while the Node_Data reflects the data to be processed by each algorithm.

Figure 5 illustrates the Directed-tree for the production of multi-period products mentioned in Figure 1.

After the adoption of the Node_Data, the original algorithm in 2.1 has been modified as follows:

(1) Create an array named ArrayData as a global variable to store all the data related to each node. Determine the root node (node D in this example), and create the corresponding array element D_Data and insert D_Data into ArrayData. Take the root node as the parent node and then execute step (2).

(2) Search though the set of two-tuples and find out those tuples that their second element are the parent node (the parent node in this example is node D and the corresponding two-tuples are (B, D) and (C, D)). And if there is any, execute step (3).

(3) Take turns to carry out the operations below on each two-tuple that is found in step (2):

   a) *Fetch out the first element in the two-tuple and create a corresponding child node ChildNode. Add ChildNode to the current parent node as a child node and see if there is any array element in ArrayData which has the same name with the element. If there exists, execute (b). Otherwise, execute (c).*

   b) *Assign the reference of the array element found in step (a) in ArrayData to the Node_Data of ChildNode to make sure that ChildNode keeps the same data with the element. Then continue step (d).*

   c) *Create a new array element NodeData and insert it into ArrayData. Make the Node_Data of ChildNode points to NodeData and continue step (d).*

   d) *Treat the first element which is fetched out from the two-tuple as the parent node and continue step (2).*

2.3   *The pruning for production process with existing products*

In the actual production, the results of some algorithm nodes may be given directly for the purpose of speeding up by skipping over the execution of these nodes and their child nodes. This kind of "pruning" would change the original algorithm flow. A specific example is illustrated in Figure 6.

Suppose that the results of C1 and D2 are already given, the gray nodes in Figure 6 need not to be executed. It means that when the result of a certain node is given, it is unnecessary to execute the algorithms related to the node and all its child nodes. It is worthwhile to note that the execution of A1 can be skipped over from the viewpoint of C1, but in order to ensure the execution of B1, A1 must be executed.

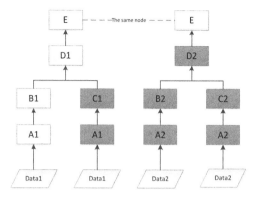

Figure 6.   The changes of the algorithm flow caused by using intermediate products.

The judgment of the execution can be determined by the algorithm below:

(1) Create an array named ArrayResult as a global variable to store the information of whether the algorithm of each node should be executed or not.
(2) Traverse every group of nodes cyclically. Begin from the root node in each group and continue step (3).
(3) According to the name of the given node, search for the Node_Data with the same name in the current set of Node_Data and carry out the operations below:

a) *Check whether the current node has been assigned any existing products or not. If it has, continue step (b). Otherwise, continue step (c).*
b) *Return directly (as the result of the algorithm related to the parent node has been provided, the execution of the algorithms related to all its child nodes can be skipped over).*
c) *If the current node is a leaf node (meaning that it has no child node), continue step (d). Otherwise, continue step (e).*
d) *Set the corresponding array element in ArrayResult as true, meaning that the algorithm related to this node should be executed. Then return to step (2) and continue.*
e) *Take turns to carry out the operations in step (3) on the child nodes of the current node. If there isn't any child node, then return.*

(4) Set the corresponding array element in ArrayResult as false, meaning that the algorithm related to the node is unnecessary to be executed. Then return to step (2) and continue.

3   USING DIRECTED-TREE TO CONTROL THE PROCESS OF THE PRODUCTION

In the actual production, a master-control program is needed to control the execution. It follows three constraints.

(1) Make sure that each node begins its production only under the condition that all the requirements it needs are met. In other words, the production of all its child nodes must be finished before the current node begins its production.
(2) In order to improve production efficiency, parallel scheduling should be applied to nodes that can be executed parallel.
(3) For the nodes provided with existing products, the production of the nodes themselves and their child nodes can be all skipped over.

Take the same example in Figure 6. In the production process above, the processing of Data1 and Data2 by single-period algorithm nodes A, B, C and D can be done parallel. In addition, the

299

algorithms related to nodes C1, D2 and its child nodes A2, B2 and C2 can be skipped over because of the use of existing products. The whole production begins with the execution of A1 and continues in order of A1-B1-D1. After the execution of E, the whole production has been completed.

In order to automate the whole production process, it is essential to find out nodes that can be processed parallel and drive them. Besides, new producible nodes should be found when the production of the current nodes has been finished. Consequently, the whole driver can be divided into two sections: a module to search for producible nodes and a parallel driver module to drive them. The searching module is responsible for searching producible nodes in the current Directed-tree and submits the nodes it has found to the parallel driver module. The parallel driver module will take the responsibility of executing the algorithms related to these nodes (the algorithms related to these nodes can be executed parallel and the granularity can be set according to the hardware resources), and informing the searching module when the execution is finished. The searching module will then set the state of the nodes accordingly and search for next producible nodes.

The searching module uses the Directed-tree established above and the searching algorithm is as follows:

(1) In the stage of initialization, create an array named ArrayNode as a global variable to store nodes that are producible.
(2) Traverse every group of nodes cyclically. Begin with the root node in each group and continue step (3).
(3) According to the name of the given node, search for the Node_Data with the same name in the current set of Node_Data and carry out the operations below:

  a) Check if the current node has been assigned any existing products or has finished its production. If so, continue step (b). Otherwise, continue step (c). Be notice that if the current node stands for a multi-period product, it is necessary to check whether the production of all its child nodes has been completed or not.
  b) Return directly without scanning any child node and continue the traversal in step (2).
  c) If the current node is a leaf node or the production of all its child nodes has been finished, continue step (d). Otherwise, continue step (e).
  d) Insert the current node into ArrayNode and return.
  e) Take turns to carry out the operations in step (3) on the child nodes of the current node.

The algorithm above can still be executed when receiving the production completion notification for a certain node. When checking whether the production of the current node and its child nodes have completed or not, the new finished nodes should be taken into account. And when the node mentioned in the completion notification is found, the state of the node should be set as completed.

4   CONCLUSIONS

A method to control the production of remote sensing products based on Directed-tree has been proposed in this paper. It resolves the three key problems in the production process of remote sensing products, which are (1) how to produce the Multi-period products; (2) how to improve the production speed with the use of some existing products; (3) the parallelization of the production process.

REFERENCES

[1] Wolf R E, Roy D P, Vermote E. MODIS land data storage, gridding, and compositing methodology: Level 2 grid [J]. IEEE Transactions on Geoscience and Remote Sensing, 1998, 36(4):1324–1338.
[2] Running S W, et al. Terrestrial remote sensing science and algorithms planned for EOS/MODIS [J]. INT. J. Remote Sensing, 1994, 15(17):3587–3620.

[3] The MODIS Science Data Support Team. SDS-TO92. MODIS Level 1A Earth Location: Algorithm Theoretical Basis Document Version3.0. [Z]. 1997.

[4] Xia Deng-wen, Wang Hong, Shi Sui-xiang, et al. Design on flow of visualization modeling system for ocean remote sensing information extraction [J]. Acta Oceanologica Sinica, 2005, 27(3): 97–103.

[5] Wei Jian-xin, Wei Dong-qi, Wu Xin-cai. Workflow-based remote sensing visualization modeling system and its distributed scheduling algorithm [J]. Arid Land Geography, 2009, 32(2):304–309.

[6] Fan Jun-tao, Li Guo-qing, Kang Lin. Data Storage in Workflow of Image Processing[J]. Computer Simulation, 2007, 24(8): 182–184, 204.

[7] Wang Cheng-yi, Zhao Zhong-ming. Design and implementation of processing flow system for remote sensing image[J]. Science of Surveying and Mapping, 2006, 31(6): 105–106.

[8] Liu Ding-sheng, Chen Yuan wei, Li Jing-shan. The Study on the Development Patterns of Remote Sensing Satellite Ground Pre-processing Systems [J]. Remote Sensing Information, 2008, 5: 87–91.

[9] Li Jing-shan, Chen Yuan-wei, Liu Ding-sheng. The Design and Implement of Multi-satellite Ground Pre-processing System Based on Workflow Technology [J]. Remote Sensing Technology and Application, 2008, 23(4): 428–433.

[10] Zhou Bing, Li Jia-guo, Wu Guan-feng, et al. A Visual Dataflow Model for Production of Remote Sensing Products [J]. Journal of Henan University (Natural Science), 2013, 43(1): 74–78.

Engineering Technology and Applications – Shao, Shu & Tian (Eds)
© *2014 Taylor & Francis Group, London, ISBN 978-1-138-02705-3*

Research on the testing technology of marine electric field sensor

Li Zhou & Ming Deng
Key Laboratory of Geo-detection, Ministry of Education, China University of Geosciences, Beijing, China

ABSTRACT: The marine electric field observation technology is widely used in the military field such as subsea oil and gas hydrate exploration as well as the electric field detection of warship. To achieve low noise detection of weak seafloor electric signals, one of its key technologies is the low noise marine electric field sensor. For a group of electrolysis technology developed electrodes, in order to detect whether they meet the expected targets, it is necessary to carry out a comparison test for its various techniques. Our test gives an evaluation mainly on signal conversion capability, difference potential stability and self-noise of the electric field sensor. Through establishing a series of test solutions, the original test results of different electrodes were gotten. The analyzed results show that the electrode fully achieve the expected goals, it also proves the effectiveness of the test solution.

Keywords: marine electric field sensor; test; difference potential stability; self-noise

1 INTRODUCTION

For a long time, people thought it was unfeasible to observe electric field of the seafloor covered by high conductive seawater. Until they found a large number of undersea sulfide, oil and gas, hydrate, geophysicists started to develop the marine electromagnetic method to study the electrical structure of the Subsea following media [1–3]; At the meantime, affected by the World War II, more and more military targets need the marine electromagnetic method as technical support, such as detecting the leak electromagnetic fields of submarine and ship, mines target identification, port security [4]. The detection by marine electromagnetic method must first achieve the ocean electromagnetic signal observation records. Driven by the western marine research institutions [1, 3], the oil service companies [2], arms manufacturers [4], the technique has made great progress. Due to the low amplitude of seafloor electric field (hundreds of μV to nV level) and frequency bandwidth (0.001 Hz to hundreds of Hz) features [5–8], a high sensitivity, low-noise marine electric field sensor must be developed to ensure the collection of precious, high-quality seafloor electric signal.

2 PRINCIPLE OF ELECTRODE PROCESS

The current mainstream electrode preparations of Ag/AgCl are powder metallurgy and electrolysis. Powder metallurgy is to purify AgCl powder and Ag powder, and through the high temperature sintering process, electrodes would have better homogeneity and stability [9]. One of its key technologies is the preparation of appropriate AgCl powder because its electrochemical characteristics directly affect the electrode's noise. The deficiencies are the complexity of process steps, high cost and difficulties to control all aspects of the parameters. They may easily lead to lower yields.

Electrolytic preparation of Ag/AgCl electrode technology is widely used in medical biotechnology, electrochemistry. Different from the Powder metallurgy method, the key step is depositing a dense layer, large-area uniform AgCl on Ag foil substrate surface by electrolysis method. Then

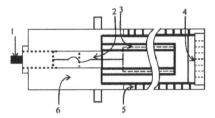

1. Watertight connectors; 2. Epoxy sealant; 3. Silver pieces; 4. Fixed cover; 5. Perforated pipe; 6. The sensor housing

Figure 1.   Structure of the marine electric field sensor.

Figure 2.   Difference potential stability test equipment connection diagram.

wrapping shield in silver foil and lead cable to solve the watertight problem. This process is proven to significantly reduce the sensor noise level and has high-yield and low cost[10–11]. Fig. 1 shows a block diagram of an electric field sensor.

3   TEST PROGRAM

In order to evaluate electrode for more angle, this test mainly focuses on testing the small-signal observation capacity of the electrode, different potential stability and self-noise. All the electrodes will be immersed in 3.5% NaCl solution for 24 hours at room temperature to remove the adverse effects of temperature fluctuations on the electrode performance.

3.1   *Small-signal observation capability test programs*

Small-signal observation capability test aims at verifying whether the electrodes can correctly convert electric signal to the electrode lead. Do self-test on multiple voltmeter and sources before test to ensure normal operation of voltmeter's each channel. Make all electrodes immersed in 3.5% NaCl solution and connect signal generator's output to two power supply electrodes. Signal generator output is set to 1 mHz, 10 mV sinusoidal signal and the electrode is connected to the switching unit's (34972A) input. Put T1 electrode as the reference electrode and set the acquisite parameters to DC voltage, 1min sampling interval, 100 mV range. Then continuously observe for about 2 hours.

3.2   *Difference potential stability test program*

Difference potential stability test is to make a short test on all channels to ensure that the noise of channel is less than 10 μVpp. There are 14 electrodes which are immersed in 3.5% NaCl solution. As shown in Fig. 2, T1 electrode is the reference electrode. Electrode lead wire is connected to the switch unit (34972A) input. Acquisition parameters are set to DC voltage, 1 min sampling interval and 100 mV range. Meanwhile record the ambient temperature and do continuous observation for about 3 days.

3.3   *Self-noise test program*

Self-noise test step is observing the self-noise through the dynamic signal analyzer (SR785) and low noise amplifier (EM A10), to ensure that the self-noise to test conditions; all electrodes are put

Figure 3. Self-noise test equipment connection diagram.

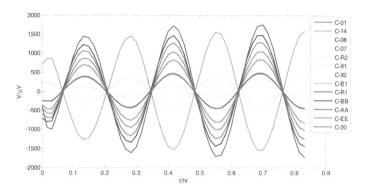

Figure 4. Figure of Small-signal test's results.

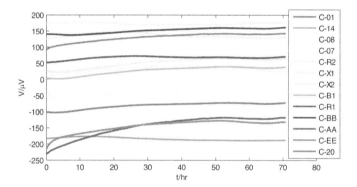

Figure 5. Difference potential stability test results.

in 3.5% NaCl solution, as shown in Fig. 3, the electrode leads access to the input of the amplifier, the amplifier gain set to 60 dB, dynamic signal analyzer parameter is set to 50 Hz bandwidth, single-ended input, and observing the noise PSD one by one.

4 TEST RESULTS

4.1 *Small-signal observation capability test results*

Fig. 4 shows the small-signal observation result. It proves that all electrodes can correctly convert electric signals. As shown in Fig. 4, the polarity, amplitude and drift of the electric field switching signal differ from another electric, and they are related to the geometric position of the electrode in the tank.

4.2 *Difference potential stability test results*

The temperature stabilized between $19 \pm 0.5°C$ within three days; Fig. 5 shows the measurement result of difference potential stability. In spite of the slightly worse performance of C-AA, C-EE,

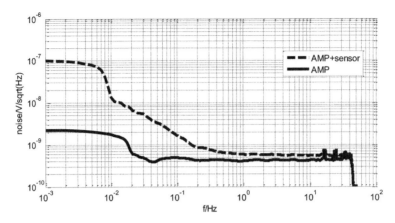

Figure 6.   The test results of CUGB's electrode self-noise.

Table 1.   7 pairs of electrode self-noise test results.

Electrode ID	The cumulative noise of electrode and amplifier (nV/sqrt(Hz))	Self-noise (nV/sqrt(Hz))
C 01-20	0.632	0.42
C 07-R2	0.687	0.50
C 08-14	0.831	0.68
C AA-R1	0.526	0.22
C B1-X1	0.666	0.47
C BB-X2	0.681	0.49
C EE-T1	0.661	0.46

performance of the remaining electrics are stable. Typical potential is about $100\,\mu V$ and typical differential potential drift is about $20\,\mu V/day$.

4.3   Self-noise test results

According to Fig. 6, self-noise test results of an electrode is shown, in which the solid black line is the result of shorting the amplifier input, and the dotted line is the cumulative noise of electrode and amplifier. As seen at 1 Hz frequency point, the amplifier noise is 0.476 nV/sqrt (Hz), the cumulative noise of electrode and amplifier is 0.687 nV/sqrt(Hz). Except for the amplifier's self-noise, the calculated noise of the electrodes is 0.50 nV/sqrt(Hz) @ 1 Hz. The results of seven pairs of electrode are in Table 1, they show the noise distribution is in 0.22~0.68 nV/sqrt (Hz) range.

4.4   Test conclusion

– By analyzing the data of small signal observation capability test, difference potential stability test and self-noise test, we knew that the electrodes worked properly. Their self-noise located in 0.22~0.68 nV/sqrt (Hz) range.
– The typical potential of electrode is about $100\,\mu V$ and typical differential potential drift is about $20\,\mu V/day$.

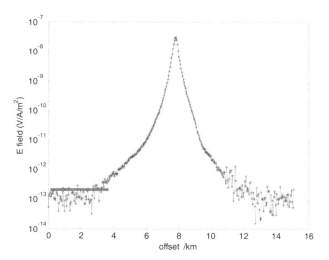

Figure 7. The test results of MVO.

5 SEA TEST

Our research group did the marine controlled source electromagnetic detection test in 2013 in somewhere of the northern South China Sea, the water depth is about 1000 m. An amplitude offset curve of the observed electric field component data was worked out after sending current normalization, navigation data calibration, attitude data calibration and azimuth rotation. Fig. 7 shows the observed MVO curves of the Ex component when 1 Hz send jobs , the maximum value of Ex component is about 4×10^{-8} V/A/m^2, local noise of the system marked by the red line is around $10-13$ V/A/m^2, it shows channel dynamic range is better than 112 dB, the effective observation range is from 2 km to 14 km.

6 CONCLUSIONS

In order to evaluate the low-noise characteristics of electrode from more aspects, we prepared a detailed test plan. We did the small signal observation capacity test, difference potential stability test, self-noise test for the 14 electrodes and got the sea trial results of partial electrodes. Test results showed the scientificalness and effectiveness of the test program, and also proved the low-noise characteristic of the electrodes.

ACKNOWLEDGMENT

We thank Guangzhou Marine Geological Survey (GMGS) for the opportunity of Sea trials and thank all the staffs of "HAIYANG LIUHAO" for their active cooperation during sea test.

REFERENCES

[1] Cox, C.S., Constable, S., Chave, A.D., et al. (1986) Controlled source electromagnetic sounding of the oceanic lithosphere. *Nature*, 320(1), 52–54.
[2] T. Eidesmo, S. Ellingsrud, L.M. MacGregor, S. Constable, M.C. Sinha, S. Johansen, F.N. Kong & H. Westerdahl (2002) Seabed logging (SBL), a new method for remote and direct identification of hydrocarbon filled layers in deepwater areas. *First Break*, (20)3.

[3] Constable, S. & Srnka, L.J. (2007) An induction to marine controlled-source electromagnetic methods for hydrocarbon exploration. *Geophysics*, 72(2), 3–12.

[4] Filloux, J.H. (1987) Instrumentation and experimental methods for oceanic studies. In: Geomagnetism, (ed. J.A. Jacobs), Academic Press, pp. 143–248.

[5] Goto, T., Kasaya, T., Machiyama, H., Takagi, R., Matsumoto, R., Okuda, Y., et al. (2008) A marine deep-towed DC resistivity survey in a methane hydrate area, Japan Sea. *Exploration Geophysics*, 39, 52–59.

[6] Constable, S., Key, K. & Lewis, L. (2009) Mapping offshore sedimentary structure using electromagnetic methods and terrain effects in marine magnetotelluric data. *Geophysical Journal International*, 176, 431–442.

[7] Dearth, L.C. (1987) *Investigation of Electrode Materials for Low Frequency Electric Antennas in Sea Water*. California: Navy and Marine Corps.

[8] Constable, S., A. Orange, G.M. Hoversten & H.F. Morrison. (1998) Marine magnetellurics for petroleum exploration Part 1. A seafloor instrument system, *Geophysics*, 63, 816–825.

[9] Deng, M., Liu, Z.G., Bai Y.C., et al. (2002) The principle and testing technology of marine electric field sensor. *Geology and Prospecting*, 38(6), 43–47.

[10] S.J. Davidson, P.G. Rawlins & H. Jones. (2006) The Choice of Sensor Type for Electric Field Measurement Applications. *MARELEC 2006. Amsterdam*.

[11] Geir Bjarte Havsgard, Hans Roger Jensen & Alexander Kurrasch. (2011) Low Noise Ag_AgCl Electric Field Sensor System for Marine CSEM and MT Applications. [Online] Available from : http://www.emgs.com.

Engineering Technology and Applications – Shao, Shu & Tian (Eds)
© 2014 Taylor & Francis Group, London, ISBN 978-1-138-02705-3

The testing technology of low noise chopper amplifier

Xinyu Shi, Kai Chen, Zhouying Lu & Yu Sun
Key Laboratory of Geo-detection, Ministry of Education, China University of Geosciences, Beijing, China

ABSTRACT: The marine controlled source electromagnetic method is used in exploring natural gas hydrate undersea and oil gas, and one of the most pivotal techniques is how to observe the electric signal from the seabed at a low noise level. This property of marine controlled source electromagnetic receivers mainly depends on the electric-field sensors and the chopper amplifiers. In order to check the technical specifications of the chopper amplifiers which were developed, we made a thorough test on their frequency response, voltage noise, CMRR and power consumption. The result indicates that our test method is scientific and effective and confirms the low-noise property of chopper amplifiers at the same time.

Keywords: marine controlled source electromagnetic method; chopper amplifier; low noise

1   INTRODUCTION

Benefited from the rapid development of semiconductor technology, in recent years, there are a lot of low noise amplifiers on the market that the voltage noise is below 10 nV/sqrt(Hz), and some even better than 1 nV/sqrt(Hz). The most of them, however, have their best working frequency band above 1 kHz, meanwhile, the low-frequency 1/f noise is obvious, and the corner frequency is between 100 Hz and 1 Hz. They can barely meet the demand of observing the seabed electric signal. Therefore, people invented chopper amplifiers to serve the need to observe low frequency weak signal (like measuring high precision temperature), these amplifiers improved the signal-to-noise ratio (SNR) of low frequency signal as well.

Table 1 gives the technical parameters of low noise amplifiers, low noise integrated chopper amplifiers and low noise discrete chopper amplifiers which are already on the market. As can be seen, the existing integrated operational amplifiers and integrated chopper amplifiers are not suitable to observe the seabed electric signal, while the discrete type chopper amplifiers made by discrete components are clearly better than the first two. Cox and some other people in America SIO built the chopper amplifier circuit for their self-made electric-field sensor with discrete components in 1985. It became a chopper amplifier with a low noise about 1.3 nV/sqrt(Hz), and it was integrated into the seafloor electromagnetic receiver Mk IV[1]. Dietmar Drung, a German, and his partners made an improvement on the chopper amplifier technology. They drove down the noise to 0.73 nV/sqrt(Hz) and applied this technique to SQUID devices made by Magnicon company[2]. The British arms manufacturer Ultra-PMES company also developed chopper amplifiers for their electric-field sensors. The biggest marine electromagnetic exploration services nowadays – Norway EMGS company integrated the electric-field sensors and amplifiers provided by Ultra-PMES at its magnetotelluric acquisition station[3]. British EM Company invented a series of low noise observation equipment, A10 stands for a type of low noise amplifiers among them, and their module performance is similar to the products made by Ultra-PMES company[4].

2   PRINCIPLE OF A CHOPPER AMPLIFIER

Electrical signals of Marine controlled source on the bottom have some features. Their energy is weak and they are at relatively low frequencies. Moreover, they have high dynamic ranges, so the

Table 1. Current foreign low noise amplifier's parameters comparison.

Manufacturer	Model	Craft	The typical noise nV/sqrt(Hz)			Corner frequency (Hz)
			@1kHz	@1 Hz	@0.1 Hz	
Linear	LT1028	BJT	0.85	1.5	5	3.5
TI	OPA211	BJT	1.1	4	About 25	100
Analog Device	AD797	BJT	0.9	About 10	About 100	100
NF	SA200F3	Unspecified	1	3	10	1
Analog Device	ADA4528-1	Integrated chopping	5.6	6	6	Unspecified
Cirrus	CS3001	Integrated chopping	6	6	6	0.08
TI	ADS1282	Integrated chopping	5	5	5	Unspecified
SIO	Unspecified	Discrete chopping	1.3	1.3	1.3	0.02
Dietmar Drung	Unspecified	Discrete chopping	0.7	0.7	0.7	0.01
Ultra-PMES	Unspecified	Discrete chopping	0.5	0.5	0.5	0.01
EM	A10	Discrete chopping	0.5	0.5	0.5	0.02

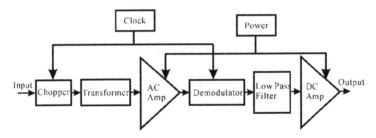

Figure 1. Principle of a chopper amplifier.

observation circuit must has low drift and high dynamic ranges, and has low noise especially at low-frequency stage[5]. Based on these requirements, the amplifier firstly chops the low frequency weak signals and modulates them to thousands of Hz, then amplifies them in low noise. As the modulation frequency is high, we can ignore the 1/f low noise that the amplifier makes. With a much better noise-signal ratio, the treated signals which are demodulated to low frequency ones would make the amplifier have an advantage in the feature of low-frequency noise.

Fig. 1 shows the principle of a chopper amplifier. It comprises a clock circuit, a chopper circuit, a transformer coupled circuit, an AC amplifier circuit, a synchronous demodulation circuit, a low pass filter circuit and a DC amplifier circuit. Clock circuit generates signals to control the chopper circuit and the synchronous demodulation circuit. Low-frequency signals of the electric-field sensor are converted into AC signals. Transformer coupled circuit does the impedance conversion. AC amplifier circuit amplifies AC signals in low noise. Synchronous demodulation circuit completes demodulation of AC signals while low pass filter circuit restrains broadband noise and DC amplifier circuit gets high-gain performance[6].

3 TEST PLAN

To get a comprehensive evaluation of the chopper amplifier we made, the test is going to execute in terms of properties like voltage noise, frequency response, input resistance and power consumption.

Testing equipment mainly contains low noise DC power supply, dynamic signal analyzer, signal source and oscilloscope. We choose Agilent E3640A desktop DC power supply, output power 30 W, optional output 20 V/1.5 A or 8 V/3 A, current resolution 1 mA with readback accuracy better than 0.15% + 5 mA, voltage resolution 10 mV with readback accuracy better than 0.05% + 5 mA, ripple

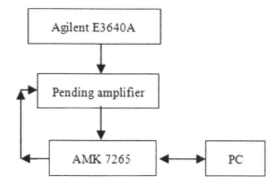

Figure 2. Schematic diagram of frequency response test equipment.

Figure 3. Schematic diagram of voltage noise test equipment.

wave smaller than 0.5 mVrms. Signal source is the AMK Signal recovery model-7265 phase-locked amplifier, output voltage range 1 μVrms to 5 Vrms, frequency range 1 mHz to 250 kHz, output impedance 50 Ω, THD smaller than −100 dB. The source is used to detect frequency response, input and output resistance kind of indices. Owing to the limited space, the paper just gives the testing plan about two specifications.

3.1 *Frequency response*

Fig. 2 shows how the devices connect. DC power supply gives power to the amplifier, and the output port of signal source is connected to the input port of the amplifier. When the treated signals are sent as input, the signal source calculates the frequency response of a frequency point according to the input and output signals, and also records its amplitude and phase. Then, the signal source constantly changes to another frequency under the control of a PC, and do the same thing again and again.

3.2 *Noise*

As shown in Fig. 3, we use Agilent E3640A DC power supply to support the amplifier. The amplifier's input port is shorted and the gain is set to 60 dB. The dynamic signal analyzer SR 785 then gets the output. We set the analyzer's bandwidth to 50 Hz and continuously observe for 10 hours. Ultimately, a noise power spectrum will be worked out.

4 TEST RESULTS

4.1 *Frequency response*

Fig. 4 shows the results of frequency response. The amplifier's passband is approximately 60 dB, −3 dB bandwidth is 0.01 Hz to 50 Hz. Band pass filter is composed of a first order high-pass filter and a second order low-pass filter.

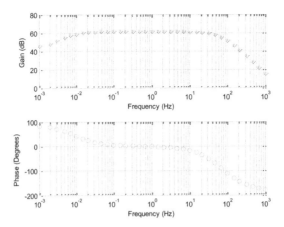

Figure 4.   Test results of frequency response.

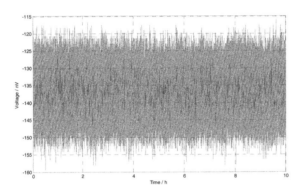

Figure 5.   Temporal noise waveforms.

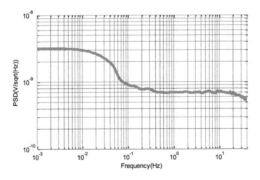

Figure 6.   PSD calculations of voltage noise.

4.2 *Noise*

With the amplifier's input shorted, the results in 10 hours can be seen in Fig. 5. The equivalent peak-peak value of input voltage noise is about 40 nVpp.

Calculating the noise data by PSD, the results are as shown in Fig. 6. Voltage noise at the frequency point is 0.7 nV/sqrt(Hz) and the corner frequency is about 0.1 Hz.

312

Table 2. Low noise chopper amplifier's test results.

Index name	Test result	Remarks
Voltage noise	0.7 nV/sqrt(Hz)@1 Hz	Corner frequency fc = 0.1 Hz
Current noise	27.5 pA/sqrt(Hz)@1 Hz	
−3 dB bandwidth	0.01 Hz–50 Hz	
Passband gain	About 60 dB	
Channel number	3	
Input resistance	About 300 Ω	
Output resistance	About 20 Ω	
Maximum undistorted output voltage	8.5 Vpp@5 Hz	
Dynamic range	Better than 110 dB	
Crosstalk	104 dB@1 Hz	
Common Mode Rejection Ratio	102 dB@1 Hz	
Offset voltage	Smaller than ±1 μV	
Input/output	Differential/single-ended	
Power source	±6 V–±9 V	
Power consumption	180 mW@±7 V	
Volume	180 mm*110 mm*30 mm	Length*width*height

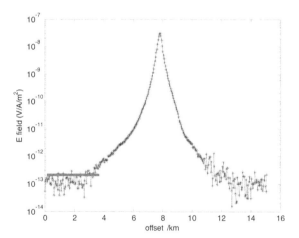

Figure 7. MVO results of sea test.

4.3 Test conclusion

Table 2 gives the detailed test results of other technical specifications. The indoor test results indicate the properties of low noise, low power consumption, broadband and high dynamic range.

5 SEA TEST

Our research group did the marine controlled source electromagnetic sounding test at somewhere in Northern South China Sea in 2013. The water there was 1000 meters deep. An amplitude offset curve of the observed electric field component data was worked out after sending current normalization, navigation data calibration, attitude data calibration and azimuth rotation. Fig. 7 shows the observed MVO curve by Ex component while 1 Hz sent operations. The maximum value of Ex component is about $4*10^{-8}$ V/A/m^2. Local noise of the system marked by the red line is around 10^{-13} V/A/m^2. From that we know the channel dynamic range is better than 112 dB and the effective observation range is from 2 km to 14 km.

6 CONCLUSION

The low noise chopper amplifier was built according to the low noise observation requirements of benthoal electric field. A series of comprehensive test proved that the schemes are feasible.

ACKNOWLEDGMENT

Our team thanks Guangzhou Marine Geological Survey for offering the sea test opportunity, and also thanks all the staffs on HAI YANG LIU HAO for their active cooperation during the experiment.

REFERENCES

[1] Marine EM Laboratory, "Webb, S.C., S.C. Constable, C.S. Cox, and T.K. Deaton, A seafloor electric field instrument, Geomag. Geoelectr., 37, pp. 1115–1129, 1985." http://marineemlab.ucsd.edu.
[2] Magnicon, http://www.magnicon.com.
[3] S.J. Davidson, P.G. Rawlins and H. Jones, The Choice of Sensor Type for Electric Field Measurement Applications, Marelec 2006, Amsterdam.
[4] EM ELECTRONICS, http://www.emelectronics.co.uk/spec/A10.html.
[5] Deng M, Du G, Zhang Q H, YU P and DENG J W, The Characteristic and Prospecting Technology of the Marine Magnetotelluric Field, Chinese Journal of Scientific Instrument, 2004, 25(6):742–746.
[6] Geir Bjarte Havsgard, Hans Roger Jensen and Alexander Kurrasch, Low Noise Ag_AgCl Electric Field Sensor System for Marine CSEM and MT Applications. http://www.emgs.com 2011.

Engineering Technology and Applications – Shao, Shu & Tian (Eds)
© 2014 Taylor & Francis Group, London, ISBN 978-1-138-02705-3

A review on biomechanics and its applications in human movement

Xiaomei Lv, Yuanyuan Hou & Jinli Yang
Hebei Institute of Physical Education, Shijiazhuang, Hebei, China

ABSTRACT: All life forms on earth, including humans, are constantly subjected to the universal force of gravitation, and thus to forces from within and surrounding the body. Through the study of the interaction of these forces and their effects, the form, function and motion of our bodies can be examined and the resulting knowledge applied to promote quality of life. Under gravity and other loads, and controlled by the nervous system, human movement is achieved through a complex and highly coordinated mechanical interaction between bones, muscles, ligaments and joints within the musculoskeletal system. Any injury to, or lesion in, any of the individual elements of the musculoskeletal system will change the mechanical interaction and cause degradation, instability or disability of movement. On the other hand, proper modification, manipulation and control of the mechanical environment can help prevent injury, correct abnormality, and speed healing and rehabilitation. Therefore, understanding the biomechanics and loading of each element during movement using motion analysis is helpful for studying disease etiology, making decisions about treatment, and evaluating treatment effects. In this article, the history and methodology of human movement biomechanics, and the theoretical and experimental methods developed for the study of human movement, are reviewed. It is suggested that further study of the biomechanics of human movement and its applications will benefit from the integration of existing engineering techniques and the continuing development of new technology.

Keywords: human movement; musculoskeletal system; biomechanics; application

1 INTRODUCTION

All life forms on earth are subjected to gravity. How the form, function and motion of these biological systems are affected directly or indirectly by gravity has been a subject of scientific study over centuries. It started with the development of mechanics that were concerned with the behavior of physical bodies when subjected to forces or displacements, and the subsequent effect of the bodies on their environment. More than 3000 years ago Babylonian astronomers studied the problem of planetary positions. The Renaissance astronomer Nicolaus Copernicus (1473–1543) published his heliocentric theory in 1543, which was in contrast to the widely accepted geocentric system model popular during the 13th–17th centuries that had been proposed by Aristotle (384–322 BC) and Claudius Ptolemy. During the early modern period (1453–1789), renowned scientists, including Galileo, Kepler and Newton, laid the foundation of classical mechanics. Galileo Galilei (1564–1642) worked theoretically and experimentally on the motions of bodies, particularly of falling bodies. Johannes Kepler (1571–1630) empirically discovered his laws of planetary motion which gave an approximate description of the motion of planets around the sun and provided one of the foundations for Isaac Newton's theory of universal gravitation. In 1687 Isaac Newton (1643–1727) used the newly developed mathematics of calculus to give a detailed mathematical explanation of mechanics and published his classic Philosophiae Naturalis Principia Mathematica. In this three-volume publication he formulated the law of universal gravitation and the three laws of motion which can be applied to the motion of planets in the heavens and all forms of movements on earth [1,2]. Since all life forms on earth are under the influence of universal gravitation, there is no doubt that mechanics not only governs the motions of lifeless objects, but also affects the form, motion and function of the biological systems on earth. The mechanical interactions within the biological systems and with the environment received attention from early scholars such as Aristotle, Leonardo da Vinci (1452–1519), Galileo Galilei (1564–1642), Johannes Kepler (1571–1630), Rene Descartes (1596–1650), and Isaac Newton, among others, as well as from scientists of the present day. Their efforts on the discovery of these mechanical interactions have come to form a discipline of research called biomechanics.

Biomechanics is the study of continuum mechanics (that is, the study of loads, motion, stress, and strain of solids and fluids) of biological systems and the mechanical effects on the body's movement, size, shape and structure. Mechanical influence on biological systems can be found at multiple levels, from molecular and cellular, all the way up to the tissue, organ and system level. Therefore, the study of biomechanics of humans ranges from the inner workings of a cell, the mechanical properties of soft and hard tissues, to the development and movement of the neuro musculoskeletal system of the body. Molecular biomechanics refers to the study of how mechanical forces and deformation affect the conformation, binding/reaction, function and transport of biomolecules, such as DNA, RNA and proteins, and how mechanobiochemistry couples inbiomolecular motors and ion channel flows, etc. Cellular biomechanics is concerned with the study of how cells sense mechanical forces or deformations, and transduce them into biological responses, especially for the study of how mechanical forces alter cell growth, differentiation, movement, signal transduction, protein secretion and transport, gene expression and regulation. The properties of living tissues are affected by applied loads and deformations, and tissue biomechanics is mainly concerned with the growth and remodeling of tissues as a response to applied mechanical stimuli. For example, the effects of elevated blood pressure on the mechanics of the arterial wall, and the behavior of cardiomyocytes within a heart with a cardiac infarct, have been widely regarded as instances in which living tissue is remodeled as a direct consequence of applied loads. Another example is Wolff's law of bone remodeling, developed by Julius Wolff (1836–1902) in the 19th century. Wolff's laws states that the internal architecture of the trabecular bone and the external cortical bone in a healthy person or animal will adapt to the loads placed on the bone and it will remodel itself over time to become stronger to resist that type of loading. The converse is also true. If the loading on a bone decreases, the bone will become weaker owing to turnover and a lack of stimulus for continued remodeling that is required to maintain bone mass [2,3].

At the system level, mechanical factors also affect the form, performance and function of the musculoskeletal system. Human movement is achieved by a complex and highly coordinated mechanical interaction between bones, muscles, ligaments and joints within the musculoskeletal system under the control of the nervous system [3]. Muscles generate tensile forces and apply moments at joints with short lever arms in order to provide static and dynamic stability of the body under gravitational and other loads while regularly performing precise limb control [3]. Any injury or lesion of any of the individual elements of the musculoskeletal system will change the mechanical interaction and cause degradation, instability or disability of movement. On the other hand, proper modification, manipulation and control of the mechanical environment can help prevent injury, correct abnormality, and speed healing and rehabilitation. Therefore, understanding the biomechanics and loading of each element is helpful for studying disease etiology, making treatment decisions and evaluating the effects of treatment. However, because of ethical considerations and technological limitations, direct measurement of the forces transmitted in the human body is possible only in exceptional circumstances, such as through instrumented implants [4–7]. A further challenge is the redundant nature of the musculoskeletal system. In the human body there are more joints and muscles than are necessary for performing our daily motor tasks. Therefore, a certain task can be achieved by more than one musculo-skeletal strategy. However, this compensatory mechanism is essential for coping with the consequences of injuries or diseases to the musculoskeletal system, but it makes it difficult to determine the internal forces noninvasively.

Currently, combining noninvasive measurements of the movement, such as the position of segments and the strain on force-measuring instruments, with computer graphics-based anatomicalmodeling is a useful approach to estimating these loadings. In this approach it is essential to integrate the techniques of motion analysis and medical imaging. They include measuring human motion and external loads, developing three-dimensional (3D) computer graphics-based biomechanical models based on medical imaging, calculating internal forces and validating the results. A validated 3D computer biomechanical model can then be applied to the simulation of various movements and surgical procedures. A recording of an electromyogram(EMG) of the active muscles can further be used to understand the muscle activity during human movement [8,9]. However, developing a precise and noninvasive method for measuring the internal force within the human body for clinical and other purposes still remains a great challenge in the field of human biomechanics and motion analysis. In this article, a brief introduction to the history and methodology of human movement biomechanics is given, followed by a review of the theoretical and experimental methods developed by the authors for the study of human movement. The clinical applications of these methods in various orthopedic and neurological pathology contexts, and in sports medicine are also described. In [10], a robust methodology to reconstruct injuries sustained in real world crashes are developed using vehicle and human body finite element models.

2 HUMAN MOTION ANALYSIS

All movements and changes in movements arise from the action of forces, both internal and external. A change in the force acting on an object is necessary for moving an object from a stationary position or for changing its velocity. The amount of change in the velocity of an object depends on the magnitude and direction of the applied force. Newton's laws of motion give a clear relationship between the changing force and the resultant change in movement, and this is applicable to all forms of movement, including human locomotion [3]. Human motion analysis is the systematic study of human motion by careful observation, augmented by instrumentation for measuring body movements, body mechanics and the activity of the muscles. It aims to gather quantitative information about the mechanics of the musculoskeletal system during the execution of a motor task [11]. A special branch of human motion analysis is gait analysis which is specific to the study of human walking, and is used to assess, plan and treat individuals with conditions affecting their ability to walk. The following is a brief account of the history of human motion analysis/gait analysis.

In pursuing mental and physical excellence, the ancient Greeks found that harmony of mind and body required athletic activity to complement the pursuit of knowledge. Their interest in sport and human movement can be seen in the predominance of kinematic representations of Greek athletics in the artistic media. With the mechanical, mathematical and anatomical paradigms developed during Greek antiquity, the great philosopher Aristotle wrote the first book on human movement (About the Movements of Animals), which is the first scientific analysis of human and animal movement in terms of observing and describing muscular action and movement [2].

The great figure from the Renaissance, Leonardo da Vinci, was the first to study human anatomy through dissections of at least 30 cadaveric bodies. He was particularly interested in the structure of the human body as it relates to performance, center of gravity, balance and center of resistance. He identified muscles and nerves in the human body and described the mechanics of the body during standing, walking up- and downhill, rising from a sitting position, jumping, and human gait. He also suggested that cords be attached to a skeleton at the points of origin and insertion of the muscles to demonstrate the progressive action and interaction of various muscles during movement [2].

Even though Leonardo da Vinci had produced very detailed descriptions of the human body, it was not until the mid-16th century when Andreas Vesalius published the first anatomy book, De Humani Corporis Fabrica, which gave him the credit of being "The Father of Modern Anatomy" [2]. Another two chief figures of the Renaissance were Galileo Galilei and Giovani Alfonso Borelli. While Galilei applied mechanical theory to study animal movement, and published a treatise De Animaliam Motibus (The movement of Animals) [2], Borelli published De Motu Animalum (On the Motion of Animals) in which successfully clarified muscular movement and body dynamics. Borelli also estimated the center of mass of the entire body by stretching the body out on a rigid platform that was supported on a knife edge and then repositioned until it was balanced [1]. Borelli is often considered the "Father of Biomechanics" [2].

During the age of enlightenment (17th–19th centuries), Wilhelm Eduard Weber and his younger brother, Eduard Friedrich Weber published the results of their collaborative study on the mechanism of walking in mankind in. Since then, the study of human motion has greatly progressed from an observatory/descriptive science to one based on quantitative measurements, for which E'tienne-Jules Marey made the most important breakthrough. He determined a series of actions of human locomotion in various forms according to measurements of the effort exerted at each moment using graphical methods, and glass-plate and celluloid film chronophotography [1,2]. Carlet added a heel and separate forefoot chambers to Marey's pressure-recording shoes, obtaining more measurements of the onset and duration of weight-bearing, and the vertical reaction force [1].

Among the noted scientists, Eadweard Muybridge began what was probably the first assessment of gait and deserves to be considered the "Father of Modern Gait Analysis" [8]. Since Muybridge found that it was not possible to capture the rapid limb movements of horses in motion by eye [12], he improved photography by creating a camera with a shutterspeed of up to 1/100 of a second and recorded the motion in men, women, children, animals and birds [1]. With the aid of computer vision and the techniques of pattern recognition and artificial intelligence, photogrammetry using photographs, radiographs, and video images continued to develop after Muybridge's invention. Stereophotogrammetry, the technique of measuring 3D landmark coordinates, was then developed by the "Father of Stereophotogrammetry", Carl Pulfrich [13]. Christian Wilhelm Braune and Otto Fischer then used analytical close-range photogrammetry, combined with the geometrical properties of central projection from multi-camera observations, to estimate the 3D position data fromdigitized and noisy image data [14]. In the 1890s, they used stereophotogrammetry and ground reaction force (GRF) measurements to study the biomechanics of human gait under loaded and unloaded conditions using their pioneering 3D mathematical technique based on Newtonian mechanics [15]. They also carefully studied the mass, volume center of mass, and body segments of three adult male cadavers and introduced the use of regression equations for estimating body segment parameters, based on the length and mass of body segments. To date, the mathematical methodology of gait analysis developed by Braune and Fischer has essentially remained unchanged in modern gait analysis [1,2]. It is noted that during the age of Enlightenment, computers had not yet been developed. Therefore, manual involvement was necessary for determining the specificmarkers on the human body in each image, which was not only laborious and time consuming, but also one of the main sources of errors. This inevitably limited the clinical application of the mainly two dimensional (2D) measurement and analysis of the human motion during that period.

In the modern era (20th century – today), human motion analysis developed rapidly as the knowledge of anatomy and mechanics, and measurement technology was progressively established. In the 1940s, Harold Eugene Edgerton (1903–1990) pioneered high-speed stroboscopic photography that was used to photograph objects in motion at a frequency of several million exposures per second. During the 1970s, video camera systems, such as infrared high-speed cameras, began their widespread application in the analysis of pathological gait, producing detailed motion analysis results within realistic cost and time constraints. With the collocation of high-speed computers and video camera systems, 3D analysis of human motion became feasible. However, it had to wait until after World War II to make its debut in clinical 3D applications.

After the war there were many retired soldiers who had sustained limb injuries and who needed orthopedic treatment, prostheses, orthoses and subsequent rehabilitation for recovery of functional activities, especially for level walking. In order to provide better medical services and achieve treatment goals, numerous investigators devoted themselves to the study of gait analysis for clinical applications. Among them was Verne Thompson Inman who began by applying the theory of mechanical engineering to clinical problems, such as designing prostheses for amputees. He studied the biomechanics of locomotion and proved the assumption that the most efficient gait pattern is achieved by minimizing vertical and lateral excursions of the body's center of gravity (COG). He also identified the so-called gait determinants for normal walking, i.e., features of the movement pattern that minimizes these COG excursions [1]. He suggested that these features be used to determine whether a movement pattern is normal or pathological. Following Inman's work, Jacquelin Perry divided the gait cycle into five stance phase periods and three swing phase periods [16]. David Sutherland refined the definition of the gait cycle to have three periods of stance, namely initial double support, single limb stance, and second double support [17]. He also carried out a comprehensive investigation on the walking patterns in a total of nearly 300 normal children aged from 1 to 7 years for the study of the development of mature gait [1,18,19].

Because of the tedious nature of processing and analyzing cine film, the need for a more scientific approach to automate the process led to the development of the 3D Vicon motion capture system by Professor John P. Paul and his PhD students, Mick Jarrett and Brian Andrews. This system was made to capture human movement data in numerical digits instead of analog data, and which is now widely used in the study of human motion [1].

Apart from kinematic measurements using video cameras, Dr. J. Robert Close used a 16-mm movie camera with a sound track for studying the phasic action of muscles in subjects after muscle transfers for poliomyelitis. Doctor Close was the first to record synchronously the kinesiological electromyography (KEMG) of one muscle and kinematic data on cine film [14]. Jacquelin Perry pioneered using fine wire electrodes to record the gait electromyogram (EMG) and used it as a primary clinical tool in determining the appropriateness of surgical procedures to correct gait deformities [14]. Because muscles are the engines for producing active movements, EMG has been a useful assessment tool for detecting the electrical activity of specific muscles and assessing their contribution to movement or gait. Between 1944 and 1947, Vern Inman and colleagues added KEMG to other measurements, i.e., 3D force and energy, in the study of walking in normal subjects and amputees, and thus significantly moved the science of gait analysis forward.

The essential aim of human motion analysis is to understand the mechanical function of the musculoskeletal system during the execution of a motor task. Since the forces for generating movement in the musculoskeletal system are too difficult to measure noninvasively, combined experimental and mathematical modeling approaches have been used. The power of mathematical modeling is that it enables the values of parameters which are difficult or impossible to measure to be calculated from the values of quantities which can be measured. An example of this approach is determining the force in a spring which cannot be measured directly. With the measureable deformation of the spring, the force in the spring can be calculated using Hooke's law that relates the deformation with the force in the spring (Fig. 1). Therefore, spring from the measured deformation of the spring (Dx) using Hooke's law. Numerous studies have used mathematical modeling in conjunction with noninvasive experimental measurements to calculate non-measurable internal forces in the musculoskeletal system through inverse dynamics analysis. The measurable values of quantities in human motion analysis are usually the motion of the musculoskeletal system defined by skin markers and measured by the motion capture systems, and the external forces applied to this musculoskeletal system measured by force plates. With the 3D trajectories of skin markers obtained using

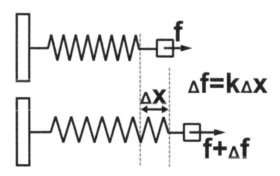

Figure 1.   Determining the non-measurable force (Df) in the spring from the measured deformation of the spring (Dx) using Hooke's law.

stereo-photogrammetry, and the GRFs and center of pressure (COP) measured using force plates, intersegmental forces and internal moments at the joints of the lower limbs are then calculated from the solution of equations based on Newton's laws of motion. This approach is called inverse dynamics analysis.

3 CONCLUSIONS

Universal gravitation affects all life forms on earth. Our body is constantly subject to forces from within and surrounding the body. Through the study of the interaction of the forces and their effects on the body, the form, function and motion of our biological body can be studied and the resulting knowledge can be applied to promoting quality of life. Using stereophotogrammetry-based human motion analysis techniques combined with measured GRFs and muscle activities, deviations from normal kinematic, kinetic or EMG patterns can be identified and then used to evaluate neuro-musculoskeletal conditions, to help with subsequent treatment planning, and to assess the efficacy of treatments in various patient groups. It can also be used to improve athletic performance and to help identify posture or movement related problems in people with injuries or diseases. Further establishment of the biomechanics of human movement and its clinical applications will benefit from the integration of existing engineering techniques and the continuing development of new technology.

REFERENCES

[1] Clinical Gait Analysis, University of Vienna, Austria. History of the study of locomotion. http://www.univie
.ac.at/cga/history/.
[2] Nigg, B.M. & Herzog, W. (2007) Biomechanics of the musculo-skeletal system. 3rd edition. Hoboken NJ: John Wiley & Sons; 2007.
[3] Lu TW, Yen HC, Chen HL, Hsu WC, Chen SC, Hong SW, et al. Symmetrical kinematic changes in highly functioning older patients post stroke during obstacle-crossing. Gait Posture, 2010; 31:511–6.
[4] Bergmann G, Deuretzbacher G, Heller M, Graichen F, Rohlmann A, Strauss J, et al. Hip contact forces and gait patterns from routine activities. J Biomech 2001; 34:859–71.
[5] Heller MO, Bergmann G, Deuretzbacher G, Durselen L, Pohl M, Claes L, et al. Musculo-skeletal loading conditions at the hip during walking and stair climbing. J Biomech 2001; 34:883–93.
[6] Lu TW, Taylor SJ, O'Connor JJ, Walker PS. Influence of muscle activity on the forces in the femur: an in vivo study. J Biomech 1997; 30:1101–6.
[7] Stansfield BW, Nicol AC, Paul JP, Kelly IG, Graichen F, Bergmann G. Direct comparison of calculated hip joint contact forces with those measured using instrumented implants. An evaluation of a three-dimensional mathematical model of the lower limb. J Biomech 2003; 36:929–36.
[8] Allard P, Stokes IAF, Blanchi JP. Three-dimensional analysis of human movement. Champaign, IL: Human Kinetics Publishers; 1995.
[9] Lu TW. Geometric and mechanical modelling of the human locomotor system. Oxford: University of Oxford; 1997.
[10] Golman A, Danelson K A, Miller L E, Stitzel, J D. Injury prediction in a side impact crash using human body model simulation. Accident Analysis and Prevention, v 64, p 1–8, 2014.
[11] Cappozzo A, Della Croce U, Leardini A, Chiari L. Human movement analysis using stereophotogram-metry. Part 1: theoretical background. Gait Posture 2005; 21:186–96.
[12] Andriacchi TP, Alexander EJ. Studies of human locomotion: past, present and future. J Biomech 2000; 33:1217–24.
[13] Center for Photogrammetric Training, Ferris State University. History of photogrammetry. Michigan: Ferris State University; 2008.
[14] Chiari L, Della Croce U, Leardini A, Cappozzo A. Human movement analysis using stereophotogram-metry. Part 2: instrumental errors. Gait Posture 2005; 21:197–211.
[15] Sutherland DH. The evolution of clinical gait analysis part l: kinesiological EMG. Gait Posture 2001; 14:61–70.

[16] Perry J. Gait analysis: normal and pathological function. Thorofare, N.J.: Slack; 1992.
[17] Wang TM, Chen HL, Hsu WC, Liu MW, Chang CF, Lu TW. Biomechanical role of the locomotor system in controlling body center of mass motion in older adults during obstructed gait. J Mech 2010; 26:195–203.
[18] Sutherland DH. The development of mature gait. Gait Posture 1997; 6:163–70.
[19] Liu MW, Hsu WC, Lu TW, Chen HL, Liu HC. Patients with type II diabetes mellitus display reduced toe-obstacle clearance with altered gait patterns during obstacle-crossing. Gait Posture 2010; 31:93–9.

Engineering Technology and Applications – Shao, Shu & Tian (Eds)
© 2014 Taylor & Francis Group, London, ISBN 978-1-138-02705-3

A generative and discriminative hybrid model for object detection and tracking

Qian Liu
Cyber Physical System R&D Center, The Third Research Institute of Ministry of Public Security, Shanghai, China

Jian Wang
Cyber Physical System R&D Center, The Third Research Institute of Ministry of Public Security, Shanghai, China
School of Electronic Information and Electrical Engineering, Shanghai Jiao Tong University, Shanghai, China
Shanghai Chenrui Information Technology Company, Shanghai, China

Qianying Hou
Cyber Physical System R&D Center, The Third Research Institute of Ministry of Public Security, Shanghai, China

Wei Zhao
Information Center, China Food and Drug Administration, Beijing, China

Lin Mei
Cyber Physical System R&D Center, The Third Research Institute of Ministry of Public Security, Shanghai, China
Shanghai Chenrui Information Technology Company, Shanghai, China

Chuanping Hu & Yi Li
The Third Research Institute of Ministry of Public Security, Shanghai, China

ABSTRACT: A novel object detection and tracking method is proposed based on a generative and discriminative model. Firstly, stable features of the object are extracted and represented by DAISY feature descriptor, which has computational efficiency and invariant property for illumination, deformation, viewpoint and scale. In this way, the object generative model is constructed. Secondly, Hough Forest classifier is adopted as discriminative model and the input patches of object are trained. Moreover, the discriminative codebook is updated by computing the similarity measurement between the detection results of the following video sequence and the codebook, and makes the codebook dynamic and adaptive. Experiment results show that the proposed algorithm, combining DAISY feature descriptor and Hough Forest, has satisfactory tracking precision and good real time performance, and works well under the condition of partial occlusions and different image resolutions.

Keywords: detection and tracking; generative model; discriminative model; DAISY; Hough Forest

1 INTRODUCTION

The chief challenge of Object detection and tracking is a difficulty in handling the appearance variability of intrinsic appearance variability such as pose variation and shape deformation and extrinsic such as illumination change, camera motion and camera viewpoint.

For object detection and tracking, there are, roughly speaking, two different main models: generative model and discriminative model [1–2]. The generative approach constructs an adaptive model for prediction or match, the adaptive model is built by extracting features of objects. This method requires a good feature description to represent the object and a state model which is able to cover as much state as possible. Kalman Filter [3] use a state transition model and measurements to update the estimated state; [4–5] represents the appearance of the object with a set of feature descriptor which has invariant property for scale, rotation and affine.

The discriminative approach treat object detection as a binary classification problem, aimed at finding the optimal hyper plane and decision function to make the difference between object and background largest. In [6], support vector machine (SVM) is used to separate the object from the background.

Generally speaking, generative model is difficult to meet real-time requirements because of large amount of calculation in feature description and match; discriminative model based approach requires extensive training samples, and it is hard to implement in practice.

To display their respective advantages, researchers try to combine generative model and discriminative model [1, 7, 8]. A generative and discriminative hybrid model for object detection and tracking is proposed in this paper. Firstly, stable features of object are extracted and represented by DAISY feature descriptor, which has computational efficiency and invariant property for illumination, deformation, viewpoint and scale. Secondly, Hough Forest classifier is adopted as discriminative model and the input patches of object are trained. Moreover, the discriminative codebook is updated by computing the similarity measurement between the detection results of the following video sequence and the codebook, and makes the codebook dynamic and adaptive. Experiment results show that the proposed algorithm, combining DAISY feature descriptor and Hough Forest, has satisfactory tracking precision and good real time performance, and works well under the condition of partial occlusions and different image resolutions.

2 GENERATIVE MODEL

To make the object model could well describe the appearance changes of the objects, firstly, feature points of the image are extracted through Difference of Gaussian and represented by DAISY feature descriptor.

2.1 Feature extraction

The main task of object model description is to extract stable feature points, which has salient structural information and invariant property for scale and viewpoint. In this paper, DoG is used to extract feature points from the image.

2.2 Feature description

Engin et al. proposed an efficient feature description methods-DAISY, which not only inherits the advantages of SIFT, but also has high efficiency, strong robustness to partial occlusions and different image resolutions. Next, we introduce DAISY feature descriptor.

DAISY computes the descriptor of every point from its neighbors. Fig 1 depicts the structure of DAISY descriptor: the weighted sums of gradient norms are computed by convolution of the gradients in different directions with several Gaussian filters.

Fig 1: the black dot represents a feature point, each circle represents a region, each red dot represents the sample locations, the radius is proportional to the standard deviations of the Gaussian kernels, it is recommended that using smaller blur kernels near the center and larger away from it. By changing the size of the Gaussian kernel can effectively remove the impact of distant location gradient. Compared with SIFT/SURF, DAISY uses a circular grid to replace the weighted sums of gradient norms by convolutions of the gradients in different direction with several Gaussian

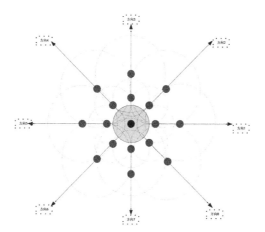

Figure 1.   The structure of DAISY descriptor.

Table 1.   Computing time comparison.

Image size	DAISY (s)	SIFT (s)	PCA-SIFT (s)	SURF (s)
800 × 600	3.8	252	243	83
1024 × 768	6.5	432	417	149
1280 × 960	9.8	651	630	217

Figure 2.   Generative model.

filters. The efficiency of DAISY comes from the fact that convolutions can be implemented very efficiently especially when using Gaussian filters.

In Table 1, we show typical computation times of different local descriptors such as SIFT, PCA-SIFT, SURF and DAISY over all the pixels of various sized images: we can conclude that the computation time of DAISY descriptors is almost 100 times faster than SIFT's, and it is also better than PCA-SIFT, SURF. The efficient computation is of great significance for real-time object detection and tracking.

2.3   Forming generative model

To properly handle occlusions, in this section, we extract several meaningful patches from the object, for each patch we compute its DAISY descriptors.

In Fig. 2, the object (car) image is divided into N patches. For each patch, we extract its feature points using DoG, then randomly extract M feature points to represent it.

$$I_n = \{k_{n1}, k_{n2}, \ldots k_{nm}, \ldots, k_{nM}\} \tag{1}$$

I_n denotes the representation of the nth patch of the object.

$$I_n^{DY} \leftarrow \text{DAISY}(k_{n1}, k_{n2}, \ldots k_{nm}, \ldots, k_{nM}) \tag{2}$$

I_n^{DY} is DAISY descriptors for the nth patch of the object. The appearance model for the object can be denoted as:

$$\{I_1^{DY}, I_2^{DY}, \ldots I_n^{DY}, \ldots, I_N^{DY}\} \tag{3}$$

3 DISCRIMINATIVE MODEL

Stable features of object are extracted and represented by DAISY feature descriptor, then generative model which is the input of the Hough Forest classifier is constructed. We train a Hough Forest classifier as the discriminative model to separate the object and background.

3.1 *HF training*

Hough Forest is proposed by J. Gall in 2009 [10], which is a random forest that directly maps the image patch appearance to Hough space.

In this paper, the training patches are sampled from the video frames, the patches sampled from the object patches (the interior of the object bounding boxes) are assigned the class label $+1$, while the patches sampled from the background (background patches) are assigned the class label -1. We train a Hough forest on a random subset of the training data. Each tree is constructed based on a set of patches, where the appearance of the patch represented by DAISY descriptor.

3.2 *Hough vote*

For random forests, the final decision is computed by averaging the results of all the decision trees, while Hough forests are used to cast probabilistic votes within the generalized Hough transform.

During object detecting, Hough forests can be used to localize the centroid of the bounding boxes of the whole object. Considering H as a Hough space, the element of H is generated by a hypotheses sets $\{h(x_c, s)\}$, x_c denotes the object centered at the location x, s is the scale. When the patch reaches the leaf L, we assume that the patch has all the attributes of leaf node L. To integrate the votes coming from different patches, we accumulate them in an additive way into a Hough image, corresponding to the maxima of the Hough image, we can find the possible location of the object centroid and confidence of the scale.

Firstly, set B_x to be the initial object bounding box, p_L is the probability of B_x belonging to the object class, D_n is the offset from the center of the object to the possible location of the object. Consider a patch $\rho(y) = (I(y), c(y), d(y))$ centered at the position y in the patch. If $y \notin B(x)$, then $I(y)$ doesn't belong to the object. So in this paper, we just consider the patch when $y \in B(x)$. We need to find a conditional probability $p(h(x_c, s)|L(y))$ based on $I(y)$, $L(y)$, $h(x_c, s)$.

$$p(h(x_c, s)|L(y)) = p((h(x_c, s), c(y) = 1)|L(y)) = p(d|(c(y) = 1, L(y)) \cdot p(c(y) = 1|L(y)) \tag{4}$$

From Eq. (4), $p(c(y) = 1|L(y))$ is the probability distribution of the object class, $p(d|(c(y) = 1, L(y)))$ is the spatial probability distribution of the object. The information is stored in the discriminative model. The first factor can then be approximated using the Parzen window estimate based on the offset vectors d collected in the leaf, while the second factor can be straightforwardly estimated as the proportion p_L of object patches at training time.

$$p(h(x_c, s)|L(y)) = p((h(x_c, s), c(y) = 1)|L(y)) =$$
$$\frac{1}{|D_n|}[\frac{1}{s_u}\sum_{d \in D_n} \exp\{\frac{\|(y - x_c) - d\|^2}{2\sigma^2}\}] \bullet p_L \tag{5}$$

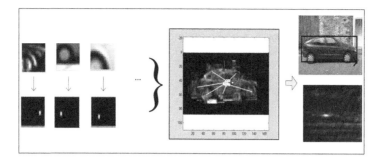

Figure 3.　The votes of every patch to bounding box.

Assuming that there are T decision trees in the forest, for the entire forest, we simply average the probabilities coming from different trees to obtain the forest-based estimate:

$$p(h(x_c,s)\,|\,I(y);\{T_t\}_1^T)=\frac{1}{T}\sum_{t=1}^{T}p(h(x_c,s)\,|\,L(y);T_t) \tag{6}$$

To integrate the votes coming from different patches, we accumulate them into a Hough image:

$$V(x_c,s)=\sum_{y\in B_x}p(h(x_c,s)\,|\,I(y);\{T_t\})_1^T) \tag{7}$$

The detection procedure simply computes the Hough image $V(x_c,s)$ and returns the set of its maxima locations and values $h(x_c,s)$ as the detection hypotheses. x_c is the centroid of the object we detect and s is the scale, $V(x_c,s)$ is the confidence of the detection hypotheses $h(x_c,s)$.

Fig. 3 shows the results that the votes of every patch to bounding box. The left figure shows the image patches which are represented by the feature descriptors, the middle figure shows the object which is constructed by the leaf node information where all image patches passed, the region where yellow dots gathered is the range of the object centroids, the yellow lines represent the offset that the image patch to the possible object centroid being voted. The first row of the right figure shows the object which is located by object centroid, next row shows the object centroids in Hough space, the higher brightness of the Hough image is, the larger possibility of the existence of the object centroid.

3.3　Model updating

In a single object tracking scene, objects appearance change a lot. It's not robust to build the object model obtained from the initial frame. So this paper proposes a updating strategy for object model during the detection period.

From Eq. (7), we can see that the class probability p_L of identification code is equal to the weighting factor of the vote of the object. After training, the p_L of every leaf node should be fixed. This paper proposes a new updating strategy in the following detecting process.

Firstly, unlabeled patches, which are extracted from following video frames, traverse all the decision trees in the forest and find the leaf node to make p_L the biggest, which satisfies $p_{max} = \max(p_L)$. We define the corresponding codebook as:

$$\text{Codebook}_{max}=\{\text{patch}_1,\text{patch}_2,...,\text{patch}_m\} \tag{8}$$

The number of the patches contained by a leaf node is denoted as m. To a newly reached patch, calculate the similarity between it and every patch in the $codebook_{max}$. The similarity of the two

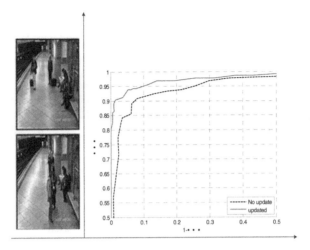

Figure 4. Comparison between No-updated and updated model.

blocks $patch_i$, $patch_j$ is defined as:

$$S(i, j) = 0.5(NCC(patch_i, patch_j) + 1) \qquad (9)$$

We set an experienced threshold $\theta = 0.6$ for $S(i,j)$. Calculate the sum N of the codebook elements in the codebooks whose similarity is bigger than 0.6. If $N > \frac{m}{2}$, then put the new patch into the codebook. To keep the balance of the number of the codebooks, we randomly discard a patch in the codebook. Thus we get the new codebook and the leaf node's p'_L and $\boldsymbol{D}'_L$ is computed correspondingly. Finally, during this updating strategy, what can be assured is that the similarity between the object model and the object status in the most recent period is the biggest.

Figure 4 is the validation experiment for the model updating mechanism, and data used in this experiment is the first 900 frames in AVSS 2007. The left 1st, 2nd figure is the visual representation of the object tracking. The right 3rd figure draws the recall-(1-precision) (recall ratio and precision ratio) performance curves based on the experiment results before and after model updating. The figure tells that the accuracy (the solid red line) with model updating has improved a lot, in comparison with the fixed model.

4 EXPERIMENTS AND EVALUATION

This paper focuses on tracking a single object. The tracking target is selected manually in the first frame.

As figure 5 shows: In the video 1, the fan is rotated in several angles and move near or far away from the camera. Figure (5.a) is the girl's face selected by the mouse as the tracking object. Figure (5.b) and (5.c) have a 90° rotation on the fan and we observe the object from different perspectives. It shows the target can be tracked successfully even the perspective is changing. Figure (5.d) shows that when the object is approaching the camera and the revolution changes a lot, the object remains being tracked still.

Figure (6.a)~(6.d), video 2 is for confirming partial occlusions on the target from different directions. From the result we can see that the algorithm can still track the object accurately, owing to the local DAISY feature and hough transformation.

Table 2 is the experimental analysis of the 4 video clips in changing scenarios (different size, resolution, light and the partial occlusions). We calculate the tracking frames and the frame rate of

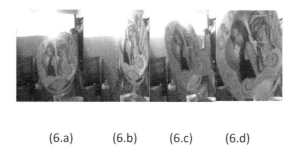

(6.a)　　　(6.b)　　　(6.c)　　　(6.d)

Figure 5.   Tracking in different views and scales.

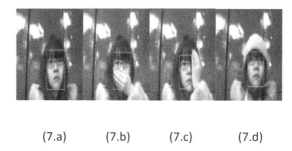

(7.a)　　　(7.b)　　　(7.c)　　　(7.d)

Figure 6.   Tracking with partial occlusion.

Table 2.   Performance analysis.

Video	Frame number	Different Scales	Different resolutions	Occlusion	Different illumination	Tracked frames	FPS
Video 1	536	√	√	—	√	536	21
Video 2	842	√	—	√	—	838	20

each video. "√" means that this dealing can be handled while "—" represents that such scenario change doesn't exist in the corresponding video clip.

5   CONCLUSIONS

We combine the scale-invariant DoG feature with DAISY feature to build the object model which has invariant property for illumination, deformation, viewpoint and scale. In addition to the random forest classifier, we introduce the Hough transformation – use Hough voting to accurately track the object. And we propose an efficient and accurate model updating strategy. It's a generative and discriminative hybrid model for object detection and tracking, this model can be widely applied to pedestrian or face detection, traffic monitoring and many other areas.

ACKNOWLEDGEMENT

Our research was sponsored by following projects:

– National High-tech R& D Program of China ("863 Program") (No. 2013AA01A603);
– National Science and Technology Support Projects of China (No. 2012BAH07B01);

– Program of Science and Technology Commission of Shanghai Municipality (No. 12510701900).
– 2012 IoT Program of Ministry of Industry and Information Technology of China.

REFERENCES

[1] Lei Y., Ding X. Q., Wang S. J. Visual Tracker Using Sequential Bayesian learning: Discriminative, Generative, and Hybrid. IEEE Transactions on System Man And Cybernetics, 2008, 38(6):1578–1591.

[2] Dinh T. B., Medioni G. Cotraining Framework of Generative and Discriminative Trackers with Partial Occlusion Handling. Proceedings of IEEE Workshop on Applications of Computer Vision. Los Angeles, USA: IEEE, 2011:642–649.

[3] Curwen R. W., Amini A. A., Duncan J. S. Tracking Vascular Motion in X-ray Image Sequences with Kalman Snakes. Proceedings of Computers in Cardiology. Los Angeles, USA: Computers in Cardiology, 1994:109–112.

[4] Lowe D. G. Object recognition from local scale-invariant features. Proceedings of IEEE International Conference on Computer Vision (ICCV), Corfu, Greece: IEEE, 1999, 2:1150–1157.

[5] Mikolajczyk K., Schmid C. A performance evaluation of local descriptors. Proceedings of IEEE Computer Society Conference on Computer Vision and Pattern Recognition (CVPR). Madison, USA: IEEE, 2003, 27(10):257–263.

[6] A. Adam, E. Rivlin, I. Shimshoni. Robust fragments-based tracking using the integral histogram. Proceedings of IEEE Computer Society Conference on Computer Vision and Pattern Recognition (CVPR). New York, USA: IEEE, 2006, 1:798–805.

[7] Avidan S. Support vector tracking. IEEE Transaction on Pattern Analysis and Machine Intelligence (PAMI), 2004, 26(8):1064–1072.

[8] Kalal Z., Mikolajczyk K., Matas J.. Tracking-Learning-Detection. IEEE Transactions on Pattern Analysis and Machine Intelligence, 2012, 34(7):1409–1422.

[9] Kalal Z., Matas J., Mikolajczyk K. P-N learning: Bootstrapping binary classifiers by structural constraints. Proceedings of IEEE Conference on Computer Vision and Pattern Recognition (CVPR). San Francisco, CA, USA:IEEE, 2010:49–56.

[10] Tola E., Lepetit V., Fua P. A Fast Local Descriptor for Dense Matching. Proceedings of IEEE Conference on Computer Vision and Pattern Recognition (CVPR). Anchorage, USA:IEEE, 2008:1–8.

[11] Gall J., Lempitsky V. Class-Specific Hough Forests for Object Detection. Proceedings of IEEE Conference on Computer Vision and Pattern Recognition (CVPR). Miami, USA: IEEE, 2009:1022–1029.

[12] Breiman L. Random forests. Machine Learning, 2001, 45(1):5–32.

[13] Categorization and segmentation. International Journal of Computer Vision, 2008, 77(1):259–289.

Author index

Printed and bound by CPI Group (UK) Ltd, Croydon, CR0 4YY

18/10/2024

01776219-0005